·

Predictive Analytics

Predictive Analytics

Parametric Models for Regression and Classification Using R

Ajit C. Tamhane

Northwestern University

with contributions from Edward C. Malthouse, Northwestern University

The right of Ajit C. Tamhane to be identified as the author of this work has been asserted in accordance with law.

Registered Office
John Wiley & Sons, Inc., 111 River Street, Hoboken, NJ 07030, USA

Editorial Office
John Wiley & Sons, Inc., 111 River Street, Hoboken, NJ 07030, USA

For details of our global editorial offices, customer services, and more information about Wiley products visit us at www.wiley.com.

Wiley also publishes its books in a variety of electronic formats and by print-on-demand. Some content that appears in standard print versions of this book may not be available in other formats.

Library of Congress Cataloging-in-Publication Data

Names: Tamhane, Ajit C., author.
Title: Predictive analytics : parametric models for regression and
    classification using R / Ajit C. Tamhane, Northwestern University ; with
    contributions from Edward C. Malthouse, Northwestern University.
Description: Hoboken, NJ : Wiley, 2020. | Series: Wiley series in
    probability and statistics | Includes index.
Identifiers: LCCN 2020034678 (print) | LCCN 2020034679 (ebook) | ISBN
    9781118948897 (cloth) | ISBN 9781118948910 (adobe pdf) | ISBN
    9781118948903 (epub)
Subjects: LCSH: Data mining – Statistical methods. | Probabilities. | R
    (Computer program language)
Classification: LCC QA76.9.D343 T349 2020 (print) | LCC QA76.9.D343
    (ebook) | DDC 006.3/12 – dc23
LC record available at https://lccn.loc.gov/2020034678
LC ebook record available at https://lccn.loc.gov/2020034679

Cover Design: Wiley
Cover Image: © draganab/Getty Images

Set in 11/13.5pt Computer Modern by SPi Global, Chennai, India

To All My Students
From Whom I Have Learned More
Than I Have Taught Them

Contents

Preface

We are living in a highly data-rich environment, popularly referred to as the world of Big Data. This book is my attempt to present the classical methods of parametric regression and classification in the context of this world. Data analytics commonly refers to modern computer intensive methods such as classification and regression trees (CART), random forests, deep learning, support vector machines, and so on. However, the knowledge of classical methods covered in this book is still essential for a sound foundation. They also provide simpler solutions in many cases. The main distinguishing and attractive feature of the book is that it covers a broad range of parametric methods, including multiple regression, logistic regression (binary and multinomial), discriminant analysis, generalized linear models, and survival analysis. The presentation is streamlined so that these topics can be covered in a single course. The book is primarily aimed at graduate students in statistics, data science, and applied fields such as marketing, engineering, and so on, as well as at practitioners in these disciplines.

Although the book covers mainly classical parametric methods, I have added short sections in several chapters briefly reviewing modern computer intensive extensions. The idea of these sections is to provide sneak previews of these methods, which students may want to learn in more detail and use them as possible alternatives to classical methods. For example, we briefly review regression trees and neural nets in the multiple regression chapter and classification trees and support vector machines in the logistic regression chapter.

There are many excellent texts on classical regression, for example, Chatterjee and Hadi (2012) and Montgomery et al. (2012) on multiple regression and Hosmer and Lemeshow (1989) on logistic regression. However, I have felt the need to modernize the presentations and use larger data sets with business applications of interest to many of our students. At the same time, the core of these classical methods needs to be presented rigorously. To balance these two somewhat conflicting needs, I have put the technical details, such as derivations and proofs of the results, in the technical notes section of each chapter. The main body of the text is devoted to real data examples with only the necessary statistical methodology details included. Frequently, I have used small data sets as they are often helpful for pedagogical purposes. I have also separated the exercises into theoretical and applied. This format allows the book to be used in both masters and PhD classes. Masters classes may skip the Technical Notes sections and most of the theoretical exercises, while PhD classes may use both theoretical and applied parts of the book, emphasizing the former. Prospective instructors may consider using the book in this manner for these two groups of students. The choice of sections and exercises to skip or include would depend upon the interests of the instructor and students.

The ideal background for this book is a good course in probability-based statistics including sampling distributions, confidence intervals, hypothesis tests, and simple linear regression. The more advanced topic of maximum likelihood estimation, which underlies fitting of logistic regression and generalized linear models, is covered in an appendix. Courses in multivariable calculus and matrix algebra are a necessary mathematical prerequisite. A review of matrix algebra is given in an appendix.

The book uses R for all data analyses. The R scripts along with their outputs for all examples are included in the book. Excel files for all the data sets used in examples and exercises in the `csv` format are posted on the book's website. A background in R is helpful but not essential. A couple of lab classes at the beginning of the term should be adequate in preparing students to start programming in R.

A team project is an integral part of a course based on this book. Three project assignments used in my courses are given in an appendix. Thus students learn to apply multiple regression and logistic regression covered in the first half of the course in a project setting with large and dirty data.

It should be possible to cover most of the book in a semester-long course. Chapter 2 on simple linear regression and correlation is mainly intended for review since most students should have had this background. In a quarter-long course at Northwestern, I need to skip some sections from most chapters. These sections are marked with asterisks. In a PhD level course, I do cover some of these starred sections or have students go through them (particularly Technical Notes sections).

Answers to selected exercises are provided at the end of the book. A solution manual giving detailed solutions to all exercises and slides for the lectures are available from the publisher to the instructors who adopt the book.

If the readers find this book as an entryway to the fascinating world of predictive analytics using statistical modeling, then my goal in writing this book will be fulfilled. Enjoy the journey.

Ajit C. Tamhane

Department of Industrial Engineering and Management Sciences

Northwestern University

Evanston, Illinois

January 2020

Acknowledgments

I am grateful to many people who contributed in various ways to this book. Above all, I am most grateful to Professor Ed Malthouse with whom I had originally planned to coauthor this book. Unfortunately, he did not have the time, but he gave many insightful comments and suggestions throughout. He also provided many interesting data sets, including the data sets for all three projects. Ed's ability to drill through complex large data sets and dig out interesting nuggets of information is simply unparalleled. He was my PhD student about 25 years ago, but I have learned more from him over the course of these years and especially during the writing of this book.

I am also very grateful to Dr. Jiangtao Gou, another ex-PhD student of mine, who helped with computations for many data examples and created figures and plots for them.

The book has gone through many drafts which have been tested over several years in my masters and PhD classes. Students in these classes were essentially guinea pigs, but they patiently suffered through many errors in these drafts pointing out them to me and providing other useful feedback. The following PhD students from the Industrial Engineering and Management Sciences (IEMS) Department served as teaching assistants for my classes: Young Woong Park, Mingyang Di, Baiyang Wang, Kungang Zhang, Yi Zhu, Xin Qian and Boyang Shang. They wrote R codes for many exercises and examples. I would like to express my sincere thanks to all of them. I also would like to acknowledge Yayu Zhou, another IEMS PhD student, who assisted in preparing the lecture slides based on this book.

My final thanks to all the students who have taken my classes over many years and a number of PhD students whom I have advised. I can honestly say that I have learned more from them than I have taught them. In trying to answer their questions, I have developed a much deeper understanding of the subject. I dedicate this book to all of them.

A.C.T.

Abbreviations

Although all abbreviations are defined in the text where they occur, they are assembled here for the convenience of the reader.

AIC:	Akaike information criterion
ANOVA:	Analysis of variance
AUC:	Area under the curve
AV:	Added variables
BIC:	Bayesian information criterion
BLUE:	Best linear unbiased estimator
CART:	Classification and regression trees
CCR:	Correct classification rate
c.d.f.:	Cumulative distribution function
CI:	Confidence interval
CR:	Components plus residuals
CV:	Cross validation
d.f.:	Degrees of freedom
EDA:	Exploratory data analysis
FM:	Full model
FNR:	False negative rate
FPR:	False positive rate
GAM:	Generalized additive model
GLM:	Generalized linear model
GLR:	Generalized likelihood ratio
GLS:	Generalized least squares
GMAT:	Graduate Management Admissions Test
GPA:	Grade point average
IRWLS:	Iteratively reweighted least squares
KM:	Kaplan-Meier
LDA:	Linear discriminant analysis
LDF:	Linear discriminant function
L.H.S.:	Left hand side
LOF:	Lack of fit
LR:	Likelihood ratio

LV:	Latent vector
LS:	Least squares
MANOVA:	Multivariate analysis of variance
MBA:	Master of Business Administration
MLE:	Maximum likelihood estimator/estimation
MLR:	Multiple linear regression
MSEP:	Mean square error of prediction
MSE:	Mean square error
MSH:	Mean square hypothesis
MSLOF:	Mean square lack of fit
MSPE:	Mean square pure error
MSR:	Mean square regression
MST:	Mean square total
NN:	Neural net
OLS:	Ordinary least squares
PC:	Principal component
PCA:	Principal components analysis
PCR:	Principal components regression
p.d.f.:	Probability density function
PH:	Proportional hazards
PI:	Prediction interval
PLS:	Partial least squares
PM:	Partial model
p.m.f.:	Probability mass function
QDA:	Quadratic discriminant analysis
QDF:	Quadratic discriminant function
Q-Q:	Quantile-quantile
r.v.:	Random variable
R.H.S.:	Right hand side
RMSE:	Root mean square error
RMSEP:	Root mean square error of prediction
ROC:	Receiver operating characteristic
SCI:	Simultaneous confidence interval
SCR:	Simultaneous confidence region
SD:	Standard deviation
SE:	Standard error
SM:	Saturated model
SS:	Sum of squares
SSE:	Error sum of squares
SSG:	Group sum of squares
SSH:	Hypothesis sum of squares
SSLOF:	Lack of fit sum of squares
SSPE:	Pure error sum of squares

SSR:	Regression sum of squares
SST:	Total sum of squares
SVM:	Support vector machine
VIF:	Variance inflation factor
WBC:	White blood cell count
WLS:	Weighted least squares
w.r.t.:	with respect to

About the companion website

This book is accompanied by a companion website:

www.wiley.com/go/tamhane/predictiveanalytics

Excel files for all the data sets used in examples and exercises in the csv format are posted on this website.

Chapter 1

Introduction

Statistical learning is the science of discovering patterns in the data and building models to make predictions and decisions. These are not new topics, and statisticians have studied them for a long time. The modern resurgence of this field using highly computer intensive methods was spurred by the big data movement. Its foundations were laid in the well-known book by Hastie et al. (2001). **Machine learning** is the related area in computer science, which develops computational structures (e.g., deep learning neural networks) and fast algorithms for performing tasks that are generally far more complex and wider in scope, e.g. voice and image recognition and text analytics. While machine learning mainly emphasizes minimizing prediction errors, statistical learning also emphasizes modeling and statistical inferences, e.g. testing the significance of the effects of predictors on the outcome variable. Both statistical and machine learning have become increasingly popular in today's business and scientific worlds driven by the abundance of data, availability of statistical tools to analyze them and advent of immense and superfast computing power.

    Predictive analytics is the part of statistical learning that is concerned with making predictions about a certain outcome based on available data. **Data mining** is the part that is concerned with discovering patterns, associations, and trends in the data. Often data mining is interpreted to cover both data exploration and modeling including predictive analytics. In the statistical learning approach to predictive analytics, the focus is on building predictive models and use them to draw inferences and make predictions. This book is devoted to classical parametric statistical models for regression and classification. In this introductory chapter, we provide an overview of this subject in the larger

Predictive Analytics: Parametric Models for Regression and Classification Using R.
First Edition. Ajit C. Tamhane.
© 2021 John Wiley & Sons, Inc. Published 2021 by John Wiley & Sons, Inc.
Companion website: www.wiley.com/go/tamhane/predictiveanalytics

landscape of statistical learning. Although the origins of predictive analytics, such as linear regression, go more than a century back, the subject is thriving today with many modern developments in computing and optimization algorithms.

1.1   Supervised versus unsupervised learning

Statistical learning covers two main broad areas: **supervised learning**, which is synonymous with predictive analytics and **unsupervised learning**. In supervised learning, one of the variables is designated as a **response**, **outcome**, **dependent**, or **output** variable. The goal is to model its relationship with the so-called **predictor**, **covariate**, **independent** or **input** variables (called **features** in machine learning). In this book, for the most part, we will use the terminology of response variable and predictor variables. Some examples of supervised learning are the following: predicting customer purchase behavior as a function of past purchase history and demographic and socioeconomic data; predicting the selling price of a house as a function of floor area, plot size, amenities, school district and economic indicators; classifying a patient's disease status based on a battery of laboratory tests and classifying a bank loan as in good standing or likely to default as a function of borrower's financial parameters such as income, debt etc.

In unsupervised learning, there is no designated response variable, all variables being of the same genre. The goal is to uncover relationships, associations, and patterns that underlie the data. Some examples are the following: determination of latent factors in multivariate data and market segmentation, e.g. clustering of customers with similar attributes.

Although the primary goal in predictive analytics is prediction, it is also generally of interest to make inferences and *interpretations* about the relationships between the predictor variables and the response variable. For example, it is of interest to know which attributes of a house are the key determinants of its price and how much they contribute to the price. What would raise the value of a house more relative to the cost–a finished basement or addition of a sun room? Going beyond mere prediction, it may be of interest to get a deeper understanding of the phenomenon at hand, especially in scientific studies, by establishing causal connections between predictor variables and the response variable. On the other hand, in a prediction problem, it is of interest to get the best prediction possible regardless of the relative importance of the predictors used. Most supervised learning problems have elements of both–prediction and inference.

Because there is a designated response variable in supervised learning, the accuracy of a predictive model can be tested by comparing its predictions against the actual observed responses. Usually, this is done by dividing the data randomly into a **training set** and a **test set**. The model is fitted on the training set and tested on the test set. Thus, the test set serves as an independent data set. If predictions are made on the training set themselves, then we get a measure of how well the model fits the data–not of its predictive power.

1.2 Parametric versus nonparametric models

As mentioned earlier, this book focuses on parametric models for supervised learning. The functional form of these models is specified; only the parameters of the model are unknown and are estimated from data. On the other hand, in nonparametric models, the functional form of the model is either not specified or is too complex to be expressed in a closed form involving a few parameters; examples of the latter are mentioned below. Thus, model fitting involves estimation of an unknown function rather than unknown parameters of a known function. The estimation problem is thus intrinsically harder.

Examples of parametric models of supervised learning include multiple regression, logistic regression, discriminant analysis, generalized linear models, and the Cox proportional hazards model for survival data, which are all covered in this book. Examples of nonparametric models of supervised learning include classification and regression trees (CART), random forests, neural nets (NN), and support vector machines (SVM). Technically, NN and SVM are complex parametric models as we will see in the brief reviews of these techniques in Sections 3.7.2 and 7.6.2, respectively, e.g. NN involves specified activation functions but unspecified weights on the edges connecting the nodes which need to be estimated. Nonetheless, the resulting models are not easily expressible in closed forms. So these models are classified as nonparametric models. Examples of parametric models of unsupervised learning include correlation analysis, principal components analysis, and factor analysis. Examples of nonparametric models of unsupervised learning include cluster analysis and association rules.

Clearly, parametric models require a lot less data for fitting them than nonparametric models do. However, if the model is misspecified, then we may end up fitting a wrong model. In nonparametric models, this risk is minimized since the data determine the form of the model. Parametric models are easier to fit if the model is correctly specified even if the data are not abundant. Also, standard statistical inference procedures are available for them. Although, nonparametric models are more flexible and data-adaptive, they require a data-rich environment and statistical inference procedures for them generally involve bootstrap or resampling, which are computer intensive methods. So parametric models should be used when the form of the model is known or can be approximated by a simple function. Nonparametric models should be used when the functional form of the model is unknown or too complex, but sufficient data are available for its estimation.

1.3   Types of data

Data are gathered on subjects (e.g. patients), objects (e.g. items or products), processes (e.g. services), or organizations (e.g. companies). We refer to the entities on which the data are gathered generically as **sampling units** even though they may not be samples drawn from larger populations. If the data are gathered on sampling units at some fixed point in time, then the data are referred to as **cross-sectional data**. Frequently, data are gathered over time on a given sampling unit (e.g. quarterly sales of a company). Such data are referred to as **time-series data**. In this book, for the most part, we deal with cross-sectional data.

Data can be classified in many other ways. One way is by the measurement scale used. There are mainly two measurement scales: **numerical** or **quantitative** and **categorical** or **qualitative**. Quantitative data are of two types: data that can be treated as **continuous** for all practical purposes such as length, height, weight and time, and **discrete**, typically measured on an integer scale, such as count or frequency data. Qualitative data are also of two types: **ordinal** where data consist of ordered labels, e.g. the grade in a course or rating on a scale of 1–5 in a customer survey, and **nominal** where data consist of unordered labels, e.g. ethnicity or party affiliation or color of eye. Any set of distinct symbols may be assigned to distinct values of a nominal variable, including numbers, but obviously no arithmetic operations, such as averaging can be done on them. Ordered numbers may be assigned to distinct values of an ordinal variable, e.g. 0–4 for grades F through A. Again, arithmetic operations do not make sense for ordinal data, although they are often performed such as when computing the grade point average by implicitly ascribing a numerical scale to ordinal data.

Another way that data can be classified is the type of study or process used to collect them. Mainly, there are two types of studies used to collect data: **observational** and **experimental**. Observational data typically come from archival sources or from observational studies in which a phenomenon is observed passively and data are recorded. On the other hand, experimental data come from designed experiments in which predictor variables (called **factors** in the design of experiments terminology) are actively manipulated to evaluate their effects on the response variable. Observational data are more likely to be subject to confounding from selection and other biases because of possible uncontrolled and unobserved variables (called **noise factors**), whereas experimental data that come from well-controlled randomized experiments are much less likely to be subject to such biases. It should be emphasized that if a research study is biased then no amount of data or fancy statistical analyses can rectify invalid conclusions that can result from such studies. In predictive analytics, we generally work with observational data, and so it is extremely important to verify the data quality before conducting any statistical analyses.

1.4 Overview of parametric predictive analytics

Generically, we will denote the response variable by y and the predictor variables by $x_1, \ldots, x_p$. We want to build a model relating y to the x's. If y is a numerical variable such as salary, sales, then we have a **regression problem**. If y is a categorical variable, such as success or failure of a treatment, a customer's choice of the brand of a detergent or letter grade in a course, then we have a **classification problem** since the response is classified into one of several categories.

In a regression problem, the predictive model is typically of the form:

$$y = f(x_1, \ldots, x_p) + \varepsilon, \tag{1.1}$$

where $f(x_1, \ldots, x_p) = E(y|x_1, \ldots, x_p)$ is referred to as the **model** part, which is the expected value of y conditioned on the x's, and ε is the **random error** part with a zero mean. The model part is sometimes referred to as **signal** and the random part is referred to as **noise**. The goal of model fitting is to extract the signal from data contaminated with noise. Model **overfitting** occurs when noise is mistakenly fitted as part of the signal, e.g. when a higher than necessary degree polynomial is fitted to a y versus x plot to account for wiggles in the data that are merely random variations around a linear or a quadratic trend.

There are two types of models: **empirical models** and **mechanistic** or **theoretical models**. In an empirical model, the true form of $f(\cdot)$ is unknown but is approximated by a relatively simple function such as a linear or a multiplicative function. Mechanistic models commonly arise in sciences and engineering and are derived from mechanistic theories underlying a given physical, chemical, or biological phenomenon under study. They are often solutions of differential equations that describe the phenomenon. Both types of models are parametric in nature, so fitting these models involves estimation of unknown parameters. In this book, we mainly focus on building empirical models from observational data.

The simplest empirical model is the **multiple regression** model:

$$y = \beta_0 + \beta_1 x_1 + \cdots + \beta_p x_p + \varepsilon,$$

where $\beta_0, \beta_1, \ldots, \beta_p$ are unknown parameters and ε is assumed to be a $N(0, \sigma^2)$ random error. This is called a **linear model** because it is linear in the β's, not necessarily in the x's. For example, x_2 can be equal to x_1^2 or $\log x_1$ and x_3 can be equal to $x_1 x_2$. If the specified model is nonlinear in the unknown parameters, then it is called a **nonlinear regression model**. Mechanistic models are often nonlinear in the β's. Some models are superficially nonlinear but can be transformed to a linear form. For example multiplicative models can be log-transformed to a linear form. Models that cannot be transformed into a linear form by such transformations are called intrinsically nonlinear. We do not discuss nonlinear models in this book. Multiple linear regression is covered in

Chapters 3–6. Simple linear regression, which is a special case of multiple linear regression for a single predictor, and correlation analysis are covered in Chapter 2.

For binary responses, the logistic regression model discussed in Chapter 7 is commonly used. If y is coded as 1 for success and 0 for failure, then $p = E(y)$ denotes the probability of success, which is modeled as a function of the x's. The logistic regression model postulates a linear model on the **logistic transform** $\ln[p/(1-p)]$, i.e.,

$$\ln\left[\frac{p}{1-p}\right] = \beta_0 + \beta_1 x_1 + \cdots + \beta_p x_p.$$

Binary responses are assumed to follow the Bernoulli distribution. For polytomous responses, the logistic regression model needs to be extended appropriately depending on whether the responses are nominal or ordinal. These topics are covered in Chapter 7.

Another approach to predicting dichotomous or polytomous responses is discriminant analysis. A binary linear discriminant function is a linear function of the predictor variables that best discriminates between two responses. For polytomous outcomes more than one linear function is needed. Discriminant analysis is covered in Chapter 8.

A generalized linear model (GLM) extends multiple regression and logistic regression models in two different ways. First, it extends the response variable distribution to any member of the so-called **exponential family** of distributions. Both the normal and the Bernoulli distributions belong to this family, but there are many other distributions including exponential, gamma, and Poisson, which belong to this family. Second, it postulates a linear model on the so-called **link function** $g(\mu)$, where $\mu = E(y)$. In the case of multiple regression, the link function is the identity function $g(\mu) = \mu$. In the case of logistic regression, the link function is the logistic transform defined above. GLM provides a powerful generalization of the linear model since it allows building predictive models for many other response distributions. Exponential and gamma distributions are used for lifetime data that arise in reliability and survival studies', and in marketing (e.g. time since last order). Poisson distribution is used for count data that arise in applications such as traffic studies and marketing (e.g. number of orders). GLMs for different distributions and associated link functions are discussed in Chapter 9.

Finally, in Chapter 10, we cover analysis of survival data. A unique feature of survival data is that they are often censored, i.e. the actual survival time is not observed because the subject (e.g. patient) withdraws from the study or the study is terminated before the event of interest occurs. We cover the Cox proportional hazards regression model to analyze the effects of possible risk factors on the lifetimes of patients.

Chapter 2

Simple linear regression and correlation

One of the simplest and yet a commonly occurring data analytic problem is exploring the relationship between two numerical variables. In many applications, one of the variables may be regarded as a **response variable** and the other as a **predictor variable** and the goal is to find the best fitting relationship between the two. For example, it may be of interest to predict the amount of sales from advertising dollars or the reduction in tumor size from the amount of radiation exposure. This is referred to as a **regression problem**. In other applications, there is no such distinction between the two variables, and it is of interest to simply assess the strength of relationship between them. For example, it may be of interest to assess the relationship between the average summer temperature and the average winter temperature in a given region of the country or the amounts of sales in two divisions of a company. This is referred to as a **correlation problem**. We study both these problems in this chapter.

We will use the following two data sets to illustrate the methods introduced in this chapter:

Example 2.1 (Bacteria Counts: Data) Chatterjee and Hadi (2012) gave the data shown in Table 2.1 on the number of surviving bacteria (in hundreds) exposed to 200 kv X-rays for 15 six-minute intervals. The main question of interest is how do the bacteria decay with time, in particular, does the exponential decay law apply and if so what is the decay rate? □

Predictive Analytics: Parametric Models for Regression and Classification Using R, First Edition. Ajit C. Tamhane.
© 2021 John Wiley & Sons, Inc. Published 2021 by John Wiley & Sons, Inc.
Companion website: www.wiley.com/go/tamhane/predictiveanalytics

Table 2.1. Surviving bacteria count (in hundreds) at time t (in six-minute intervals)

Time (t)	Bacteria count (N_t)	Time (t)	Bacteria count (N_t)
1	355	9	56
2	211	10	38
3	197	11	36
4	166	12	32
5	142	13	21
6	106	14	19
7	104	15	15
8	60		

Source: Adapted from Chatterjee and Hadi (2012).

Table 2.2. Cardiac output measurements (liters per minute)

	Method			Method	
Patient	Invasive (x)	Noninvasive (y)	Patient	Invasive (x)	Noninvasive (y)
1	6.3	5.2	14	7.7	7.4
2	6.3	6.6	15	7.4	7.4
3	3.5	2.3	16	5.6	4.9
4	5.1	4.4	17	6.3	5.4
5	5.5	4.1	18	8.4	8.4
6	7.7	6.4	19	5.6	5.1
7	6.3	5.7	20	4.8	4.4
8	2.8	2.3	21	4.3	4.3
9	3.4	3.2	22	4.2	4.1
10	5.7	5.5	23	3.3	2.2
11	5.6	4.9	24	3.8	4.0
12	6.2	6.1	25	5.7	5.8
13	6.6	6.3	26	4.1	4.0

Source: Private data.

Example 2.2 *(Cardiac Output Measurements: Data)* Frequently, we have an accurate measurement method (often called a gold standard) that is expensive or difficult to use in practice. Hence, a less accurate but cheaper and a more practical method is used. By making measurements using both methods, we can correlate the two and calibrate the less accurate method against the more accurate method. A medical device company compared two methods of cardiac output measurement: an accurate, but invasive method, and a noninvasive, but less accurate method. Cardiac outputs of 26 patients were measured using both methods. The data are shown in Table 2.2. Two questions to ask are (i) how well are the measurements of the two methods correlated and (ii) how accurately can we predict the actual cardiac outputs from the noninvasive method by calibrating it against the more accurate invasive method? ☐

2.1 Fitting a straight line

2.1.1 Least squares method

The simplest equation one can fit to bivariate numerical data is a straight line. Denote the response variable by y and the predictor variable by x. Let $\{(x_i, y_i),\ i = 1, \ldots, n\}$ denote a data set of size n. Before fitting a straight line (or more generally any curve) to data, first we should make a scatter plot and see whether it displays any pattern, and if so, what sort of pattern. Of course, it would be foolhardy to fit a straight line if the scatter plot displays a nonlinear pattern or displays no pattern at all, i.e. if the plot is just a random scatter. Some nonlinear trends can be linearized by making suitable transformations as we shall see in Section 2.1.2.

In this section, we consider the problem of fitting a straight line $y = \beta_0 + \beta_1 x$ to the data, where β_0 and β_1 are unknown **intercept** and **slope** parameters of the straight line. The **least squares (LS) estimates** of β_0 and β_1, denoted by $\widehat{\beta}_0$ and $\widehat{\beta}_1$, minimize the sum of the squared differences between the observed y_i's and their predicted values from the straight line:

$$Q = \sum_{i=1}^{n} [y_i - (\beta_0 + \beta_1 x_i)]^2. \tag{2.1}$$

We refer to Q as the **LS criterion**. The minimum of Q can be found by setting the partial derivatives, $\partial Q / \partial \beta_0$ and $\partial Q / \partial \beta_1$, equal to zero. The resulting equations are called the **normal equations** (see (2.28) and (2.29)). Their solutions are

$$\widehat{\beta}_0 = \overline{y} - \widehat{\beta}_1 \overline{x} \quad \text{and} \quad \widehat{\beta}_1 = \frac{S_{xy}}{S_{xx}}, \tag{2.2}$$

which are the least squares estimates of β_0 and β_1. Here $\bar{x}$ and $\bar{y}$ are the sample means of the x's and y's, respectively, and where, for compactness of notation, we have defined

$$S_{xy} = \sum_{i=1}^{n}(x_i - \bar{x})(y_i - \bar{y}), \quad S_{xx} = \sum_{i=1}^{n}(x_i - \bar{x})^2, \quad \text{and} \quad S_{yy} = \sum_{i=1}^{n}(y_i - \bar{y})^2. \tag{2.3}$$

We refer to

$$\hat{y} = \hat{\beta}_0 + \hat{\beta}_1 x \tag{2.4}$$

as the **LS line**, which is the "best fitting" straight line that minimizes the LS criterion. This line is used to predict the values of y for given x's. In particular, if $x = \bar{x}$, then it is easy to see from the formula for $\hat{\beta}_0$ that $\hat{y} = \bar{y}$. Thus, the LS line passes through the midpoint $(\bar{x}, \bar{y})$ of the scatter plot.

An alternative useful form for the LS line

The correlation coefficient r between x and y is defined in (2.21) in Section 2.3. Using that formula, we can write

$$r = \frac{S_{xy}}{\sqrt{S_{xx}S_{yy}}} = \hat{\beta}_1 \sqrt{\frac{S_{xx}}{S_{yy}}} = \hat{\beta}_1 \left(\frac{s_x}{s_y}\right), \tag{2.5}$$

where s_x and s_y are the standard deviations of the x's and the y's given by (2.22). Now Equation (2.4) for the LS line can be written as $\hat{y} - \bar{y} = \hat{\beta}_1(x - \bar{x})$. Then substituting for $\hat{\beta}_1$ from above and rearranging the terms, we get

$$\frac{\hat{y} - \bar{y}}{s_y} = r\left(\frac{x - \bar{x}}{s_x}\right). \tag{2.6}$$

This equation has a nice interpretation that one standard deviation change in x from its mean results in r standard deviation units change in y from its mean. The sign of change in y is the same as the sign of change in x if $r > 0$ and the sign is opposite if $r < 0$. Thus, the change in x is modulated by the magnitude of r; if $r = 0$ then a change in x results in no change in y.

Example 2.3 *(Galton Data: Regression to the Mean)* In a classic study of correlation between fathers' heights (x) and sons' heights (y) based on a sample of 1078 father–son pairs, the British scientist Sir Francis Galton (1822–1911) found that the average height of fathers is $\bar{x} = 68''$, the average height of sons is $\bar{y} = 69''$; thus, on the average, sons are $1''$ taller than their fathers. Furthermore, the correlation between x and y is 0.5 and the standard deviations of x and y are roughly equal to $2.7''$.

The line $y = x + 1$ and the regression line $y - 69 = 0.5(x - 68)$ calculated using (2.6) are shown in Figure 2.1. Notice that the regression line predicts y-values below the $y = x + 1$ line for $x > \bar{x} = 68''$ and above the $y = x + 1$ line for $x < \bar{x} = 68''$. As an example, consider two fathers, one of whom is $4''$ taller and the other is $4''$ shorter than average, i.e. $72''$ and $64''$ in height. Then we see that their sons will be predicted to be $71''$ and $67''$ tall, respectively. Note that they will not be predicted to be $1''$ taller than their fathers. The tall fathers' sons will be taller than average but not as tall as their fathers, while short fathers' sons will be shorter than average but not as short as their fathers. Thus, the sons heights regress toward the mean, hence referred to as the **regression to the mean** phenomenon.

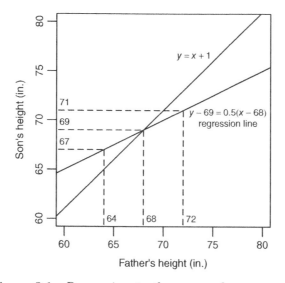

Figure 2.1. *Regression to the mean phenomenon.*   □

2.1.2 Linearizing transformations

An example of a nonlinear relationship that can be linearized by **log-transformation** of the data is the so-called **power law**, $y = ax^b$. The log-transformed equation is $\ln y = \ln a + b \ln x$, which is a straight line on the log–log scale. By making transformations $y \to \ln y$ and $x \to \ln x$, we get a straight line with intercept $\beta_0 = \ln a$ and slope $\beta_1 = b$.

There are many empirical applications of the power law. Here we consider its application to model the demand–price relationship. Suppose y is the demand and x is the price of a product. Differentiating both sides of the equation $\ln y = \ln a + b \ln x$, we get $dy/y = b(dx/x)$ or $b = (dy/y)/(dx/x)$. Thus, b represents the relative change in demand due to a unit relative change in price (e.g. 1 percent). The parameter b is known as the **price elasticity** of the product. Generally, $b < 0$ so that as the price increases, the demand decreases. If $b = 0$, then the demand is said to be inelastic with respect to the price.

Another such relationship is the so-called **exponential law**, $y = a \exp(bx)$ (where x is typically time), which is used to model exponential growth (for $b > 0$) or exponential decay (for $b < 0$). In this case, the log-transformed equation is $\ln y = \ln a + bx$. By making transformations $y \to \ln y, x \to x$, we get a straight line with intercept $\beta_0 = \ln a$ and slope $\beta_1 = b$. By fitting a straight line to the transformed data, we can get the LS estimates $\widehat{\beta}_0$ and $\widehat{\beta}_1$ from which the estimates of a and b can be obtained.

Example 2.4 *(Bacteria Counts: LS Estimates and Scatter Plots)* Consider the bacteria decay data in Table 2.1. The left panel of Figure 2.2 shows the scatter plot of bacteria count (N_t) versus time (t). The plot is negatively curved with a large outlier at $t = 1$. The theory suggests the exponential decay model

$$N_t = N_0 e^{bt} \quad \text{for } t \geq 0,$$

where $b < 0$. We make the logarithmic transformation yielding a straight line model: $\ln N_t = \ln N_0 + bt = \beta_0 + \beta_1 t$. The right panel of Figure 2.2 shows the scatter plot of $\ln N_t$ versus t, which shows a clear linear trend.

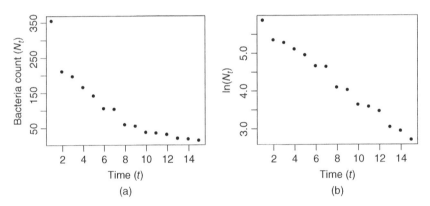

Figure 2.2. *Plots of N_t versus t (a) and $\ln N_t$ versus t (b).*

For the log-transformed data with $x = t$ and $y = \ln N_t$, we can calculate the following statistics:

$$\overline{x} = 8.0, \quad \overline{y} = 4.226, \quad S_{xx} = 280, \quad S_{yy} = 13.516, \quad S_{xy} = -61.159.$$

So

$$\widehat{\beta}_1 = \frac{-61.159}{280} = -0.218 \quad \text{and} \quad \widehat{\beta}_0 = 4.226 + (0.218)(8.0) = 5.973.$$

Thus, the fitted LS line is $\widehat{\ln N_t} = 5.973 - 0.218t$. So $\widehat{N}_0 = e^{5.973} = 392.68$ and the fitted exponential decay model is

$$\widehat{N}_t = 392.68 e^{-0.218t} \quad \text{for} \quad t \geq 0.$$

In radioactivity applications, it is of interest to estimate the half-life, denoted by $t_{0.5}$, which is the time at which the intensity of radioactivity, or in the present case, the count of surviving bacteria, will be 50% of the initial count. From the above model, we see that $N_t/N_0 = 0.5$ when $\ln 0.5 = bt$, so $t_{0.5} = \ln 0.5/b$. Therefore, the estimated half-life is $\widehat{t}_{0.5} = \ln 0.5/(-0.218) = (-0.693)/(-0.218) = 3.180$ time units, which is $3.180 \times 6 = 19.08$ minutes. $\qquad\square$

2.1.3 Fitted values and residuals

To check the model assumptions and detecting outliers, it is useful to calculate the **fitted values** and **residuals** given by

$$\widehat{y}_i = \widehat{\beta}_0 + \widehat{\beta}_1 x_i \quad \text{and} \quad e_i = y_i - \widehat{y}_i \quad (i = 1, \ldots, n), \tag{2.7}$$

respectively. Figure 2.3 shows these two quantities. It can be shown that the residuals satisfy the following two linear constraints:

$$\sum_{i=1}^n e_i = 0 \quad \text{and} \quad \sum_{i=1}^n x_i e_i = 0. \tag{2.8}$$

Thus, only $n-2$ of the n residuals are linearly independent. Hence, the **error degrees of freedom (error d.f.)** associated with the residuals is defined to be $n-2$. From the first equation above, it follows that $\sum_{i=1}^n y_i = \sum_{i=1}^n \widehat{y}_i$; thus, the average of the fitted values equals $\overline{y}$.

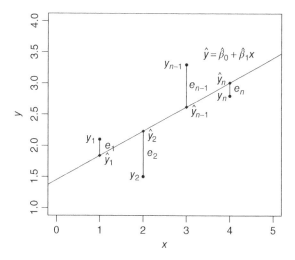

Figure 2.3. *LS straight line fit.*

Residuals are useful for checking the goodness of fit and diagnose model violations. We will study model diagnostic uses of residuals in Chapter 4.

2.1.4   Assessing goodness of fit

If the straight line model is correct, then the residuals obtained after filtering out the fitted straight line should be randomly distributed around zero. This can be assessed by plotting the residuals versus x_i's as shown in the following example.

Example 2.5 *(Bacteria Counts: Residual Plots)* We saw in Figure 2.2 that the plot of N_t versus t is curved, while the plot of $\ln N_t$ versus t is linear, both with negative slopes. The corresponding residual plots are shown in Figure 2.4. Notice a highly curved residual plot in Figure 2.4a and a random residual plot in Figure 2.4b. A lesson to draw here is that the residual plot shows departures from the fitted straight line much more vividly than the scatter plot of y versus x since the linear part of the relationship has been filtered out.

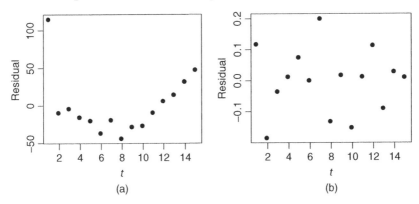

Figure 2.4. *Plots of residuals from LS fits of N_t versus t (a) and $\ln N_t$ vs. t (b).*

□

A numerical measure of goodness of fit of the LS line is obtained by comparing the residual variation of the y_i's around the LS line, referred to as the **error sum of squares (SSE)** with the total variation in the y_i's around their mean $\overline{y}$, referred to as the **total sum of squares** (SST). The difference between SST and SSE is called the **regression sum of squares (SSR)**, as it represents the part of the total variation of the y_i's that is accounted for by regression of the y_i's on the x_i's.

To see this relationship, we write the deviation of each y_i around $\overline{y}$ as the sum of two parts:

$$y_i - \overline{y} = (y_i - \widehat{y}_i) + (\widehat{y}_i - \overline{y}) = (\widehat{y}_i - \overline{y}) + e_i \quad (i = 1, \dots, n). \qquad (2.9)$$

If we square both sides and sum over $i = 1, \dots, n$, it turns out that the cross-product term, $2\sum_{i=1}^{n}(\widehat{y}_i - \overline{y})e_i$ cancels. So

$$\underbrace{\sum_{i=1}^{n}(y_i - \overline{y})^2}_{\text{SST}} = \underbrace{\sum_{i=1}^{n}(\widehat{y}_i - \overline{y})^2}_{\text{SSR}} + \underbrace{\sum_{i=1}^{n}e_i^2}_{\text{SSE}}. \qquad (2.10)$$

Note that SSE is just the minimum value of the LS criterion Q in (2.1). We refer to (2.10) as the **analysis of variance (ANOVA) identity**.

The proportion of variation accounted for by the LS line is given by

$$R^2 = \frac{\text{SSR}}{\text{SST}} = 1 - \frac{\text{SSE}}{\text{SST}}, \tag{2.11}$$

called the **coefficient of determination**. Note that $0 \le R^2 \le 1$ and a higher value indicates a better fit.

In the Technical Notes section, we derive the formula $\text{SSR} = \widehat{\beta}_1^2 S_{xx}$. Then SSE can be computed from $\text{SST} - \text{SSR} = S_{yy} - \widehat{\beta}_1^2 S_{xx}$. These calculations are illustrated in the following example.

Example 2.6 (Bacteria Counts: Goodness of Fit) Using the calculations from Example 2.4, we get

$$\text{SST} = S_{yy} = 13.516, \quad \text{SSR} = \widehat{\beta}_1^2 S_{xx} = (-0.218)^2(280) = 13.359.$$

Hence, $R^2 = \text{SSR}/\text{SST} = 13.359/13.516 = 98.8\%$. For a straight line fit of N_t versus t, we can similarly compute $R^2 = 82.3\%$. Thus, the log model provides a significantly better fit than the straight line model.   □

Table 2.3. Anscombe data sets

No.	Data Set I		Data Set II		Data Set III		Data Set IV	
	x	y	x	y	x	y	x	y
1	10	8.04	10	9.14	10	7.46	8	6.58
2	8	6.95	8	8.14	8	6.77	8	5.76
3	13	7.58	13	8.74	13	12.74	8	7.71
4	9	8.81	9	8.77	9	7.11	8	8.84
5	11	8.33	11	9.26	11	7.81	8	8.47
6	14	9.96	14	8.10	14	8.84	8	7.04
7	6	7.24	6	6.13	6	6.08	8	5.25
8	12	10.84	12	9.13	12	8.15	8	5.56
9	7	4.82	7	7.26	7	6.42	8	7.91
10	5	5.68	5	4.74	5	5.73	8	6.89
11	4	4.26	4	3.10	4	5.39	19	12.50

Source: Anscombe (1973).

Example 2.7 *(Anscombe Data: Scatter Plots)* Anscombe (1973) constructed four different bivariate data sets shown in Table 2.3 such that the same LS line fits all four data sets. The scatter plots are given in Figure 2.5 along with their LS fitted lines. The scatter plot for Data Set I shows a straight line trend. The scatter plot for Data Set II is a parabola. All points in the scatter plot for Data Set III follow an almost perfect straight line except one large outlier. Finally, all observations in Data Set IV are at $x = 8$ except one at $x = 19$. Yet, the same LS line, namely $\widehat{y} = 3.0 + 0.5x$, fits all four data sets. Not only is the LS line the same, but all statistics discussed in the following sections, e.g. the t-statistics for the regression coefficients, are the same for all four data sets. Thus we would not be able to tell these data sets apart if we just fit the straight lines to them without looking at the scatter plots.

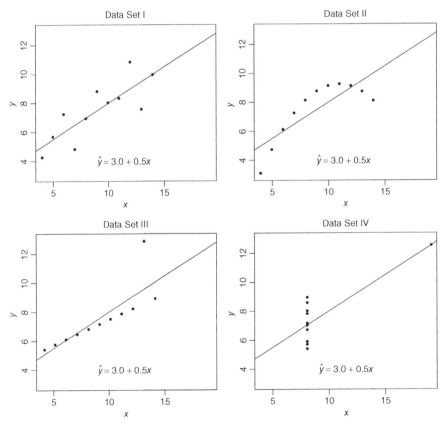

Figure 2.5. *Scatter plots for Anscombe data sets with LS fitted lines.*

What is the explanation of this seemingly bizarre result? We will see in what follows that all simple linear regression statistics depend on the raw data only through the following five summary statistics, whose values are the same for the four data sets:

$$\overline{x} = 9.00, \quad \overline{y} = 7.50, \quad S_{xx} = 110.0, \quad S_{yy} = 41.27, \quad S_{xy} = 55.00.$$

Therefore it is important to visualize the data first by making appropriate plots before computing various statistics.   □

2.2 Statistical inferences for simple linear regression

2.2.1 Simple linear regression model

To test hypotheses or compute confidence intervals (CIs) on β_0 and β_1, we need to assume a probability model for the data. The standard normal theory model for simple linear regression is

$$y_i = \beta_0 + \beta_1 x_i + \varepsilon_i \quad (i = 1, \dots, n), \tag{2.12}$$

where the ε_i's are assumed to be independent and identically distributed (i.i.d.) $N(0, \sigma^2)$ **random errors** and σ^2 is an unknown **error variance**. This model is shown graphically in Figure 2.6. It follows that the y_i's are independent normally distributed random variables with means and variance given by

$$E(y_i) = \mu_i = \beta_0 + \beta_1 x_i \quad \text{and} \quad \text{Var}(y_i) = \sigma^2 \quad (i = 1, \dots, n). \tag{2.13}$$

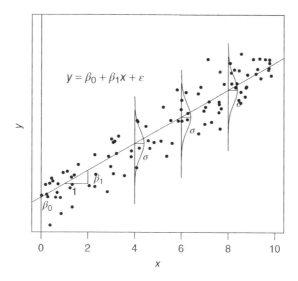

Figure 2.6. *Simple linear regression model.*

There are four assumptions implicit in this model. They are **normality**, constant variance (called **homoscedasticity**), **independence** and **linearity** of $E(y)$ with respect to x. In addition, we assume that the x_i's are nonrandom. In Chapter 4, we will discuss residual plots and other methods to check these assumptions and how to deal with any violations. The inferential methods given in the following sections are strictly valid only under these assumptions.

2.2.2   Inferences on β_0 and β_1

The sampling distributions of the LS estimators $\widehat{\beta}_0$ and $\widehat{\beta}_1$ are normal with $E(\widehat{\beta}_0) = \beta_0$ and $E(\widehat{\beta}_1) = \beta_1$ (i.e. they are unbiased), and

$$\mathrm{Var}(\widehat{\beta}_0) = \frac{\sigma^2 \sum_{i=1}^n x_i^2}{n S_{xx}} \quad \text{and} \quad \mathrm{Var}(\widehat{\beta}_1) = \frac{\sigma^2}{S_{xx}}.$$

To estimate the error variance σ^2, we compute the sample variance of the residuals, which have $n - 2$ d.f. as noted before. The mean of the residuals is zero, so the estimate of σ^2 equals

$$s^2 = \frac{\sum_{i=1}^n e_i^2}{n - 2} = \frac{\mathrm{SSE}}{n - 2} = \mathrm{MSE}, \qquad (2.14)$$

where MSE stands for **mean square error**. Thus, $s = \sqrt{\mathrm{MSE}}$ is the **root mean square error (RMSE)** of the residuals, which is used to estimate σ. Replacing σ^2 by its estimate s^2 in the formulae for the variances of $\widehat{\beta}_0$ and $\widehat{\beta}_1$ and taking the square roots, we get their estimated standard deviations, called the **estimated standard errors** or simply the **standard errors (SEs)**:

$$\mathrm{SE}(\widehat{\beta}_0) = s\sqrt{\frac{\sum x_i^2}{n S_{xx}}} \quad \text{and} \quad \mathrm{SE}(\widehat{\beta}_1) = \frac{s}{\sqrt{S_{xx}}}.$$

It then follows that

$$\frac{\widehat{\beta}_0 - \beta_0}{\mathrm{SE}(\widehat{\beta}_0)} \sim t_{n-2} \quad \text{and} \quad \frac{\widehat{\beta}_1 - \beta_1}{\mathrm{SE}(\widehat{\beta}_1)} \sim t_{n-2},$$

where t_{n-2} denotes Student's t-distribution with $n - 2$ d.f. Thus, $100(1 - \alpha)\%$ **confidence intervals (CIs)** on β_0 and β_1 are given by

$$\widehat{\beta}_0 \pm t_{n-2,\alpha/2} \mathrm{SE}(\widehat{\beta}_0) \quad \text{and} \quad \widehat{\beta}_1 \pm t_{n-2,\alpha/2} \mathrm{SE}(\widehat{\beta}_1),$$

where $t_{\nu,\alpha/2}$ is the $100(1 - \alpha/2)$th percentile (also called the upper $\alpha/2$ critical point) of the t-distribution with ν d.f. These critical points are tabulated in Table D.2.

The significance of the linear component of the relationship between y and x can be assessed by testing $H_0 : \beta_1 = 0$ versus $H_1 : \beta_1 \neq 0$. The test statistic is

$$t = \frac{\widehat{\beta}_1}{\text{SE}(\widehat{\beta}_1)} = \frac{\widehat{\beta}_1 \sqrt{S_{xx}}}{s}. \tag{2.15}$$

The α-level test rejects H_0 if

$$|t| > t_{n-2,\alpha/2}. \tag{2.16}$$

Equivalently, we reject H_0 if the P-value of the test statistic is less than α.

Example 2.8 (Bacteria Counts: Inferences on Intercept and Slope Coefficients) To calculate the standard errors of $\widehat{\beta}_0$ and $\widehat{\beta}_1$, we first calculate $s = \text{RMSE} = \sqrt{0.0121} = 0.110$. Also, $S_{xx} = 280$ as calculated before. Hence,

$$\text{SE}(\widehat{\beta}_0) = 0.110\sqrt{\frac{1240}{15 \times 280}} = 0.0598 \quad \text{and} \quad \text{SE}(\widehat{\beta}_1) = \frac{0.110}{\sqrt{280}} = 0.0066.$$

So the t-statistics for β_0 and β_1 are

$$\beta_0 : t = \frac{5.973}{0.0598} = 99.92 \quad \text{and} \quad \beta_1 : t = \frac{-0.218}{0.0066} = -33.22,$$

both of which are highly significant $(P < 0.001)$.

A 95% CI on β_1 is given by

$$\widehat{\beta}_1 \pm t_{13,0.025}\text{SE}(\widehat{\beta}_1) = -0.218 \pm (2.160)(0.0066) = [-0.232, -0.204].$$

From this, we get the following 95% CI on the half-life $t_{0.5} = (\ln 0.5)/\beta_1$:

$$\left[\frac{\ln 0.5}{-0.232}, \frac{\ln 0.5}{-0.204}\right] = \left[\frac{-0.693}{-0.232}, \frac{-0.693}{-0.204}\right] = [2.987, 3.397].$$

$\square$

2.2.3 Analysis of variance for simple linear regression

The purpose of the ANOVA is to partition the total variation in the response variable y into independent components so that each component can be attributed to a separate source of variation. In the case of simple linear regression, there are only two sources of variation, namely the variation caused by different x-values through their linear relationship with y and the residual or error variation. The ANOVA identity (2.10) gives this decomposition.

The total d.f. is always $n-1$ since SST measures the variation of the y_i's around their mean $\overline{y}$, just as we use $n-1$ d.f. to calculate the sample variance of a sample of size n around its mean. Corresponding to the ANOVA identity (2.10), the partitioning of this total d.f. is as follows:

$$\underbrace{n-1}_{\text{Total d.f.}} = \underbrace{1}_{\text{Regression d.f.}} + \underbrace{n-2}_{\text{Error d.f.}},$$

where the regression d.f. $= 1$ because the regression equation has one predictor variable and the error d.f. $= n - 2$ as explained before. A **sum of squares (SS)** divided by its d.f. is referred to as a **mean square (MS)**. Thus, the **mean square regression (MSR)** equals SSR/1 and the **mean square error (MSE)** equals SSE/$(n - 2)$ as defined in (2.14).

It can be shown that under the null hypothesis $H_0 : \beta_1 = 0$, the ratio $F = $ MSR/MSE has an F-distribution with 1 and $n - 2$ d.f. Thus, an α-level test rejects H_0 if

$$F = \frac{\text{MSR}}{\text{MSE}} > f_{1,n-2,\alpha}, \tag{2.17}$$

where $f_{1,n-2,\alpha}$ is the upper α critical point of this F-distribution. These critical points are tabulated in Table D.4. The calculations of the sums of squares and mean squares are presented in the **ANOVA** shown in Table 2.4. The t-test given in (2.16) to test $H_0 : \beta_1 = 0$ is equivalent to the F-test because $F = t^2$ (since MSR = SSR = $\widehat{\beta}_1^2 S_{xx}$ and MSE = s^2) and $f_{1,\nu,\alpha} = t_{\nu,\alpha/2}^2$. See the example below.

Table 2.4. ANOVA table for simple linear regression

Source	SS	d.f.	MS	F
Regression	SSR	1	MSR	$\frac{\text{MSR}}{\text{MSE}}$
Error	SSE	$n - 2$	MSE	
Total	SST	$n - 1$		

Example 2.9 *(**Bacteria Counts: Analysis of Variance**)* The ANOVA for the straight line fit to the log-transformed bacteria data is shown in Table 2.5. We see that $F = 1103.70$ with 1 and 13 d.f. is highly significant ($P < 0.001$). Thus, there is a significant linear component to the fit. Note that $F = 1103.70 = t^2 = (-33.22)^2$ and $f_{1,13,0.01} = 9.07 = t_{13,0.005}^2 = (3.012)^2$. $\square$

2.2.4   Pure error versus model error

We have used MSE as an estimator of σ^2, but it is unbiased only if the model is correctly specified. Otherwise, the misspecified part of the model contaminates and inflates this estimator. Therefore, we refer to MSE as the **model error**

Table 2.5. ANOVA table for regression of ln(Bacteria count) on time

Source	SS	d.f.	MS	F	P
Regression	13.359	1	13.359	1103.70	0.000
Error	0.157	13	0.0121		
Total	13.516	14			

estimator. We can rarely be certain that the specified model is correct, especially when it is an empirical model. Therefore, the only way to obtain an unbiased estimator of σ^2 (called the **pure error estimator**) is by taking independent repeat observations at each x_i and then pooling the sample variances among these repeat observations (assuming homoscedasticity). This is generally possible only in designed experiments where repeat observations should be planned for this purpose. The pure error estimator can also be used to do a **lack of fit (LOF) test** to check if the model is misspecified. Exercise 2.13 gives the details of the LOF test.

2.2.5 Prediction of future observations

Having fitted the LS line $\widehat{y} = \widehat{\beta}_0 + \widehat{\beta}_1 x$, we often want to use it to predict a future value y_{new} for a given x_{new}. For example, suppose a straight line is fitted to data on tread wear (y) as a function of mileage (x) for a particular brand and make of a car tire. A consumer may want to use this LS line to know the amount of tread wear at 50,000 miles for the tire that she purchases so that if she will need to replace the tire before its 50,000 mile warranty. This is a **prediction problem**. There is a related **estimation problem** that the tire manufacturer faces, namely estimate the mean amount of tread wear at 50,000 miles for *all* tires of that brand and make. The two problems are different because in the prediction problem we want to predict a future *random* outcome y_{new}, the actual amount of tread wear for a *random* tire, while in the estimation problem, we want to estimate an unknown *fixed mean* $\mu_{\text{new}} = E(y_{\text{new}}) = \beta_0 + \beta_1 x_{\text{new}}$, the mean amount of tread wear for *all* tires. The **calibration problem** is the inverse of the estimation problem in which we want to calculate a confidence interval for x_{new} for a given μ_{new}; this is discussed in Exercise 2.13.

In both prediction and estimation problems, we use the same formula for the predictor or the estimator:

$$\widehat{y}_{\text{new}} = \widehat{\mu}_{\text{new}} = \widehat{\beta}_0 + \widehat{\beta}_1 x_{\text{new}}. \tag{2.18}$$

The difference arises when we want to calculate an interval around it. For estimating a fixed parameter we use a CI, while for predicting a random outcome, we use a **prediction interval (PI)**. A $100(1 - \alpha)\%$ CI for μ_{new} is given by

$$\widehat{\mu}_{\text{new}} \pm t_{n-2,\alpha/2} \text{SE}(\widehat{\mu}_{\text{new}}), \tag{2.19}$$

where

$$\text{SE}(\widehat{\mu}_{\text{new}}) = s\sqrt{\frac{1}{n} + \frac{(x_{\text{new}} - \overline{x})^2}{S_{xx}}}.$$

A $100(1 - \alpha)\%$ PI for y_{new} is given by

$$\widehat{y}_{\text{new}} \pm t_{n-2,\alpha/2}\sqrt{s^2 + \text{SE}^2(\widehat{\mu}_{\text{new}})} = \widehat{y}_{\text{new}} \pm t_{n-2,\alpha/2}s\sqrt{1 + \frac{1}{n} + \frac{(x_{\text{new}} - \overline{x})^2}{S_{xx}}}. \tag{2.20}$$

The extra s^2 term added to $\text{SE}^2(\widehat{\mu}_{\text{new}})$ is the estimate of the additional variance σ^2 due to the random outcome y_{new}. As we shall see in the example below, a PI is generally much wider than a CI. Also notice that both the PI and the CI become increasingly wider as x_{new} moves away from $\overline{x}$ reflecting the fact that they become increasingly less precise. Additional error may be introduced if we try to predict y_{new} for x_{new} outside the range of the observed x's since the fitted model may not hold. This is called **extrapolation**, which should be generally avoided.

Example 2.10 *(Cardiac Output Measurements: Prediction and Confidence Intervals)* The scatter plot of the cardiac output data is shown in Figure 2.7 along with the LS fitted line $\widehat{y} = -0.528 + 1.014x$. The estimated standard deviation equals $s = 0.495$ with 24 d.f. and $R^2 = 90.6\%$, which indicates a strong linear relationship.

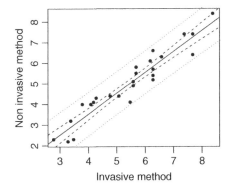

Figure 2.7. *Scatter plot of cardiac outputs measured with invasive and noninvasive methods with the LS line superimposed. The dashed curves around the LS line give pointwise CIs, while the dotted curves give pointwise PIs. Notice that the PIs are much wider than the CIs.*

Now, suppose that we want to estimate the range of noninvasive method y_{new}-values that are likely to be observed for $x_{new} = 6$ liters per minute using the invasive method. Then, we would need to calculate a PI. The predicted value $\widehat{y}_{new}$ for $x_{new} = 6$ equals

$$\widehat{y}_{new} = -0.528 + 1.014 \times 6 = 5.556.$$

To compute a 95% PI for y_{new}, we first calculate $\bar{x} = 5.469$ and $S_{xx} = 55.225$ and note that $t_{24,0.025} = 2.064$. Then a 95% PI can be calculated as

$$5.556 \pm 2.064 \times 0.495 \sqrt{1 + \frac{1}{26} + \frac{(6 - 5.469)^2}{55.225}}$$
$$= [5.556 \pm 1.044] = [4.512, 6.600].$$

If the true cardiac output using the invasive method is 6 liters per minute, then the noninvasive method reading would fall in this interval with 95% confidence.

For the sake of comparison, we calculate a 95% CI for μ_{new} if the expected cardiac output measured by the noninvasive method if $x_{new} = 6$ liters per minute. This CI is given by

$$5.556 \pm 2.064 \times 0.495 \sqrt{\frac{1}{26} + \frac{(6 - 5.469)^2}{55.225}}$$
$$= [5.556 \pm 0.213] = [5.343, 5.769].$$

Note that the confidence interval is much narrower than the prediction interval. The R output below gives the same results.

```
> cardiac=read.csv("c:/data/cardiac.csv")
> fit=lm(Noninvasive~Invasive,cardiac)
> summary(fit)
Call: lm(formula = Noninvasive ~ Invasive, data = cardiac)

Coefficients:
            Estimate Std. Error t value Pr(>|t|)
(Intercept) -0.52783    0.37685  -1.401    0.174
Invasive     1.01353    0.06658  15.222 7.88e-14 ***
---
Signif. codes:  0 '***' 0.001 '**' 0.01 '*' 0.05 '.'
0.1 ' ' 1

Residual standard error: 0.4947 on 24 degrees of freedom Multiple
R-squared:  0.9061,    Adjusted R-squared:  0.9022 F-statistic:
231.7 on 1 and 24 DF,   p-value: 7.876e-14

> predict(fit,newdata=data.frame(Invasive=6.0),interval="predict")
       fit        lwr        upr
```

```
1 5.553334 4.510229 6.596439
> predict(fit,newdata=data.frame(Invasive=6.0),interval="confidence")
       fit      lwr      upr
1 5.553334 5.340211 5.766457
```

$\square$

2.3   Correlation analysis

In correlation analysis, we assume that both x and y are random variables (in contrast to regression analysis where x is assumed to be nonrandom) with a joint distribution. The population correlation coefficient ρ of this joint distribution is defined as

$$\rho = \frac{\sigma_{xy}}{\sigma_x \sigma_y},$$

where σ_{xy} is the population covariance[1] between x and y, and σ_x and σ_y are the population standard deviations of x and y. It can be shown that ρ is between -1 and $+1$.

The sample estimate of ρ, called the **Pearson correlation coefficient** and denoted by r, can be obtained by replacing σ_{xy}, σ_x and σ_y by their following sample estimates:

$$s_{xy} = \frac{\sum_{i=1}^{n}(x_i - \overline{x})(y_i - \overline{y})}{n-1} = \frac{S_{xy}}{n-1},$$

$$s_x = \sqrt{\frac{\sum_{i=1}^{n}(x_i - \overline{x})^2}{n-1}} = \sqrt{\frac{S_{xx}}{n-1}},$$

$$s_y = \sqrt{\frac{\sum_{i=1}^{n}(y_i - \overline{y})^2}{n-1}} = \sqrt{\frac{S_{yy}}{n-1}}.$$

Thus, the Pearson correlation coefficient is given by

$$r = \frac{S_{xy}}{\sqrt{S_{xx}S_{yy}}}. \tag{2.21}$$

It can be shown that $-1 \le r \le +1$ and the sign of r is that of S_{xy} or equivalently that of $\widehat{\beta}_1 = S_{xy}/S_{xx}$. A positive sign indicates an increasing relationship while a negative sign indicates a decreasing relationship.

The R^2 defined in (2.11) is directly related to r as $R^2 = r^2$ which follows from

$$r^2 = \frac{S_{xy}^2}{S_{xx}S_{yy}} = \frac{\widehat{\beta}_1^2 S_{xx}}{S_{yy}} = \frac{\text{SSR}}{\text{SST}} = R^2.$$

---

[1]The population covariance is defined as $\sigma_{xy} = E[(x - \mu_x)(y - \mu_y)]$, where μ_x and μ_y are the expected values of x and y.

Keep in mind that both ρ and its sample estimate r are measures only of the linear association between x and y and not of any other type of association (e.g. general monotone association). Another point to keep in mind is that correlation does not equal causation. Thus, two variables being correlated does not necessarily mean that one variable is causing the other variable. The correlation between two variables may be spurious, caused by a third variable, called a **lurking variable**. The following example illustrates this phenomenon.

Example 2.11 *(Chocolate Consumption and Nobel Laureates)* Messerli (2012) showed a highly significant correlation $(r = 0.791, P < 0.001)$ between per capita chocolate consumption and per capita Nobel laureates for 23 countries. Media jumped on this study quite uncritically with headlines such as "Eat Chocolate, Win the Nobel Prize," in *Reuters*, "Study Links Eating Chocolates to Winning Nobel Prizes," in *USA Today* and "Chocolate and Nobel Prizes Linked in Study," in *Forbes*. Besides many flaws in the study (see, e.g. McClintock et al. 2014) such as the data on chocolate consumption and Nobel prizes are from different time periods, and there is no evidence that the Nobel laureates themselves ate a lot of chocolates, there is a key lurking variable, namely how affluent the countries are. Obviously, developed and affluent countries can afford to spend on a nonstaple food such as chocolates and also invest in research which leads to Nobel prizes. □

Often, false causal effect is claimed when the observed effect is the result of regression to the mean phenomenon discussed in Example 2.3. A classic example of this was given by Tversky and Kahnemann (1973) in which the instructors in a flight school adopted a policy of positive reinforcement (e.g. praise) after each successful flight maneuver. Unfortunately, the flight instructors found that the performance typically declined at the next flight maneuver. From this they concluded that high praise has a negative effect on fliers' performance. However, Tversky and Kahnemann showed that the explanation lies in regression to the mean phenomenon, which results in an outcome variable regressing toward the mean, thus a high value is often followed by a low value. Such changes are not caused by any policy change. Another example in the same vein is when one is trying to predict the final exam score from the midterm exam score of a student. If the correlation between the two scores is about 0.5, then as can be seen from Example 2.3, students who score higher than average on the midterm will tend to score higher than average on the final but not as high. Similarly, students who score lower than average on the midterm will tend to score lower than average on the final but not as low. To conclude from this that students who score higher than average on the midterm tend to slack off and students who score lower than average on the midterm work harder would be wrong. Such an argument is called **regression fallacy**.

2.3.1   Bivariate normal distribution*

Next, we shall discuss statistical inference on ρ. First, we assume a probability model for the joint distribution of (x, y). Generalizing the bell-shaped curve of the univariate normal p.d.f. to a bivariate normal p.d.f. gives a bell-shaped surface plotted over the (x, y)-plane. Whereas the univariate normal distribution has two parameters, the mean μ and variance σ^2, the bivariate normal distribution has five parameters, the means μ_x and μ_y, the variances σ_x^2 and σ_y^2, and the covariance σ_{xy} or equivalently the correlation coefficient ρ between x and y.

The bivariate normal p.d.f. is given by

$$f(x, y) = \frac{1}{2\pi\sigma_x\sigma_y\sqrt{1-\rho^2}} \exp\left\{-\frac{1}{2(1-\rho^2)}[u^2 + v^2 - 2\rho uv]\right\} \qquad (2.22)$$

for $-\infty < u, v < +\infty$, where

$$u = \frac{x - \mu_x}{\sigma_x} \quad \text{and} \quad v = \frac{y - \mu_y}{\sigma_y}.$$

Note that this p.d.f. becomes degenerate when $\rho = \pm 1$, i.e. when y is a deterministic linear function of x, say $y = a + bx$ for some constants a and $b \neq 0$. In that case the p.d.f. is concentrated on that line instead of being distributed over the (x, y)-plane. Hence, we restrict the range of ρ to $-1 < \rho < 1$.

The following are some useful properties of the bivariate normal distribution.

1. The marginal distributions of x and y are $N(\mu_x, \sigma_x^2)$ and $N(\mu_y, \sigma_y^2)$, respectively.

2. The conditional distribution of y conditioned on x, which is obtained by dividing the joint distribution (2.22) by the $N(\mu_x, \sigma_x^2)$ p.d.f., is normal with mean and variance given by

$$E(y|x) = \mu_y + \frac{\rho\sigma_y}{\sigma_x}(x - \mu_x) \quad \text{and} \quad \text{Var}(y|x) = \sigma_y^2(1 - \rho^2).$$

3. If we put

$$\beta_0 = \mu_y - \frac{\rho\sigma_y}{\sigma_x}\mu_x, \quad \beta_1 = \frac{\rho\sigma_y}{\sigma_x} \quad \text{and,} \quad \sigma^2 = \sigma_y^2(1 - \rho^2),, \qquad (2.23)$$

then we see that the conditional distribution of y is normal with conditional mean $\beta_0 + \beta_1 x$ and conditional variance σ^2. This is exactly the simple linear regression model (2.13), which can be derived from the bivariate normal distribution of (x, y) as the conditional distribution of y conditioned on x.

4. The conditional variance σ^2 of y is smaller than the unconditional variance σ_y^2 by a factor of $(1 - \rho^2)$. If $\rho = \pm 1$, then the conditional variance of y is zero because in that case, if we fix x, then y is fixed according to the deterministic linear relationship: $y = a + bx$.

2.3.2 Inferences on correlation coefficient*

Inferences on ρ are based on the sample correlation coefficient r. When $\rho = 0$, it can be shown that

$$t = \frac{r\sqrt{n-2}}{\sqrt{1-r^2}} \sim t_{n-2}. \tag{2.24}$$

This can be used as a t-statistic to test $H_0 : \rho = 0$. In fact, as can be easily verified, this t-statistic is algebraically identical to the t-statistic (2.15) used to test $H_0 : \beta_1 = 0$. This is not surprising since from (2.23) we know that β_1 is proportional to ρ and so $\beta_1 = 0$ if and only if $\rho = 0$.

To test a more general hypothesis such as $H_0 : \rho = \rho_0$, where $\rho_0 \neq 0$, or to obtain a CI on ρ, we need the so-called noncentral distribution of r when $\rho \neq 0$. This distribution is complicated and not amenable to easy manipulation. To get around this difficulty, Sir R.A. Fisher (1890–1962) suggested the following transformation of ρ and its sample estimate,

$$\psi = \frac{1}{2} \ln \left(\frac{1+\rho}{1-\rho} \right) \quad \text{and} \quad \widehat{\psi} = \frac{1}{2} \ln \left(\frac{1+r}{1-r} \right).$$

He showed that $\widehat{\psi}$ is asymptotically normal with

$$E(\widehat{\psi}) \approx \psi = \frac{1}{2} \ln \left(\frac{1+\rho}{1-\rho} \right) \quad \text{and} \quad \text{Var}(\widehat{\psi}) \approx \frac{1}{n-3}. \tag{2.25}$$

So an approximate large sample $100(1 - \alpha)\%$ CI on ψ is given by

$$\widehat{\psi} \pm z_{\alpha/2} \frac{1}{\sqrt{n-3}} = [L, U] \quad \text{(say)}. \tag{2.26}$$

From (2.25), we have $\rho = [e^{2\psi} - 1]/[e^{2\psi} + 1]$, which is a monotone function of ψ. Hence, by substituting the lower and upper confidence limits on ψ, namely L and U, in the formula for ρ yields the following confidence limits on ρ:

$$\left[\frac{e^{2L} - 1}{e^{2L} + 1}, \frac{e^{2U} - 1}{e^{2U} + 1} \right]. \tag{2.27}$$

Example 2.12 *(Cardiac Output Measurements: Inference on Correlation Coefficient)* Suppose that for the noninvasive method to be acceptable, its correlation coefficient ρ with the invasive method must be greater than 0.90. The sample correlation coefficient for these data can be computed to be $r = 0.952$, which exceeds 0.90, but we need to test whether it is significantly greater than 0.90. In other words, we need to test the one-sided hypothesis $H_0 : \rho \leq 0.90$ versus $H_1 : \rho > 0.90$ or equivalently $H_0 : \psi \leq \psi_0$ versus $H_1 : \psi > \psi_0$, where

$$\psi_0 = \frac{1}{2} \ln \left(\frac{1 + 0.90}{1 - 0.90} \right) = 1.472.$$

The test will be based on

$$\widehat{\psi} = \frac{1}{2} \ln \left(\frac{1 + 0.952}{1 - 0.952} \right) = 1.853.$$

The test statistic equals

$$z = \frac{\widehat{\psi} - \psi_0}{\sqrt{1/(n-3)}} = \frac{1.853 - 1.472}{\sqrt{1/(26-3)}} = 1.827.$$

The one-sided P-value of this statistic equals 0.034, which shows that H_0 can be rejected at the 0.05 level, and hence, we can conclude that $\rho > 0.90$.

We obtain the same result by calculating a lower 95% confidence bound on ρ. First we calculate a lower 95% confidence bound on ψ:

$$L = \widehat{\psi} - z_{0.05} \frac{1}{\sqrt{n-3}} = 1.853 - 1.645 \frac{1}{\sqrt{26-3}} = 1.510.$$

Hence, the corresponding lower confidence bound on ρ is

$$\frac{e^{2 \times 1.510} - 1}{e^{2 \times 1.510} + 1} = 0.907.$$

Since this lower confidence bound exceeds 0.90, we can reject H_0 at the 0.05 level and conclude that $\rho > 0.90$. So the noninvasive method is acceptable.   □

2.4   Modern extensions*

We have seen that the simple linear regression model considered in this chapter can be applied to exponential and power families of models by using the log transformation. Polynomial regression models of the type $y = \beta_0 + \beta_1 x + \cdots + \beta_p x^p + \varepsilon$ are special cases of the multiple regression model studied in the next chapter. Usually, we limit to second or third degree (quadratic or cubic) polynomials. More generally, one can specify p functions, $f_1(x), \ldots, f_p(x)$, called the **basis functions**, and fit the model $y = \beta_0 + \beta_1 f_1(x) + \cdots + \beta_p f_p(x) + \varepsilon$.

More complex regression models $y = f(x) + \varepsilon$ involve $f(x)$ to be piecewise linear or piecewise polynomial. In this case, the range of x is divided into a number of intervals and a separate linear or polynomial regression model is fitted over each interval. The endpoints of the intervals where adjoining fitted functions meet are called **knots**. These knots may be pre-specified or may be estimated from data. When fitting piecewise polynomials another problem needs to be addressed, namely the adjoining polynomials must be constrained to join smoothly at the knots. For this purpose, the technique of **regression splines** is used. **Cubic splines** are particularly common in use. **Smoothing splines** is a related technique which minimizes the LS criterion subject to a smoothness penalty. These approaches provide more flexibility but are somewhat more difficult to apply.

2.5 Technical notes*

Many of the following derivations can be obtained as special cases of the corresponding derivations for multiple regression in Chapter 3 using matrix methods. Nonetheless, it is useful to give them here without using matrix methods.

2.5.1 Derivation of the LS estimators

To minimize the LS criterion Q with respect to β_0 and β_1, we take the first partial derivatives of Q and set them equal to 0 resulting in the following **normal equations**:

$$\frac{\partial Q}{\partial \beta_0} = -2 \sum_{i=1}^{n} [y_i - (\beta_0 + \beta_1 x_i)] = 0 \Rightarrow \beta_0 n + \beta_1 \sum_{i=1}^{n} x_i = \sum_{i=1}^{n} y_i \qquad (2.28)$$

and

$$\frac{\partial Q}{\partial \beta_1} = -2 \sum_{i=1}^{n} x_i [y_i - (\beta_0 + \beta_1 x_i)] = 0 \Rightarrow \beta_0 \sum_{i=1}^{n} x_i + \beta_1 \sum_{i=1}^{n} x_i^2 = \sum_{i=1}^{n} x_i y_i. \qquad (2.29)$$

An easy way to solve these equations is to rewrite the simple linear regression model (2.12) by centering the x_i's as

$$y_i = (\beta_0 + \beta_1 \overline{x}) + \beta_1 (x_i - \overline{x}) + \varepsilon_i = \beta_0' + \beta_1' x_i' + \varepsilon_i \quad (i = 1, \ldots, n),$$

where

$$\beta_0' = \beta_0 + \beta_1 \overline{x}, \beta_1' = \beta_1 \text{ and } x_i' = x_i - \overline{x}.$$

Then we can apply the above normal equations to estimate β_0' and β_1'. Note that the equations simplify since $\sum_{i=1}^{n} x_i' = \sum_{i=1}^{n}(x_i - \overline{x}) = 0$. Thus, from (2.28), we get

$$\beta_0' n = \sum_{i=1}^{n} y_i \;\Rightarrow\; \widehat{\beta}_0' = \overline{y} \;\Rightarrow\; \widehat{\beta}_0 = \overline{y} - \widehat{\beta}_1 \overline{x}.$$

Similarly, from (2.29), we get

$$\widehat{\beta}_1' \sum_{i=1}^{n} x_i'^2 = \sum_{i=1}^{n}(x_i - \overline{x})y_i \;\Rightarrow\; \widehat{\beta}_1 = \frac{\sum_{i=1}^{n}(x_i - \overline{x})(y_i - \overline{y})}{\sum_{i=1}^{n}(x_i - \overline{x})^2} = \frac{S_{xy}}{S_{xx}}.$$

In the above, we have used the fact that $\sum_{i=1}^{n}(x_i - \overline{x})y_i = \sum_{i=1}^{n}(x_i - \overline{x})(y_i - \overline{y})$ since $\overline{y}\sum_{i=1}^{n}(x_i - \overline{x}) = 0$. These are the formulae (2.2) for $\widehat{\beta}_0$ and $\widehat{\beta}_1$.

2.5.2   Sums of squares

First note that SST $= \sum_{i=1}^{n}(y_i - \overline{y})^2 = S_{yy}$. Next note that

$$\begin{aligned}
\text{SSE} &= \sum_{i=1}^{n}(y_i - \widehat{y}_i)^2 \\
&= \sum_{i=1}^{n}(y_i - \widehat{\beta}_0 - \widehat{\beta}_1 x_i)^2 \\
&= \sum_{i=1}^{n}[y_i - \overline{y} - \widehat{\beta}_1(x_i - \overline{x})]^2 \quad (\text{by putting } \widehat{\beta}_0 = \overline{y} - \widehat{\beta}_1 \overline{x}) \\
&= \sum_{i=1}^{n}(y_i - \overline{y})^2 - 2\widehat{\beta}_1 \sum_{i=1}^{n}(x_i - \overline{x})(y_i - \overline{y}) + \widehat{\beta}_1^2 \sum_{i=1}^{n}(x_i - \overline{x})^2 \\
&= S_{yy} - 2\widehat{\beta}_1 S_{xy} + \widehat{\beta}_1^2 S_{xx} \\
&= S_{yy} - \widehat{\beta}_1^2 S_{xx},
\end{aligned}$$

where we have used the formula $\widehat{\beta}_1 = S_{xy}/S_{xx}$. Finally,

$$\text{SSR} = \text{SST} - \text{SSE} = S_{yy} - S_{yy} + \widehat{\beta}_1^2 S_{xx} = \widehat{\beta}_1^2 S_{xx} = \frac{S_{xy}^2}{S_{xx}}.$$

2.5.3   Distribution of the LS estimators

Both $\widehat{\beta}_0$ and $\widehat{\beta}_1$ are linear functions of the y_i's and hence, are normally distributed since the y_i's are normally distributed. Using the fact that the x_i's are nonrandom, the expected values and variances of $\widehat{\beta}_0$ and $\widehat{\beta}_1$ can be calculated as follows. First note that $\widehat{\beta}_1$ can be written as

$$\widehat{\beta}_1 = \sum_{i=1}^{n} c_i y_i \quad \text{where} \quad c_i = \frac{x_i - \overline{x}}{S_{xx}} \quad (1 \leq i \leq n).$$

The c_i's satisfy

$$\sum_{i=1}^{n} c_i = 0 \quad \text{and} \quad \sum_{i=1}^{n} c_i^2 = \frac{\sum_{i=1}^{n} (x_i - \overline{x})^2}{S_{xx}^2} = \frac{1}{S_{xx}}.$$

Hence,

$$\sum_{i=1}^{n} c_i x_i = \frac{1}{S_{xx}} \sum_{i=1}^{n} (x_i - \overline{x}) x_i = \frac{1}{S_{xx}} \sum_{i=1}^{n} (x_i - \overline{x})(x_i - \overline{x}) = \frac{1}{S_{xx}} \sum_{i=1}^{n} (x_i - \overline{x})^2 = 1.$$

It follows that

$$E(\widehat{\beta}_1) = \sum_{i=1}^{n} c_i E(y_i) = \sum_{i=1}^{n} c_i (\beta_0 + \beta_1 x_i) = \beta_0 \sum_{i=1}^{n} c_i + \beta_1 \sum_{i=1}^{n} c_i x_i = \beta_1.$$

Substituting this result in the formula for $\widehat{\beta}_0$, we get

$$E(\widehat{\beta}_0) = E(\overline{y} - \widehat{\beta}_1 \overline{x}) = E(\overline{y}) - E(\widehat{\beta}_1)\overline{x} = \beta_0 + \beta_1 \overline{x} - \beta_1 \overline{x} = \beta_0.$$

Next,

$$\mathrm{Var}(\widehat{\beta}_1) = \sum_{i=1}^{n} c_i^2 \mathrm{Var}(y_i) = \sigma^2 \sum_{i=1}^{n} c_i^2 = \frac{\sigma^2}{S_{xx}}.$$

Finally, using a not-so-difficult-to-prove result (see Exercise 2.5) that $\overline{y}$ and $\widehat{\beta}_1$ are independent, we can write

$$\mathrm{Var}(\widehat{\beta}_0) = \mathrm{Var}(\overline{y}) + \overline{x}^2 \mathrm{Var}(\widehat{\beta}_1) = \frac{\sigma^2}{n} + \frac{\sigma^2 \overline{x}^2}{S_{xx}} = \sigma^2 \left(\frac{1}{n} + \frac{\overline{x}^2}{S_{xx}} \right) = \frac{\sigma^2 \sum_{i=1}^{n} x_i^2}{n S_{xx}}.$$

Therefore, it follows that

$$\frac{\widehat{\beta}_0 - \beta_0}{\sigma \sqrt{\sum_{i=1}^{n} x_i^2 / n S_{xx}}} \sim N(0,1) \quad \text{and} \quad \frac{\widehat{\beta}_1 - \beta_1}{\sigma \sqrt{1/S_{xx}}} \sim N(0,1).$$

Replacing σ by s results in t_{n-2} random variables.

2.5.4 Prediction interval

Note that $\widehat{y}_{\text{new}} = \widehat{\mu}_{\text{new}} = \widehat{\beta}_0 + \widehat{\beta}_1 x_{\text{new}}$ is normally distributed since it is a linear combination of $\widehat{\beta}_0$ and $\widehat{\beta}_1$, which are normally distributed. Furthermore,

$$E(\widehat{y}_{\text{new}}) = E(\widehat{\mu}_{\text{new}}) = \beta_0 + \beta_1 x_{\text{new}} = \mu_{\text{new}}$$

since $E(\widehat{\beta}_0) = \beta_0$ and $E(\widehat{\beta}_1) = \beta_1$. Next, again using the fact that $\overline{y}$ and $\widehat{\beta}_1$ are independent, we can write

$$\begin{aligned}
\text{Var}(\widehat{y}_{\text{new}}) = \text{Var}(\widehat{\mu}_{\text{new}}) &= \text{Var}(\widehat{\beta}_0 + \widehat{\beta}_1 x_{\text{new}}) \\
&= \text{Var}[\overline{y} + \widehat{\beta}_1(x_{\text{new}} - \overline{x})] \\
&= \text{Var}(\overline{y}) + (x_{\text{new}} - \overline{x})^2 \text{Var}(\widehat{\beta}_1) \\
&= \sigma^2 \left[\frac{1}{n} + \frac{(x_{\text{new}} - \overline{x})^2}{S_{xx}} \right].
\end{aligned}$$

Therefore,

$$\frac{\widehat{\mu}_{\text{new}} - \mu_{\text{new}}}{\sigma \sqrt{\frac{1}{n} + \frac{(x_{\text{new}} - \overline{x})^2}{S_{xx}}}} \sim N(0, 1).$$

Replacing σ by s results in a t_{n-2} random variable from which the CI (2.19) for μ_{new} follows.

To obtain the PI (2.20) for y_{new}, we note that

$$E(\widehat{y}_{\text{new}} - y_{\text{new}}) = \mu_{\text{new}} - \mu_{\text{new}} = 0 \text{ and } \text{Var}(\widehat{y}_{\text{new}} - y_{\text{new}})$$

$$= \text{Var}(\widehat{y}_{\text{new}}) + \text{Var}(y_{\text{new}}) = \sigma^2 \left[\frac{1}{n} + \frac{(x_{\text{new}} - \overline{x})^2}{S_{xx}} \right] + \sigma^2.$$

Therefore,

$$\frac{\widehat{y}_{\text{new}} - y_{\text{new}}}{\sigma \sqrt{1 + \frac{1}{n} + \frac{(x_{\text{new}} - \overline{x})^2}{S_{xx}}}} \sim N(0, 1),$$

and the rest follows as before.

Exercises

Theoretical Exercises

2.1 **(Regression through the origin)** In some applications, the LS line is required to pass through the origin, e.g. fuel consumption as a function of the weight of the car. So we can assume the intercept to be zero. Show that the LS estimator of the slope β, obtained by minimizing the LS criterion $Q = \sum_{i=1}^{n} (y_i - \beta x_i)^2$, equals

$$\widehat{\beta} = \frac{\sum_{i=1}^{n} x_i y_i}{\sum_{i=1}^{n} x_i^2}.$$

2.2 (Properties of residuals) Show that the residuals satisfy the two constraints (2.8). Further show that in the case of regression through the origin, the first constraint $\sum_{i=1}^{n} e_i = 0$ is not necessarily satisfied.

2.3 (Weighted least squares) Suppose that the observations y_i have different precisions and so we would like to weight them using different weights $w_i > 0$. For example, each y_i may be a sample mean of n_i i.i.d. observations so that their variances are inversely proportional to the n_i. Hence, we use the n_i as the weights. Show that the **weighted least squares (WLS)** estimator of the slope β for regression through the origin, obtained by minimizing the WLS criterion $Q = \sum_{i=1}^{n} w_i(y_i - \beta x_i)^2$, equals

$$\widehat{\beta} = \frac{\sum_{i=1}^{n} w_i x_i y_i}{\sum_{i=1}^{n} w_i x_i^2}.$$

2.4 (Omitted variables) Suppose that there are two predictor variables, x_1 and x_2, but we fit the straight line model $y = \beta_0 + \beta_1 x_1 + \varepsilon$ omitting x_2. If, in fact, the true model is $y = \beta_0 + \beta_1 x_1 + \beta_2 x_2 + \varepsilon$, show that

$$E(\widehat{\beta}_1) = \beta_1 + \beta_2 \sum_{i=1}^{n} c_{i1} x_{i2} = \beta_1 + \beta_2 r_{12}\left(\frac{s_2}{s_1}\right),$$

where $c_{i1} = (x_{i1} - \overline{x}_1)/S_{11}$, $S_{11} = \sum_{i=1}^{n} (x_{i1} - \overline{x}_1)^2$, r_{12} is the sample correlation coefficient between x_1 and x_2 and s_1, s_2 are the sample SDs of x_1, x_2, respectively. Thus, $\widehat{\beta}_1$ is biased with the bias given by the second term in the above expression. Under what condition is this bias zero?

2.5 (Independence between $\overline{y}$ and $\widehat{\beta}_1$) Show that $\overline{y}$ and $\widehat{\beta}_1$ are independent by showing that

$$\mathrm{Cov}(\overline{y}, \widehat{\beta}_1) = \frac{1}{S_{xx}} \sum_{i=1}^{n} (x_i - \overline{x})\mathrm{Cov}(\overline{y}, y_i) = 0.$$

Independence between $\overline{y}$ and $\widehat{\beta}_1$ follows since they are jointly normally distributed. (Two uncorrelated random variables are independent if they are jointly normally distributed.)

2.6 (Calibration problem) Suppose that in the tire wear problem mentioned in Section 2.2.5, the tire manufacturer wants to estimate the mileage for a given mean tread wear, e.g. the mileage which corresponds to the minimum acceptable tread depth. Thus, we are given μ_{new} and we want to estimate the corresponding x_{new}.

(a) Give a natural point estimate $\widehat{x}_{\mathrm{new}}$ of x_{new}.

(b) Construct a CI for x_{new} using **Fieller's method** by following the steps below. Define a random variable

$$t = \frac{\widehat{\beta}_0 + \widehat{\beta}_1 x_{\mathrm{new}} - \mu_{\mathrm{new}}}{s\sqrt{1/n + (x_{\mathrm{new}} - \overline{x})^2/S_{xx}}},$$

which is t_{n-2} distributed. Simplify the inequality $t^2 \le t^2_{\nu,\alpha/2} = f_{1,\nu,\alpha}$, to show that a $(1 - \alpha)$-level CI for x_{new} is given by the two roots of the quadratic equation:

$$Ax_{\text{new}}^2 + Bx_{\text{new}} + C = 0,$$

where

$$A = \widehat{\beta}_1^2 - \frac{u^2}{S_{xx}}, \quad B = 2\left(\frac{u^2 \bar{x}}{S_{xx}} - \widehat{\beta}_1^2 \widehat{x}_{\text{new}}\right),$$

$$C = (\widehat{\beta}_1 \widehat{x}_{\text{new}})^2 - u^2\left(\frac{1}{n} + \frac{\bar{x}^2}{S_{xx}}\right),$$

and $u = s t_{n-2,\alpha/2}$.

(c) For the cardiac output data in Table 2.2, calculate $\widehat{x}_{\text{new}}$ and a 95% CI for x_{new} if $\mu_{\text{new}} = 6$ liters per minute. Use any values needed for this calculation from Example 2.10.

2.7 (*t*-**Statistic for testing** $\rho = 0$) Show that the *t*-statistic (2.24) for testing $\rho = 0$ is algebraically identical to the *t*-statistic (2.15) for testing $\beta_1 = 0$.

Applied Exercises

2.8 (**Regression to the mean**) Refer to Example 2.3. For the tall and short fathers considered in that example, calculate the expected heights of their sons for $r = 0.25$ and $r = 0.75$. What do you conclude?

2.9 (**Beta coefficients of stocks**) The β of a stock is a coefficient that describes how the return on that stock is related to the return on a diversified stock portfolio. It is the slope coefficient in the simple linear regression model, $y = \alpha + \beta x$, where y is the return on that stock and x is the return on a benchmark stock market index representing a diversified portfolio. In this exercise, we want to compare the β's of IBM and Apple with reference to S&P 500.

The file `IBM-Apple-SP500 RR Data.csv` contains data on percentage monthly rates of return (adjusted for dividends and stock splits) from February 2005 until September 2013 for IBM, Apple and S&P 500. These rates were calculated by downloading historical monthly prices from the Yahoo Finance website (http://finance.yahoo.com/). (You may download the current data instead.)

(a) Make scatter plots of rates of return of IBM versus S&P 500 and of Apple versus S&P 500 and comment on them.

(b) Calculate the β's for IBM and Apple with reference to S&P 500. Comment on the relative magnitudes of the β's. Which stock had a higher expected return relative to S&P 500?

(c) Calculate the sample standard deviations (SDs) of the rates of return for S&P 500, IBM and Apple. Also calculate the correlation matrix. Check that $\hat{\beta} = rs_y/s_x$ for each stock where r is the correlation coefficient between S&P 500 and the given stock, s_x is the sample SD of S&P 500 and s_y is the sample SD of the given stock.

(d) Explain based on the statistics calculated how a higher expected return is accompanied by higher volatility of the Apple stock.

2.10 (Price elasticities of steaks) Data file `steakprices.csv` gives time series data on the prices and quantities sold of three types of beef steaks, chuck, porterhouse and rib eye, (*Source*: http://www.aabri.com/manuscripts/08118.pdf).

(a) Estimate the price elasticities of all three steaks. Given that chuck is the least expensive cut and porter house is the most expensive cut of beef among these three cuts, are their price elasticities in the expected order?

(b) Estimate how much the demand will change if the price is increased by 10% for each cut.

2.11 (Smoking versus cancer) Data file `smoking-cancer.csv` contains data from 43 states and Washington, D.C., on the average number of cigarettes smoked (hundreds/capita) and number of deaths per 100,000 population due to four types of cancer (bladder, lung, kidney, and leukemia).

(a) Make scatter plots of the number of deaths due to each type of cancer versus cigarettes smoked to see what types of relationships (linear, nonlinear) exist and if there are any outliers.

(b) Perform tests on the correlations to see which type of cancer deaths are most significantly correlated with cigarette smoking.

2.12 (Spearman rank correlation coefficient) The Pearson correlation coefficient measures only the extent of linear association between x and y; it does not measure the degree of monotone (increasing or decreasing) nonlinear association. Spearman's rank correlation coefficient (denoted by r_S) measures monotone association. It is simply the Pearson correlation coefficient between the ranks assigned to the original data. Let $u_i = \text{rank}(x_i)$ and $v_i = \text{rank}(y_i)$ $(i = 1, \ldots, n)$, where average ranks are assigned to tied observations. Then $r_S = r_{uv}$. Compute r_S for the cardiac output data in Table 2.2 and compare it with the Pearson correlation coefficient calculated for the same data in Example 2.12.

2.13 **(Lack of fit test)** Suppose that the data are collected at $m \geq 2$ distinct x-values, $x_1, \ldots, x_m$ and at least at one x_i there are $n_i \geq 2$ repeat independent observations, $y_{i1}, \ldots, y_{in_i}$. The total number of observations is $\sum_{i=1}^{m} n_i = n$. From each group of repeat observations, we can compute the **pure error sum of squares (SSPE)** as $\text{SSPE} = \sum_{i=1}^{m} \sum_{j=1}^{n_i} (y_{ij} - \overline{y}_i)^2$. The pure error d.f equal $\sum_{i=1}^{m}(n_i - 1) = n - m$. The pure error estimator is given by $s^2 = \text{MSPE} = \text{SSPE}/(n - m)$. This is an unbiased estimator of σ^2 regardless of which particular model is fitted. The total SSE can be partitioned into $\text{SSE} = \text{SSPE} + \text{SSLOF}$. The lack of fit d.f. is $= (n - 2) - (n - m) = m - 2$ and so $\text{MSLOF} = \text{SSLOF}/(m - 2)$. Then $F = \text{MSLOF}/\text{MSPE}$ can be used to perform an F-test of lack of fit. If $F > f_{m-2,n-m,\alpha}$, we conclude that there is a significant at lack of fit.

Test the lack of fit of a straight line to the data given in Table 2.6 on weight loss due to corrosion as a function of iron content in 90/10 Cu-Ni alloy specimens.

Table 2.6. Corrosion weight loss data

Iron content (%)	Weight loss	Iron content (%)	Weight loss
0.01	127.6	0.95	103.9
0.01	130.1	1.19	101.5
0.01	128.0	1.44	92.3
0.48	124.0	1.44	91.4
0.48	122.0	1.96	83.7
0.71	110.8	1.96	86.2
0.71	113.1		

Source: Draper and Smith (1998). Reproduced with permission of John Wiley & Sons.

Chapter 3

Multiple linear regression: basics

Multiple regression is a basic and essential tool of predictive modeling and serves as a foundation for more advanced techniques. It takes into account the effects of multiple predictors simultaneously. Running separate simple linear regressions on each predictor can give misleading results since each such regression accounts only for the direct bivariate relationship between the response variable and that predictor. Thus, a simple linear regression may show strong relationship, but it may be because of other intervening variables. When these other variables are included in the model via multiple regression, the apparently strong relationship may be shown to be actually weak or nonexistent or even with an opposite sign.

To illustrate the various concepts and techniques associated with multiple regression, we will use one small data set with only two predictors and 40 data points and the other relatively large data set with more than a dozen predictors and over 800 data points. The first example involves both prediction and inference, while the second example mainly involves prediction.

Example 3.1 *(College GPA and Entrance Test Scores: Data)* College admission committees are faced with the task of who to admit from thousands of applicants. There are many considerations including their college entrance test scores, high school grade point average (GPA) and rank, essays, extracurricular activities, recommendation letters, and so on. In this example, we will look at predicting just one measure of academic success in college, namely graduating GPA.

Predictive Analytics: Parametric Models for Regression and Classification Using R.
First Edition. Ajit C. Tamhane.
© 2021 John Wiley & Sons, Inc. Published 2021 by John Wiley & Sons, Inc.
Companion website: www.wiley.com/go/tamhane/predictiveanalytics

We will consider two predictors of that success, namely, college entrance Verbal and Math test scores. Data on 40 students are shown in Table 3.1 and are stored in the file GPA.csv. Some questions of interest are the following: (i) what is the relative importance of each test score in predicting the GPA and (ii) what is a good prediction model for GPA in terms of these two predictors. □

Table 3.1. Entrance test scores and graduating GPA

Verbal	Math	GPA	Verbal	Math	GPA	Verbal	Math	GPA
81	87	3.49	83	76	3.75	97	80	3.27
68	99	2.89	64	66	2.70	77	90	3.47
57	86	2.73	83	72	3.15	49	54	1.3
100	49	1.54	93	54	2.28	39	81	1.22
54	83	2.56	74	59	2.92	87	69	3.23
82	86	3.43	51	75	2.48	70	95	3.82
75	74	3.59	79	75	3.45	57	89	2.93
58	98	2.86	81	62	2.76	74	67	2.83
55	54	1.46	50	69	1.90	87	93	3.84
49	81	2.11	72	70	3.01	90	65	3.01
64	76	2.69	54	52	1.48	81	76	3.33
66	59	2.16	65	79	2.98	84	69	3.06
80	61	2.60	56	78	2.58			
100	85	3.30	98	67	2.73			

Source: McClave and Dietrich (1994, p. 811).

Example 3.2 *(Used Car Prices: Data)* This example deals with building a prediction model for used car prices from a number of predictors listed in Table 3.2. Retail prices of 2005 General Motors (GM) cars were calculated from the data provided in the 2005 Central Edition of the *Kelly Blue Book*. All cars in this data set were less than one year old and in excellent condition. Data are in file usedcarprices.csv. Questions of interest are the following: (i) which are the best predictors of the used car price and (ii) what are their relative contributions to the price. □

Table 3.2. Variables for the used car prices example

Variable	Description
Price	Suggested retail price (response variable)
Mileage	Odometer reading in thousands of miles
Make	Division of GM (Buick, Cadillac, Chevrolet, Pontiac, SAAB, Saturn)
Model	Specific model of a given make
Trim	Variations on body types (e.g. Sedan 4D, LS Sedan 4D, Coupe 2D)
Type	Body type (convertible, coupe, hatchback, sedan, wagon)
Cylinders	Number of cylinders
Liters	Size of engine
Doors	Number of doors
Cruise	Indicator variable for cruise control (1 = cruise control)
Sound	Indicator variable for upgraded speakers (1 = upgraded)
Leather	Indicator variable for leather seats (1 = leather)

Source: Kuiper (2008).

3.1 Multiple linear regression model

3.1.1 Model in scalar notation

Denote the response variable by y and the predictor variables by $x_1, \ldots, x_p$. Suppose that we have n complete data vectors $(x_{i1}, \ldots, x_{ip}, y_i)$ $(i = 1, \ldots, n)$ on these variables from which we want to fit the model:

$$y_i = \beta_0 + \beta_1 x_{i1} + \cdots + \beta_p x_{ip} + \varepsilon_i \quad (i = 1, \ldots, n). \tag{3.1}$$

Here $\beta_0, \beta_1, \ldots, \beta_p$ are unknown **parameters**, β_0 is the **intercept** or the **constant term**, $\beta_1, \ldots, \beta_p$ are the **regression coefficients** and the ε_i are independent and identically distributed (i.i.d.) $N(0, \sigma^2)$ **random errors**. From this model, it follows that the y_i are independent $N(\mu_i, \sigma^2)$, where

$$\mu_i = E(y_i) = \beta_0 + \beta_1 x_{i1} + \cdots + \beta_p x_{ip} \quad (i = 1, \ldots, n). \tag{3.2}$$

Note the following points about this model:

- This is called a **linear model** because it is linear in the β parameters – not necessarily in the x's. Any nonlinear functions of the x's, e.g. x^2, $\log x$, or $x_1 x_2$, may be used as predictors, and the model is still regarded as linear. The reason that linearity in the β's is critical is that it makes the equations for finding their **least squares (LS) estimators** linear, which makes them easy to solve with closed form solutions. Furthermore, these LS estimators are then linear functions of the responses y_i's, which makes their sampling distributions simple and so inferences on them straightforward.

- Some nonlinear models can be transformed into linear models. We will see some examples in Section 3.2.5. An intrinsically **nonlinear model** is nonlinear in the β's and cannot be transformed into a linear form. An example from chemical process kinetics is the model

$$y = \frac{\beta_1}{\beta_1 - \beta_2} [e^{-\beta_2 t} - e^{-\beta_1 t}] + \varepsilon,$$

where y is the percent reaction completed, t is the reaction time, and β_1 and β_2 are reaction rate constants. No transformation can linearize this model and nonlinear regression techniques must be used to estimate the rate constants.

3.1.2   Model in matrix notation

By introducing the matrix notation below, the multiple regression model and the results associated with it (e.g. the LS estimators of the β's and their sampling distributions) can be written in a compact form and these results can be derived more easily and elegantly. We denote the vectors and matrices by bold letters (vectors by lower case letters and matrices by upper case letters); their elements are denoted by the corresponding unbolded lower case letters with a single subscript in case of vectors and double subscripts in case of matrices. All vectors are assumed to be column vectors. A prime on a vector or a matrix denotes its transpose. Generally, we suppress the dimensions of any vector or a matrix if they are clear from the context. When necessary, we indicate the dimension by a subscript.

Let

$$\boldsymbol{y} = \begin{bmatrix} y_1 \\ y_2 \\ \vdots \\ y_n \end{bmatrix}, \quad \boldsymbol{X} = \begin{bmatrix} 1 & x_{11} & \cdots & x_{1p} \\ 1 & x_{21} & \cdots & x_{2p} \\ \vdots & \vdots & \ddots & \vdots \\ 1 & x_{n1} & \cdots & x_{np} \end{bmatrix}, \quad \boldsymbol{\beta} = \begin{bmatrix} \beta_0 \\ \beta_1 \\ \vdots \\ \beta_p \end{bmatrix}, \quad \boldsymbol{\varepsilon} = \begin{bmatrix} \varepsilon_1 \\ \varepsilon_2 \\ \vdots \\ \varepsilon_n \end{bmatrix}.$$

Here $\boldsymbol{y}$ is the **response vector**, $\boldsymbol{X}$ is called the **model matrix**, $\boldsymbol{\beta}$ is the **parameter vector**, and $\boldsymbol{\varepsilon}$ is the **random error vector**. From (3.2), we can write $E(y_i) = \mu_i = \boldsymbol{x}_i' \boldsymbol{\beta}$ where $\boldsymbol{x}_i' = (1, x_{i1}, \dots, x_{ip})$ is the ith row vector of $\boldsymbol{X}$. Hence, it follows that $E(\boldsymbol{y}) = \boldsymbol{\mu} = \boldsymbol{X}\boldsymbol{\beta}$ where $\boldsymbol{\mu} = (\mu_1, \mu_2, \dots, \mu_n)'$ and the model (3.1) can be written in matrix notation as

$$\boldsymbol{y} = \boldsymbol{X}\boldsymbol{\beta} + \boldsymbol{\varepsilon}. \tag{3.3}$$

Another way of saying that the ε_i are i.i.d. $N(0, \sigma^2)$ is that the random vector $\boldsymbol{\varepsilon}$ has an n-variate normal distribution with mean vector $\boldsymbol{0}$ (the null vector of all 0's) and covariance matrix $\sigma^2 \boldsymbol{I}$, where $\boldsymbol{I}$ is an $n \times n$ identity matrix. The diagonal entries of $\sigma^2 \boldsymbol{I}$ are the variances σ^2 of the y_i's and the off-diagonal entries are the covariances between the pairs y_i and y_j, which are all 0 under the assumption of independence among the random errors ε_i's. The distribution of the response vector $\boldsymbol{y}$ is then the n-variate normal distribution with mean vector $\boldsymbol{\mu} = \boldsymbol{X}\boldsymbol{\beta}$ and covariance matrix $\sigma^2 \boldsymbol{I}$.

3.2 Fitting a multiple regression model

3.2.1 Least squares (LS) method

We extend the LS method for simple linear regression to multiple regression by minimizing the **LS criterion**:

$$Q = \sum_{i=1}^{n} [y_i - (\beta_0 + \beta_1 x_{i1} + \dots + \beta_p x_{ip})]^2. \tag{3.4}$$

w.r.t. $\beta_0, \beta_1, \dots, \beta_p$. We can set the partial derivatives of Q w.r.t. the β_j equal to zero resulting in $p + 1$ simultaneous linear equations (called the **normal equations**):

$$\frac{\partial Q}{\partial \beta_j} = -2 \sum_{i=1}^{n} x_{ij} [y_i - (\beta_0 + \beta_1 x_{i1} + \dots + \beta_p x_{ip})] = 0 \quad (j = 0, \dots, p), \tag{3.5}$$

where we have set $x_{i0} = 1$ for all i. We solve these equations for $\beta_0, \beta_1, \dots, \beta_p$. Unique solutions to these equations exist under certain conditions on the $\boldsymbol{X}$ matrix (discussed below). The resulting LS estimators can be more easily expressed in a closed form by writing the above LS criterion in matrix notation as follows:

$$Q = \sum_{i=1}^{n} \varepsilon_i^2 = \boldsymbol{\varepsilon}' \boldsymbol{\varepsilon} = (\boldsymbol{y} - \boldsymbol{X}\boldsymbol{\beta})'(\boldsymbol{y} - \boldsymbol{X}\boldsymbol{\beta}).$$

We can set the vector derivative of Q w.r.t. $\boldsymbol{\beta}$ equal to $\mathbf{0}$ (see Section 3.8) to obtain the normal equations in a compact form below:

$$(\boldsymbol{X'X})\boldsymbol{\beta} = \boldsymbol{X'y}. \tag{3.6}$$

If the inverse of $\boldsymbol{X'X}$ exists, then this equation has a unique solution $\widehat{\boldsymbol{\beta}} = (\widehat{\beta}_0, \widehat{\beta}_1, \ldots, \widehat{\beta}_p)'$ given by

$$\widehat{\boldsymbol{\beta}} = (\boldsymbol{X'X})^{-1}\boldsymbol{X'y}, \tag{3.7}$$

which is the LS estimator of the parameter vector $\boldsymbol{\beta}$. It can be shown that the $\widehat{\beta}_j$'s satisfy the equation

$$\overline{y} = \widehat{\beta}_0 + \widehat{\beta}_1\overline{x}_1 + \cdots + \widehat{\beta}_p\overline{x}_p. \tag{3.8}$$

In other words, the fitted regression plane passes through the centroid $(\overline{x}_1, \ldots, \overline{x}_p; \overline{y})$ of the data.

From linear algebra, it is known that the inverse of $\boldsymbol{X'X}$ exists if and only if the columns of $\boldsymbol{X}$ are linearly independent. These column vectors are the data vectors of the predictor variables. They are linearly independent if there are no linear relationships among the predictors. Otherwise, one or more of them can be expressed in terms of the others, and so we can reduce them to an independent set by eliminating the extra ones. For example, suppose that a data set has three predictors: income, expenditure, and saving. Since saving equals income minus expenditure, we do not need to keep all three; any two will suffice. When there are approximate or exact linear dependencies among the columns of $\boldsymbol{X}$, difficulties arise in the computation of $(\boldsymbol{X'X})^{-1}$ and hence of $\widehat{\boldsymbol{\beta}}$. This is called the **multicollinearity problem**, which we shall study in Chapter 4. In this chapter, we assume that the columns of $\boldsymbol{X}$ are linearly independent, so $(\boldsymbol{X'X})^{-1}$ exists and $\widehat{\boldsymbol{\beta}}$ is unique.

Are the LS estimators optimal in some sense? **Gauss–Markov theorem** stated and proved in Section 3.8 provides an answer to this question.

Example 3.3 (*College GPA and Entrance Test Scores: Multiple Regression*) Before running a regression of GPA on Verbal and Math scores, it is useful to study relationships among them by making scatter plots and computing the correlation matrix. Scatter plots between all three of them can be combined into a single composite plot called the **matrix scatter plot** shown in Figure 3.1. This plot and the correlation matrix below are obtained by using the following R commands.

```
> gpa = read.csv("c:/data/GPA.csv")
> plot(gpa)
> cor(gpa)
```

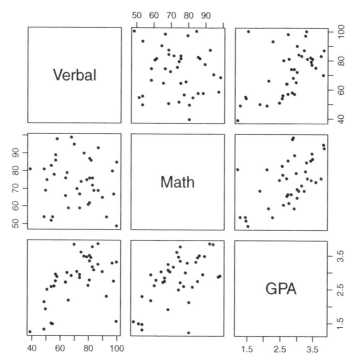

Figure 3.1. *Matrix scatter plot between GPA, Verbal score and Math score.*

We can see from this plot that Verbal and Math scores are fairly uncorrelated with each other while both are moderately correlated with GPA. Thus, the two predictors make relatively independent and roughly equal contributions to the GPA and it makes sense to fit a linear model for GPA that includes both of them.

The above visual impressions based on the matrix scatter plot are confirmed by the correlation matrix below.

$$\begin{array}{c} \\ \text{Verbal} \\ \text{Math} \\ \text{GPA} \end{array} \begin{array}{ccc} \text{Verbal} & \text{Math} & \text{GPA} \\ \left[\begin{array}{ccc} 1 & -0.107 & 0.529 \\ -0.107 & 1 & 0.573 \\ 0.529 & 0.573 & 1 \end{array}\right] \end{array}.$$

Using the lm function in R, we get the fitted equation given in the following output.

```
> lmfit = lm(GPA ~ Verbal + Math, data = gpa)
> summary(lmfit)

Coefficients:
            Estimate Std. Error t value Pr(>|t|)
(Intercept) -1.570537   0.493749  -3.181  0.00297 **
```

```
Verbal        0.025732    0.004024    6.395 1.83e-07 ***
Math          0.033615    0.004928    6.822 4.90e-08 ***
---
Signif. codes:  0 '***' 0.001 '**' 0.01 '*' 0.05 '.'
0.1 ' ' 1

Residual standard error: 0.4023 on 37 degrees of freedom Multiple
R-squared:  0.6811,    Adjusted R-squared:  0.6638 F-statistic:
39.51 on 2 and 37 DF,  p-value: 6.585e-10
```

Thus the regression equation fitted using R is

$$\widehat{\text{GPA}} = -1.5705 + 0.0257\text{Verbal} + 0.0336\text{Math}.$$

Next, we fit this model by hand calculation to illustrate the formula (3.7). The $\boldsymbol{X}$ matrix and the $\boldsymbol{y}$ vector have 40 rows corresponding to 40 observations:

$$\boldsymbol{X} = \begin{bmatrix} 1 & 81 & 87 \\ \vdots & \vdots & \vdots \\ 1 & 84 & 69 \end{bmatrix} \quad \text{and} \quad \boldsymbol{y} = \begin{bmatrix} 3.49 \\ \vdots \\ 3.06 \end{bmatrix}$$

from which we can compute

$$\boldsymbol{X}'\boldsymbol{X} = \begin{bmatrix} 40 & 2884 & 2960 \\ 2884 & 218\,048 & 212\,533 \\ 2960 & 212\,533 & 225\,782 \end{bmatrix}$$

and

$$\boldsymbol{X}'\boldsymbol{y} = \begin{bmatrix} 1 & \cdots & 1 \\ 81 & \cdots & 84 \\ 87 & \cdots & 69 \end{bmatrix} \begin{bmatrix} 3.49 \\ \vdots \\ 3.06 \end{bmatrix} = \begin{bmatrix} 110.89 \\ 8225.68 \\ 8409.77 \end{bmatrix}.$$

Next, we compute

$$(\boldsymbol{X}'\boldsymbol{X})^{-1} = \begin{bmatrix} 1.506 & -8.182 \times 10^{-3} & -1.205 \times 10^{-2} \\ -8.182 \times 10^{-3} & 1.000 \times 10^{-4} & 1.310 \times 10^{-5} \\ -1.205 \times 10^{-2} & 1.310 \times 10^{-5} & 1.500 \times 10^{-4} \end{bmatrix}.$$

Finally,

$$\widehat{\boldsymbol{\beta}} = \begin{bmatrix} 1.506 & -8.182 \times 10^{-3} & -1.205 \times 10^{-2} \\ -8.182 \times 10^{-3} & 1.000 \times 10^{-4} & 1.310 \times 10^{-5} \\ -1.205 \times 10^{-2} & 1.310 \times 10^{-5} & 1.500 \times 10^{-4} \end{bmatrix} \begin{bmatrix} 110.89 \\ 8225.68 \\ 8409.77 \end{bmatrix}$$

$$= \begin{bmatrix} -1.5705 \\ 0.0257 \\ 0.0336 \end{bmatrix}.$$

Thus $\widehat{\beta}_0 = -1.5705$, $\widehat{\beta}_1 = 0.0257$, and $\widehat{\beta}_2 = 0.0336$.   □

3.2.2 Interpretation of regression coefficients

The following points are important to keep in mind when interpreting the regression coefficients.

1. In general, the $\widehat{\beta}_j$'s depend on which other predictors are included in the model. If the other variables in the model change, then the $\widehat{\beta}_j$'s change, too. This is because the $\widehat{\beta}_j$'s measure the marginal contributions of the x_j's to y conditional on other variables in the model.

2. A variable can have an apparently large effect in presence of some variables but not in presence of other variables. Also, the signs of the $\widehat{\beta}_j$'s may change and thus the apparent direction of the effect of x_j on y may change if other variables in the model change.

3. The only exception to the above two items is when the columns of the $\boldsymbol{X}$ matrix are mutually orthogonal so that $\boldsymbol{X}'\boldsymbol{X}$ and its inverse are diagonal matrices. In this case, adding or deleting the variables from the model does not change the regression coefficients of the other variables; thus their contributions are independent of each other. This case typically occurs only in designed experiments when the so-called **orthogonal designs** are used; see Exercise 3.4.

4. The $\widehat{\beta}_j$'s have units, namely the units of y divided by the units of the x_j's. Therefore, the magnitudes of the $\widehat{\beta}_j$'s depend on the units of the x_j's and cannot be directly compared to each other. For example, suppose we fit a model for the gas mileage (mpg) of a car on two variables: engine size (l) and car weight (lb). Then the coefficients have the units of mpg per liter and mpg per lb, respectively. Comparing them is like comparing apples and oranges.

5. To address the problem of units, **standardized regression coefficients** may be used by fitting a regression model to standardized x's and y; see Section 3.6.3. Although the resulting coefficients are unitless, they inherit some of the same issues as the unstandardized regression coefficients. For instance, they also depend on other predictors included in the model.

3.2.3 Fitted values and residuals

Fitted values and residuals are useful to assess the goodness of the LS fit of the model to the observed data as well as for model diagnostics. This latter use will be discussed in Chapter 4. Here we focus on their use for assessing the goodness of fit.

The **fitted values** are defined as

$$\widehat{y}_i = \widehat{\beta}_0 + \widehat{\beta}_1 x_{i1} + \cdots + \widehat{\beta}_p x_{ip} = \boldsymbol{x}_i'\widehat{\boldsymbol{\beta}} \quad (i = 1, \ldots, n),$$

where $\boldsymbol{x}_i' = (1, x_{i1}, \ldots, x_{ip})$ is the ith row vector of $\boldsymbol{X}$ and the **fitted values vector** is defined as

$$\widehat{\boldsymbol{y}} = \begin{bmatrix} \widehat{y}_1 \\ \widehat{y}_2 \\ \vdots \\ \widehat{y}_n \end{bmatrix} = \begin{bmatrix} \boldsymbol{x}_1'\widehat{\boldsymbol{\beta}} \\ \boldsymbol{x}_2'\widehat{\boldsymbol{\beta}} \\ \vdots \\ \boldsymbol{x}_n'\widehat{\boldsymbol{\beta}} \end{bmatrix} = \begin{bmatrix} \boldsymbol{x}_1' \\ \boldsymbol{x}_2' \\ \vdots \\ \boldsymbol{x}_n' \end{bmatrix} \widehat{\boldsymbol{\beta}} = \boldsymbol{X}\widehat{\boldsymbol{\beta}}.$$

The **residuals** are defined as

$$e_i = y_i - \widehat{y}_i \quad (i = 1, \ldots, n),$$

and the **residual vector** is defined as:

$$\boldsymbol{e} = \begin{bmatrix} e_1 \\ e_2 \\ \vdots \\ e_n \end{bmatrix} = \begin{bmatrix} y_1 - \widehat{y}_1 \\ y_2 - \widehat{y}_2 \\ \vdots \\ y_n - \widehat{y}_n \end{bmatrix} = \boldsymbol{y} - \widehat{\boldsymbol{y}}.$$

Some students get confused between the random errors ε_i and the residuals e_i. The ε_i are *unobservable* random differences between the y_i and $E(y_i) = \mu_i = \boldsymbol{x}_i'\boldsymbol{\beta}$ (unobservable because $E(y_i)$ is unknown), while the e_i are observable (because they are computable) differences between the observed y_i and the fitted $\widehat{y}_i$, where $\widehat{y}_i = \widehat{\mu}_i = \boldsymbol{x}_i'\widehat{\boldsymbol{\beta}}$.

In matrix notation, we can write the random error vector and the residual vector as

$$\boldsymbol{\varepsilon} = \boldsymbol{y} - \boldsymbol{X}\boldsymbol{\beta} \quad \text{and} \quad \boldsymbol{e} = \boldsymbol{y} - \boldsymbol{X}\widehat{\boldsymbol{\beta}}.$$

Substituting $\widehat{\boldsymbol{\beta}} = (\boldsymbol{X}'\boldsymbol{X})^{-1}\boldsymbol{X}'\boldsymbol{y}$ from (3.7) in $\widehat{\boldsymbol{y}} = \boldsymbol{X}\widehat{\boldsymbol{\beta}}$, we get

$$\widehat{\boldsymbol{y}} = \boldsymbol{X}(\boldsymbol{X}'\boldsymbol{X})^{-1}\boldsymbol{X}'\boldsymbol{y} = \boldsymbol{H}\boldsymbol{y}, \tag{3.9}$$

where

$$\boldsymbol{H} = \boldsymbol{X}(\boldsymbol{X}'\boldsymbol{X})^{-1}\boldsymbol{X}' \tag{3.10}$$

is called the **hat matrix**. Hence, the residual vector equals

$$\boldsymbol{e} = \boldsymbol{y} - \widehat{\boldsymbol{y}} = (\boldsymbol{I} - \boldsymbol{H})\boldsymbol{y}, \tag{3.11}$$

where $\boldsymbol{I}$ is the $n \times n$ identity matrix. Thus, both $\widehat{\boldsymbol{y}}$ and $\boldsymbol{e}$ are linear transforms of the observed vector $\boldsymbol{y}$.

It can be checked that $\boldsymbol{H}$ and $(\boldsymbol{I} - \boldsymbol{H})$ are symmetric and satisfy $\boldsymbol{H}\boldsymbol{H} = \boldsymbol{H}$ and $(\boldsymbol{I} - \boldsymbol{H})(\boldsymbol{I} - \boldsymbol{H}) = \boldsymbol{I} - \boldsymbol{H}$. Such matrices are called **projection matrices**.

The $\boldsymbol{H}$ matrix projects $\boldsymbol{y}$ into $\widehat{\boldsymbol{y}}$ and the $\boldsymbol{I} - \boldsymbol{H}$ matrix projects $\boldsymbol{y}$ into $\boldsymbol{e}$. These two projections are orthogonal to each other since $\boldsymbol{H}(\boldsymbol{I} - \boldsymbol{H}) = \boldsymbol{O}$, where $\boldsymbol{O}$ is the null matrix and so $\widehat{\boldsymbol{y}}'\boldsymbol{e} = \boldsymbol{y}'\boldsymbol{H}(\boldsymbol{I} - \boldsymbol{H})\boldsymbol{y} = 0$. Geometric interpretation of LS estimation is given in Section 3.8.5 along with Figure 3.8 showing the vectors $\boldsymbol{y}, \widehat{\boldsymbol{y}}$ and $\boldsymbol{e}$.

Next note that

$$\boldsymbol{X}'\boldsymbol{e} = \boldsymbol{X}'(\boldsymbol{I} - \boldsymbol{H})\boldsymbol{y} = [\boldsymbol{X}' - \boldsymbol{X}'\boldsymbol{X}(\boldsymbol{X}'\boldsymbol{X})^{-1}\boldsymbol{X}']\boldsymbol{y} = \boldsymbol{O}\boldsymbol{y} = \boldsymbol{0},$$

where $\boldsymbol{0}$ is a $(p+1)$-vector of all 0's. So the residual vector $\boldsymbol{e}$ is orthogonal to every column of $\boldsymbol{X}$ or their dot products are all zero, i.e. $\sum_{i=1}^{n} x_{ij}e_i = 0$ for $j = 0, 1, \ldots, p$. In particular, the first column of $\boldsymbol{X}$ consists of all 1's and so $\sum_{i=1}^{n} e_i = 0$, i.e. the residuals sum to 0 if the intercept term is included in the regression model. There are $p + 1$ independent linear restrictions on $\boldsymbol{e}$ and therefore, the error degrees of freedom (d.f.) equals $n - (p+1)$.

The residuals are used to estimate the error variance σ^2 by

$$s^2 = \frac{\sum_{i=1}^{n} e_i^2}{n - (p+1)} = \frac{\text{SSE}}{n - (p+1)} = \text{MSE}, \tag{3.12}$$

where $\text{SSE} = \sum_{i=1}^{n} e_i^2$ is the **error sum of squares**, $\text{MSE} = \text{SSE}/[n - (p+1)]$ is the **mean square error** and $n - (p+1)$ is the **error degrees of freedom (d.f.)**. It can be shown that s^2 is an unbiased estimate of σ^2, i.e. $E(s^2) = \sigma^2$. So σ is estimated by the **root mean square error (RMSE)**: $s = \text{RMSE} = \sqrt{\text{MSE}}$.

3.2.4 Measures of goodness of fit

For measuring the goodness of the LS fit, we follow the same approach as in Chapter 2. Define $\overline{\boldsymbol{y}}$ as a vector of dimension n, all of whose entries are $\overline{y}$ or $\overline{\boldsymbol{y}} = \overline{y}\boldsymbol{1}$, where $\boldsymbol{1}$ is a vector of all 1's. Then we can write (2.9) in vector notation by expressing the deviation of the observation vector $\boldsymbol{y}$ from $\overline{\boldsymbol{y}}$ as the sum of two vectors:

$$\boldsymbol{y} - \overline{\boldsymbol{y}} = (\widehat{\boldsymbol{y}} - \overline{\boldsymbol{y}}) + (\boldsymbol{y} - \widehat{\boldsymbol{y}}) = (\widehat{\boldsymbol{y}} - \overline{\boldsymbol{y}}) + \boldsymbol{e}.$$

It can be shown that $(\widehat{\boldsymbol{y}} - \overline{\boldsymbol{y}})$ and $\boldsymbol{e}$ are orthogonal to each other (see Section 3.8). From the Pythagoras theorem, we get the analysis of variance (ANOVA) identity:

$$\underbrace{\|\boldsymbol{y} - \overline{\boldsymbol{y}}\|^2}_{\text{SST}} = \underbrace{\|\widehat{\boldsymbol{y}} - \overline{\boldsymbol{y}}\|^2}_{\text{SSR}} + \underbrace{\|\boldsymbol{e}\|^2}_{\text{SSE}}, \tag{3.13}$$

where, e.g. $\|\boldsymbol{e}\|^2 = \boldsymbol{e}'\boldsymbol{e} = \sum_{i=1}^{n} e_i^2$ is the squared length (norm) of the vector $\boldsymbol{e}$. As in Chapter 2, $\text{SST} = \|\boldsymbol{y} - \overline{\boldsymbol{y}}\|^2 = \sum_{i=1}^{n} (y_i - \overline{y})^2$ is the **total sum of squares (SST)** and $\text{SSR} = \|\widehat{\boldsymbol{y}} - \overline{\boldsymbol{y}}\|^2 = \sum_{i=1}^{n} (\widehat{y}_i - \overline{y})^2$ is the **regression sum of squares**

(SSR). Note that error sum of squares (SSE) is just the minimum value of the LS criterion Q in (3.4).

The proportion of variation in y accounted for by its regression on the x's is given by

$$R^2 = \frac{\text{SSR}}{\text{SST}} = 1 - \frac{\text{SSE}}{\text{SST}}, \qquad (3.14)$$

called the **multiple coefficient of determination**. Its positive square root R ($0 \leq R \leq 1$) is called the **multiple correlation coefficient**. Since R measures the extent of linear association between y and multiple x's, it cannot be assigned a sign as in the case of the bivariate correlation coefficient r; by convention we assign a positive sign to R. It can be shown that R is the Pearson correlation coefficient between the observed y_i's and the fitted $\widehat{y}_i$'s.

It should be pointed out that a high R^2 does not necessarily mean a better predictive model. To see this, observe that SSE can only decrease and hence R^2 can only increase when more predictors are added to the model, whether they are related to the response variable or not. The goal of a model is not to fit it to the data as closely as possible as it may result in **overfitting**. The goal is to capture the overall trend in the data with as simple a model as possible. Such a model is useful for prediction.

To obviate the drawback of R^2 that it increases with the number of predictors in the model, we use **adjusted** R^2, defined as

$$R_{\text{adj}}^2 = 1 - \frac{\text{SSE}/[n - (p + 1)]}{\text{SST}/(n - 1)}. \qquad (3.15)$$

Note that $\text{SST}/(n - 1)$ does not depend on p while both SSE and $n - (p + 1)$ decrease with p. So their ratio, which is the MSE, may increase or decrease as p increases. Typically, R_{adj}^2 initially increases with p but then decreases as additional predictors are added to the model as their marginal contributions diminish in magnitude. So maximizing R_{adj}^2 (or minimizing MSE) is a valid criterion for model selection. However, R_{adj}^2 does not have a simple interpretation like R^2 has as the proportion of variation in y explained by the fitted model, and it may even become negative.

3.2.5   Linearizing transformations

As we saw in Chapter 2, many nonlinear models can be transformed to linear models. In addition to the power and exponential laws mentioned there, **multiplicative laws** can also be linearized and made additive by using the log-transformation.

The **Cobb–Douglas production function** in economics used to model the output of a firm (O) as a function of the capital input (K) and labor input (L) provides an example of a multiplicative law. This function has the form

$O = \gamma K^\alpha L^\beta \varepsilon$, where α, β, and γ are unknown parameters to be estimated and ε is a multiplicative random error term. Make the log-transformation:

$$\ln O = \ln \gamma + \alpha \ln K + \beta \ln L + \ln \varepsilon$$

and note that this is in the linear model form.

Similar to the price elasticity of demand defined in Section 2.1.2,

$$\alpha = \frac{(\partial O/O)}{(\partial K/K)} \quad \text{and} \quad \beta = \frac{(\partial O/O)}{(\partial L/L)}$$

are called the capital and labor elasticities of output, respectively. Thus, α and β are the relative changes in the output due to unit relative changes in the corresponding inputs.

Log-transformation faces a difficulty when some responses are zero or negative. A standard practice in such cases is to add a common small positive constant to all responses so that they all become positive and then take their logs. In database marketing models for predicting customer purchases, a large majority of the customers who are offered a sales incentive (e.g. a promotional offer) do not respond and so their purchase amounts are zero. On the other hand, the distribution of the purchase amounts for customers who make purchases is highly right-skewed with some very large purchases. For such kinds of data, we can add 1 to all the purchase amounts and then take the logs, which transforms all zero purchase amounts back to zeros.

3.3 Statistical inferences for multiple regression

3.3.1 Analysis of variance for multiple regression

There are two sources of variation in y, namely the variation caused by the x's through their linear relationships with y and the random error or residual variation. The ANOVA identity in (3.13) gives this decomposition. Unless the column vectors of $\boldsymbol{X}$ are mutually orthogonal (i.e. the design is orthogonal), the contributions of the predictors to the variation in y are not independent of each other and cannot be partitioned into additive components, i.e. SSR cannot be expressed as the sum of the SSs due to the individual x's.

As in the case of simple linear regression, we can partition the total d.f. as follows:

$$\underbrace{n-1}_{\text{Total d.f.}} = \underbrace{p}_{\text{Regression d.f.}} + \underbrace{[n-(p+1)]}_{\text{Error d.f.}}.$$

Regression d.f. equals the number of predictor variables in the linear model ($p = 1$ for simple linear regression). The error d.f. equals $n - (p+1)$ because the n residuals are subject to $p + 1$ linear constraints as noted in Section 3.2.3.

The error d.f. $n - (p + 1)$ implies that n must be greater than $p + 1$, i.e. the number of observations must be greater than the number of the unknown β's to be estimated, in order for any d.f. to be available for estimating the error variance σ^2. If $n = p + 1$, then the error d.f. $= 0$ and SSE $= 0$ since we obtain an exact fit, i.e. all fitted $\widehat{y}_i$'s equal to the observed y_i's, and so all residuals e_i's are equal to 0. Such a model is called a **saturated model**. For example, for bivariate data $\{(x_i, y_i),\ i = 1,\ \ldots, n\}$ with no repeat observations on any x_i, we can fit an $(n - 1)$th degree polynomial in x which will exactly pass through all the n points giving an exact fit with SSE $= 0$, and hence $R^2 = 100\%$, but such a fit simply follows every wiggle in the data without giving any idea of the nature of relationship between y and x and the extent of random variation around that relationship.

How large must n be relative to p in order to have sufficient error d.f. available for estimating σ^2 accurately? Opinions vary on this question, but a rough rule of thumb is that the error d.f. should be ideally 30 or more.

A sum of squares (SS) divided by its d.f. is called a **mean square (MS)**. Thus, **mean square regression** (MSR) equals SSR/p and **mean square error (MSE)** equals SSE/$[n - (p + 1)]$, which is used to estimate σ^2 as defined in (3.12). It can be shown that under the null hypothesis $H_0 : \beta_1 = \cdots = \beta_p = 0$, the ratio $F = \text{MSR}/\text{MSE}$ has an F-distribution with p and $n - (p + 1)$ d.f. Thus, an α-level test rejects H_0 if

$$F = \frac{\text{MSR}}{\text{MSE}} > f_{p, n-(p+1), \alpha}, \tag{3.16}$$

where $f_{p, n-(p+1), \alpha}$ is the upper α critical point of this F-distribution. These calculations are presented in the form of an ANOVA in Table 3.3.

Table 3.3. ANOVA table for multiple linear regression

Source	SS	d.f.	MS	F
Regression	SSR	p	MSR	$\frac{\text{MSR}}{\text{MSE}}$
Error	SSE	$n - (p + 1)$	MSE	
Total	SST	$n - 1$		

Why do we test the overall null hypothesis $H_0 : \beta_1 = \cdots = \beta_p = 0$? Shouldn't we be testing the individual β_j's directly? The problem with testing the individual β_j's directly without an overall F-test is that when there are many estimated regression coefficients, some of them may turn out to be significant simply by random chance. For example, if 100 estimated regression coefficients are tested individually at 5% significance level then, on the average, five of them will turn out to be significant even when all the true β_j's are zero. The probability that at least one $\widehat{\beta}_j$ will turn out significant is almost 1. **Multiple testing procedures** are designed to control the probability of occurrence of false positives when testing multiple hypotheses. The book by Hochberg and Tamhane (1987) covers this area in detail.

Example 3.4 (College GPA and Entrance Test Scores: ANOVA for Linear Regression) The ANOVA for the regression of GPA versus Verbal and Math test scores from Example 3.3 is shown in Table 3.4. We see that the F-statistic equals 39.51 with 2 and 37 d.f., which is highly significant with a P-value < 0.001. Also, $R^2 = \text{SSR}/\text{SST} = 12.7859/18.7735 = 0.681$. Thus, about 2/3rd of the variation in GPA is accounted for by linear regression on Verbal and Math test scores. The estimated standard deviation is $s = \sqrt{0.1618} = 0.4023$. We will use this estimate to test hypotheses and make confidence intervals on the β_j's. $\square$

Table 3.4. ANOVA table for linear regression of GPA on entrance test scores

Source	SS	d.f.	MS	F	P
Regression	12.7859	2	6.3930	39.51	0.000
Error	5.9876	37	0.1618		
Total	18.7735	39			

3.3.2　Inferences on regression coefficients

The ANOVA F-test is a test of the global null hypothesis $H_0 : \beta_1 = \cdots = \beta_p = 0$. Having rejected this null hypothesis, we would like to know which individual β_j are different from zero or estimate their magnitudes via CIs. These inferences are based on the sampling distributions of the $\widehat{\beta}_j$'s.

　　The LS estimator vector $\widehat{\boldsymbol{\beta}}$ can be shown (see Section 3.8) to have a $(p+1)$-variate normal distribution with mean vector $E(\widehat{\boldsymbol{\beta}}) = \boldsymbol{\beta}$ and covariance matrix $\text{Cov}(\widehat{\boldsymbol{\beta}}) = \sigma^2 \boldsymbol{V}$, where $\boldsymbol{V} = (\boldsymbol{X}'\boldsymbol{X})^{-1}$. Therefore, the individual $\widehat{\beta}_j$ are normally distributed with means β_j and variances $\sigma^2 v_{jj}$, where v_{jj} is the jth diagonal entry $(0 \le j \le p)$ of $\boldsymbol{V}$. Also, it can be shown that $[n - (p+1)]s^2/\sigma^2 = \text{SSE}/\sigma^2$ is distributed as χ^2 with $n - (p+1)$ d.f. independent of $\widehat{\boldsymbol{\beta}}$.

From these results, it follows that

$$\frac{\widehat{\beta}_j - \beta_j}{\mathrm{SE}(\widehat{\beta}_j)} \sim t_{n-(p+1)} \quad (0 \le j \le p),$$

where $\mathrm{SE}(\widehat{\beta}_j) = s\sqrt{v_{jj}}$. So a $100(1-\alpha)\%$ CI for any β_j is given by

$$\widehat{\beta}_j \pm t_{n-(p+1),\alpha/2}\mathrm{SE}(\widehat{\beta}_j) \quad (0 \le j \le p).$$

A two-sided α-level test of $H_{0j} : \beta_j = 0$ rejects if

$$|t_j| = \frac{|\widehat{\beta}_j|}{\mathrm{SE}(\widehat{\beta}_j)} > t_{n-(p+1),\alpha/2} \quad (0 \le j \le p). \tag{3.17}$$

Example 3.5 (College GPA and Entrance Test Scores: Inferences on Regression Coefficients) The 95% CIs on the two coefficients can be calculated as (using $t_{37,0.025} = 2.0262$ and standard errors of the regression coefficients for Verbal and Math from the R output in Example 3.3):

$$\text{Verbal}:\ 0.02573 \pm 2.0262 \times 0.004\,02 = [0.0176, 0.0339]$$

and

$$\text{Math}:\ 0.03362 \pm 2.0262 \times 0.004\,93 = [0.0236, 0.0436].$$

Note that 0 is well outside both the intervals. This is in accord with the result from Example 3.3 that both Verbal and Math are highly significant predictors of GPA. However, this only means that there are significant linear components of Verbal and Math in their relationship with GPA, but there could be quadratic components as well. In Example 3.8, we will investigate if the quadratic terms should be included in the model. □

3.3.3 Confidence ellipsoid for the β vector*

The previous section gave separate CIs for the individual β's. As will be seen later in some applications, it is useful to have a simultaneous (or joint) $100(1-\alpha)\%$ **confidence region** for all β's. This region turns out to be an ellipsoid centered at $\widehat{\beta}$, and is given by

$$\left\{ \beta : \frac{(\widehat{\beta} - \beta)' X' X (\widehat{\beta} - \beta)}{(p+1)s^2} \le f_{p+1,n-(p+1),\alpha} \right\}. \tag{3.18}$$

This region can be extended to any subset of the β vector or more generally to any linearly independent set of $r \le p+1$ linear parametric functions

$\theta_i = c_{i0}\beta_0 + \cdots + c_{ip}\beta_p\ (1 \leq i \leq r)$.   Let $\boldsymbol{\theta} = (\theta_1, \ldots, \theta_r)' = \boldsymbol{C}\boldsymbol{\beta}$,   where $\boldsymbol{C} = \{c_{ij}\}$ is an $r \times (p+1)$ matrix with linearly independent rows. Then the LS estimator of $\boldsymbol{\theta}$ is $\widehat{\boldsymbol{\theta}} = \boldsymbol{C}\widehat{\boldsymbol{\beta}}$ and $\mathrm{Cov}(\widehat{\boldsymbol{\theta}}) = \sigma^2 \boldsymbol{C}\boldsymbol{V}\boldsymbol{C}'$ (using the sandwich formula (A.7)), where $\boldsymbol{V} = (\boldsymbol{X}'\boldsymbol{X})^{-1}$. The extension of the above confidence ellipsoid for $\boldsymbol{\theta}$ is

$$\left\{ \boldsymbol{\theta} : \frac{(\widehat{\boldsymbol{\theta}} - \boldsymbol{\theta})'(\boldsymbol{C}\boldsymbol{V}\boldsymbol{C}')^{-1}(\widehat{\boldsymbol{\theta}} - \boldsymbol{\theta})}{rs^2} \leq f_{r,n-(p+1),\alpha} \right\}. \qquad (3.19)$$

Using this confidence ellipsoid, we can reject $H_0 : \boldsymbol{\theta} = \boldsymbol{C}\boldsymbol{\beta} = \boldsymbol{0}$ at level α if $\boldsymbol{0}$ falls outside the ellipsoid, i.e. if

$$F = \frac{\widehat{\boldsymbol{\theta}}'(\boldsymbol{C}\boldsymbol{V}\boldsymbol{C}')^{-1}\widehat{\boldsymbol{\theta}}}{rs^2} > f_{r,n-(p+1),\alpha}.$$

One application of interest of this formula is for $\boldsymbol{\theta} = (\beta_1, \ldots, \beta_p)'$. In that case $r = p$, the $\boldsymbol{C}\boldsymbol{V}\boldsymbol{C}'$ matrix is simply the $p \times p$ submatrix of $\boldsymbol{V}$ obtained by deleting the first row and the first column of $\boldsymbol{V}$ corresponding to $\widehat{\beta}_0$. In that case, the null vector $\boldsymbol{0}$ falling outside this confidence ellipsoid is equivalent to the α-level ANOVA F-test of the overall $H_0 : \beta_1 = \cdots = \beta_p = 0$. We will see more examples of this in the next section.

Example 3.6 (College GPA and Entrance Test Scores: Confidence Ellipsoid) For the linear model fitted to the GPA data in Example 3.3, the following covariance matrix of the regression coefficients of $(\widehat{\beta}_1, \widehat{\beta}_2)$ can be obtained by using the function vcov(lmfit) in R:

$$s^2 \boldsymbol{C}\boldsymbol{V}\boldsymbol{C}' = 10^{-5} \begin{bmatrix} 1.619 & 0.212 \\ 0.212 & 2.428 \end{bmatrix}.$$

The inverse of this matrix equals

$$\frac{1}{s^2}(\boldsymbol{C}\boldsymbol{V}\boldsymbol{C}')^{-1} = 10^5 \begin{bmatrix} 0.625 & -0.055 \\ -0.055 & 0.417 \end{bmatrix}.$$

Using $\widehat{\beta}_1 = 0.0257$ and $\widehat{\beta}_2 = 0.0336$, the equation for the confidence ellipse is given by

$$\frac{1}{2}[\beta_1 - 0.0257, \beta_2 - 0.0336](10^5) \begin{bmatrix} 0.625 & -0.055 \\ -0.055 & 0.417 \end{bmatrix} \begin{bmatrix} \beta_1 - 0.0257 \\ \beta_2 - 0.0336 \end{bmatrix}$$

$$= 10^5[0.312(\beta_1 - 0.0257)^2 - 0.055(\beta_1 - 0.0257)(\beta_2 - 0.0336)$$

$$+ 0.209(\beta_2 - 0.0336)^2].$$

To test $H_0 : \beta_1 = \beta_2 = 0$, substitute these values in the above equation to obtain the F-statistic as

$$10^5[0.312(0.0257)^2 - 0.055(0.0257)(0.0336) + 0.209(0.0336)^2] = 39.44,$$

which agrees with the F-statistic value $= 39.51$ in the R output in Example 3.3 except for round-off errors. Since it exceeds $f_{2,37,0.05} = 3.252$, which implies that the point $\beta_1 = \beta_2 = 0$ falls outside the 95% confidence ellipse for the coefficients β_1 and β_2, as can be seen from Figure 3.2. So we can confidently reject H_0.   □

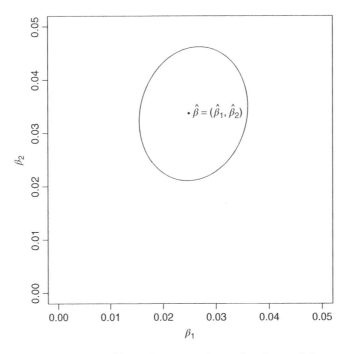

Figure 3.2. *95% confidence ellipse for β_1 and β_2.*

3.3.4   Extra sum of squares method

The **extra sum of squares (extra SS) method** is used when we want to test hypotheses on multiple β_j's simultaneously. One example of this is when we want to test whether the **full model** with all predictors fits the data significantly better than a **partial** or **reduced model** consisting of a subset of predictors. In other words, whether adding the extra predictors improves the fit significantly. The partial model is said to be **nested** under the full model.

As an example, consider comparing a first-degree model versus a second-degree model in two variables:

$$E(y) = \beta_0 + \beta_1 x_1 + \beta_2 x_2 \quad \text{versus} \quad E(y) = \beta_0 + \beta_1 x_1 + \beta_2 x_2 + \beta_3 x_1^2 + \beta_4 x_2^2$$

$$(3.20)$$

by testing $H_0 : \beta_3 = \beta_4 = 0$. The general problem of testing a partial model versus a full model can be stated as

$$E(y) = \beta_0 + \beta_1 x_1 + \cdots + \beta_q x_q \quad \text{versus} \quad E(y) = \beta_0 + \beta_1 x_1 + \cdots + \beta_p x_p, \quad (3.21)$$

where $q < p$. This problem is equivalent to testing $H_0 : \beta_{q+1} = \cdots = \beta_p = 0$, which imposes $r = p - q$ linearly independent constraints on the β's. A special case of interest is testing the so-called **null model** (which has only the intercept term) versus the full model, i.e.

$$E(y) = \beta_0 \quad \text{versus} \quad E(y) = \beta_0 + \beta_1 x_1 + \cdots + \beta_p x_p \quad (3.22)$$

by testing the **overall null hypothesis** $H_0 : \beta_1 = \cdots = \beta_p = 0$ against the alternative that at least one $\beta_j \neq 0$ $(1 \leq j \leq p)$.

Still another example of application of the extra SS method is testing equalities among subsets of β's. For example, suppose that in a regression model for predicting annual credit card purchases, one of the variables is the age of the customer classified into three categories, young, middle-age, and old. Let $\beta_1, \beta_2,$ and β_3 be the regression coefficients for the three categories. (Actually, only two categories suffice as we shall see in Section 3.6.1 on dummy variables, but that is not germane to the discussion here.) Suppose we want to test the null hypothesis that the average credit card purchases are the same for the three age groups of the customers, i.e. test $H_0 : \beta_1 = \beta_2 = \beta_3$ by testing the following two hypotheses simultaneously:

$$H_{01} : \beta_1 - \beta_2 = 0 \quad \text{and} \quad H_{02} : \beta_1 - \beta_3 = 0.$$

There are infinitely many equivalent ways of writing these hypotheses, e.g. H_{02} can be written as $\beta_2 - \beta_3 = 0$ or $(1/2)(\beta_1 + \beta_2) - \beta_3 = 0$. It can be shown that all such representations of H_0 lead to the same test.

Both the above examples are special cases of the so-called **general linear hypothesis**, namely, that the β's are subject to a set of r linearly independent constraints $H_0 : \boldsymbol{\theta} = \boldsymbol{C\beta} = \boldsymbol{0}$; r is called the **rank** or the degrees of freedom (d.f.) of the hypothesis. This is the same hypothesis that was considered in the previous section and the F-test based on the simultaneous confidence ellipsoid can be shown to be the extra SS test.

The steps in the extra SS test are as follows:

1. Fit the full model and compute its SSE and SSR.

2. Fit the partial model under H_0 and compute its SSE, denoted by SSE_0, where $\text{SSE}_0 \geq \text{SSE}$.

3. Compute the **hypothesis sum of squares**, $\text{SSH}_0 = \text{SSE}_0 - \text{SSE}$, and the **hypothesis mean square**, $\text{MSH}_0 = \text{SSH}_0/r$.

4. Compute the F-statistic

$$F = \frac{\text{MSH}_0}{\text{MSE}},$$

which has r and $n - (p + 1)$ d.f.

5. Reject H_0 at level α if $F > f_{r,n-(p+1),\alpha}$.

This test is based on the result that under the general linear hypothesis $H_0 : \boldsymbol{C\beta} = \mathbf{0}$ of rank r, the F-statistic has an F-distribution with r and $n - (p + 1)$ d.f.

As an application of the extra SS method, consider the overall null hypothesis testing problem (3.22). We know how to compute SSE under the full model. To compute SSE_0, note that the LS estimator of β_0 under $H_0 : E(y) = \beta_0$ is just $\bar{y}$. Therefore, under H_0, we have $\hat{y}_i = \bar{y}$ for all i and $\text{SSE}_0 = \sum_{i=1}^{n} (y_i - \bar{y})^2 = \text{SST}$. Using the ANOVA identity (3.13), we get

$$\text{SSH}_0 = \text{SSE}_0 - \text{SSE} = \text{SST} - \text{SSE} = \text{SSR}.$$

The hypothesis d.f. equals $r = p$ and $\text{MSH}_0 = \text{SSR}/p = \text{MSR}$, namely the regression mean square. Thus, H_0 is rejected at level α if $F = \text{MSR}/\text{MSE} > f_{p,n-(p+1),\alpha}$, which is the ANOVA F-test (3.16). The two-sided t-test (3.17) on a single β_j: reject $H_{0j} : \beta_j = 0$ if $F_j = t_j^2 > f_{1,n-(p+1),\alpha}$, can also be derived using the extra SS method.

Example 3.7 (College GPA and Entrance Test Scores: ANOVA for the Linear Model) The ANOVA in Table 3.4 for the linear model gives SSR = 12.7859. A question that is often of interest in such regression problems is what proportion of this SS can be attributed to each predictor, in this case to Verbal and Math scores. Using the extra SS method, we can compute the SS due to each predictor by taking the difference between SSR for the full model and SSR_0 for the partial model, which omits the predictor of interest. In the present problem, the full model includes both Verbal and Math, while one partial model includes only Math (to find the SS due to Verbal) and the other partial model includes only Verbal (to find the SS due to Math). In R, this calculation is performed by using the `drop1` function. The result is as follows:

```
       Df Sum of Sq    RSS     AIC F value    Pr(>F)
<none>                5.9876 -69.968
Verbal  1    6.6187 12.6063 -42.187  40.900 1.834e-07 ***
Math    1    7.5311 13.5186 -39.392  46.538 4.902e-08 ***
```

Note two things about this ANOVA table.

1. The F-test for each predictor is exactly equivalent to the t-test given in the R output from Example 3.5 in that $F = t^2$ and the P-values are identical (the former calculated from the two tails of the t_{37} distribution and the latter calculated from the upper tail of the $F_{1,37}$ distribution).

2. Further note that the SSs for Verbal and Math do not add up to SSR as can be checked from

$$\text{SS}_{\text{Verbal}} + \text{SS}_{\text{Math}} = 6.6187 + 7.5311 = 14.2098 \neq \text{SSR} = 12.7859.$$

This will be the case whenever the predictors are not orthogonal to each other and so their contributions to SSR are not mutually exclusive. Therefore, it is not possible to apportion SSR into two independent components corresponding to Verbal and Math scores.

If we use the `anova` function in R, then we get the following output:

```
          Df Sum Sq Mean Sq F value    Pr(>F)
Verbal     1 5.2549  5.2549  32.472 1.608e-06 ***
Math       1 7.5311  7.5311  46.538 4.902e-08 ***
Residuals 37 5.9876  0.1618
```

In this table, the SSs for Verbal and Math do add up to SSR as can be checked from

$$\text{SS}_{\text{Verbal}} + \text{SS}_{\text{Math}} = 5.2549 + 7.5311 = 12.7860.$$

However, these SSs are calculated by adding the predictors sequentially in the order in which the predictors are specified in the `lm` function, first Verbal and then Math in the present example, and attributing the increase in SSR (or equivalently decrease in SSE) to the last added predictor. These SSs are referred to as **sequential sums of squares** or **type I sums of squares**, and it is easy to see that they always add up to SSR. But they depend on the order in which the predictors are entered into the regression model; if the order is changed, then they change as well. Only if the predictors are orthogonal, then the order of entry does not matter and the SSs of the predictors are the same as those obtained from the `drop1` function based on the extra SS method. These latter SSs are referred to as **adjusted sums of squares** or **type III sums of squares**. Only these latter SSs should be used for testing purposes and not the sequential sums of squares. □

Example 3.8 (College GPA and Entrance Test Scores: Quadratic Model) In Example 3.3, we fitted a model to the GPA data that was linear in both Verbal and Math scores. In this example, we fit the quadratic model:

$$\text{GPA} = \beta_0 + \beta_1 \text{Verbal} + \beta_2 \text{Math} + \beta_3 \text{Verbal}^2 + \beta_4 \text{Math}^2 + \varepsilon,$$

to see if it provides a significantly better fit. The fitted quadratic model is

$$\widehat{\text{GPA}} = -11.458 + 0.189\text{Verbal} + 0.159\text{Math} - 0.0011\text{Verbal}^2 - 0.0009\text{Math}^2.$$

To check if the two quadratic terms significantly improve the fit, we can do the extra SS test as follows. Taking the linear model as the partial model and the quadratic model as the full model, we get $\text{SSE}_0 = 5.9876$ and $\text{SSE} = 1.2890$ from Tables 3.4 and 3.5, respectively. So the F-statistic for testing $\beta_3 = \beta_4 = 0$ in the quadratic model equals

$$F = \frac{(5.9876 - 1.2890)/2}{0.0368} = 63.79,$$

with 2 and 35 d.f., which is clearly highly significant ($P < 0.001$).

Table 3.5. ANOVA table for GPA versus entrance test scores: quadratic regression

Source	SS	d.f.	MS	F	P
Regression	17.4845	4	4.371	118.69	0.000
Residual error	1.2890	35	0.037		
Total	18.7735	39			

This extra SS test to compare the linear model versus the quadratic model can be carried out in R by fitting the two models and then doing the test using the **anova** function as shown below.

```
> gpa = read.csv("c:/data/GPA.csv")
> lmfit = lm(GPA ~ Verbal+Math , data = gpa)
> qmfit = lm(GPA ~ Verbal + Math +  I(Verbal^2) + I(Math^2), data = gpa)
> anova(lmfit,qmfit)
Analysis of Variance Table

Model 1: GPA ~ Verbal + Math
Model 2: GPA ~ Verbal + Math + I(Verbal^2) + I(Math^2)
  Res.Df RSS Df Sum of Sq F Pr(>F)
1  37 5.9876
2  35 1.2890  2  4.6986  63.792  2.125e-12 ***
```

□

3.3.5 Prediction of future observations

In this section, we consider the problem of predicting the response for a given set of predictors, e.g. predict an entering freshman's college graduating GPA from his Verbal and Math scores. Denote the vector of the given set of predictors by $\boldsymbol{x}_{\text{new}} = (1, x_{\text{new},1}, \ldots, x_{\text{new},p})'$, the corresponding value of y by y_{new} and $E(y_{\text{new}})$ by $\mu_{\text{new}} = \boldsymbol{x}'_{\text{new}}\boldsymbol{\beta} = \beta_0 + \beta_1 x_{\text{new},1} + \cdots + \beta_p x_{\text{new},p}$. As in Section 2.2.5, we will give the formulae for the CI for μ_{new} and the prediction interval (PI) for y_{new}.

The unbiased estimate of both y_{new} and μ_{new} is given by

$$\widehat{y}_{\text{new}} = \widehat{\mu}_{\text{new}} = \boldsymbol{x}'_{\text{new}}\widehat{\boldsymbol{\beta}} = \widehat{\beta}_0 + \widehat{\beta}_1 x_{\text{new},1} + \cdots + \widehat{\beta}_p x_{\text{new},p}.$$

By using the formula for $\text{Cov}(\widehat{\boldsymbol{\beta}}) = \sigma^2 (\boldsymbol{X}'\boldsymbol{X})^{-1}$ and a special case (A.5) of the sandwich formula, we get

$$\text{Var}(\widehat{y}_{\text{new}}) = \text{Var}(\widehat{\mu}_{\text{new}}) = \boldsymbol{x}'_{\text{new}}\text{Cov}(\widehat{\boldsymbol{\beta}})\boldsymbol{x}_{\text{new}} = \sigma^2 \boldsymbol{x}'_{\text{new}}(\boldsymbol{X}'\boldsymbol{X})^{-1}\boldsymbol{x}_{\text{new}}.$$

So

$$\text{SE}(\widehat{\mu}_{\text{new}}) = s\sqrt{\boldsymbol{x}'_{\text{new}}(\boldsymbol{X}'\boldsymbol{X})^{-1}\boldsymbol{x}_{\text{new}}}.$$

Then analogous to (2.19) and (2.20), we get the following formulae for the CI for μ_{new}:

$$\widehat{\mu}_{\text{new}} \pm t_{n-(p+1),\alpha/2} s\sqrt{\boldsymbol{x}'_{\text{new}}(\boldsymbol{X}'\boldsymbol{X})^{-1}\boldsymbol{x}_{\text{new}}} \tag{3.23}$$

and for the PI for y_{new}:

$$\widehat{y}_{\text{new}} \pm t_{n-(p+1),\alpha/2} s\sqrt{1 + \boldsymbol{x}'_{\text{new}}(\boldsymbol{X}'\boldsymbol{X})^{-1}\boldsymbol{x}_{\text{new}}}. \tag{3.24}$$

Example 3.9 *(College GPA and Entrance Test Scores: Confidence and Prediction Intervals)* Suppose we want to estimate the graduating GPA of a typical freshman with 80 on Verbal and 90 on Math using the quadratic model fitted in Example 3.8. The estimated GPA is

$$\widehat{\text{GPA}} = -11.458 + 0.189(80) + 0.159(90) - 0.0011(80)^2 - 0.0009(90)^2 = 3.584.$$

To calculate the standard error of this estimate, note from the ANOVA in Table 3.5 that $s = \sqrt{0.0368} = 0.1918$, $\boldsymbol{x}_{\text{new}} = (1, 80, 90, 80^2, 90^2)'$ and $(\boldsymbol{X}'\boldsymbol{X})^{-1}$ is a 5×5 matrix not shown here. Substituting these values in the formula for $\text{SE}(\widehat{\mu}_{\text{new}})$, we get $\text{SE}(\widehat{\mu}_{\text{new}}) = 0.0566$. So the 95% CI for the average graduating GPA of all freshmen with Verbal score = 80 and Math score = 90 is (using $t_{35,0.025} = 2.030$)

$$3.584 \pm (2.030)(0.0566) = [3.469, 3.699].$$

The 95% PI for the GPA of a randomly chosen freshman is

$$3.584 \pm (2.030)\sqrt{1 + (0.0566)^2} = [3.178, 3.990].$$

Note that the PI is much wider than the CI.

In R, these calculations can be done by using the `predict` function as shown below.

```
> predict(qmfit,newdata=data.frame(Verbal=80,Math=90),interval="confidence")
       fit      lwr      upr
1 3.583893 3.469074 3.698712
> predict(qmfit,newdata=data.frame(Verbal=80,Math=90),interval="predict")
       fit      lwr      upr
1 3.583893 3.177738 3.990048
```

$\square$

3.4   Weighted and generalized least squares

3.4.1   Weighted least squares

In this section, we generalize the method of weighted least squares (WLS) introduced in Exercise 2.3. In the case of multiple regression, the **WLS criterion** is

$$Q = \sum_{i=1}^{n} w_i [y_i - (\beta_0 + \beta_1 x_{i1} + \cdots + \beta_p x_{ip})]^2, \tag{3.25}$$

where $w_i > 0$ $(i = 1, \ldots, n)$ are the weights. This criterion can be expressed in matrix notation as

$$Q = (\boldsymbol{y} - \boldsymbol{X}\boldsymbol{\beta})' \boldsymbol{W} (\boldsymbol{y} - \boldsymbol{X}\boldsymbol{\beta}), \tag{3.26}$$

where $\boldsymbol{W} = \text{diag}\{w_1, \ldots, w_n\}$ is an $n \times n$ diagonal matrix. The weights may be chosen to reflect the relative importance of the observations. A common choice is $w_i \propto [\text{Var}(y_i)]^{-1}$ if the $\text{Var}(y_i)$ are not constant, i.e. if they are heteroscedastic. This case is discussed below.

Let $\boldsymbol{W}^{1/2} = \text{diag}\{w_1^{1/2}, \ldots, w_n^{1/2}\}$ so that $\boldsymbol{W}^{1/2}\boldsymbol{W}^{1/2} = \boldsymbol{W}$. Then the WLS criterion can be written as

$$\begin{aligned} Q &= (\boldsymbol{y} - \boldsymbol{X}\boldsymbol{\beta})' \boldsymbol{W}^{1/2} \boldsymbol{W}^{1/2} (\boldsymbol{y} - \boldsymbol{X}\boldsymbol{\beta}) \\ &= (\boldsymbol{W}^{1/2}\boldsymbol{y} - \boldsymbol{W}^{1/2}\boldsymbol{X}\boldsymbol{\beta})'(\boldsymbol{W}^{1/2}\boldsymbol{y} - \boldsymbol{W}^{1/2}\boldsymbol{X}\boldsymbol{\beta}) \\ &= (\boldsymbol{y}^* - \boldsymbol{X}^*\boldsymbol{\beta})'(\boldsymbol{y}^* - \boldsymbol{X}^*\boldsymbol{\beta}), \end{aligned}$$

where $\boldsymbol{y}^* = \boldsymbol{W}^{1/2}\boldsymbol{y}$ and $\boldsymbol{X}^* = \boldsymbol{W}^{1/2}\boldsymbol{X}$. This representation shows that the WLS criterion is the same as the LS criterion (3.4) with $\boldsymbol{y}$ replaced by $\boldsymbol{y}^*$ and $\boldsymbol{X}$

replaced by $\boldsymbol{X}^*$. Therefore, using formula'(3.7) for the LS estimator, the **WLS estimator** of $\boldsymbol{\beta}$ can be written as

$$
\begin{aligned}
\widehat{\boldsymbol{\beta}}_{\text{WLS}} &= (\boldsymbol{X}^{*\prime}\boldsymbol{X}^*)^{-1}\boldsymbol{X}^{*\prime}\boldsymbol{y}^* \\
&= (\boldsymbol{X}'\boldsymbol{W}^{1/2}\boldsymbol{W}^{1/2}\boldsymbol{X})^{-1}\boldsymbol{X}'\boldsymbol{W}^{1/2}\boldsymbol{W}^{1/2}\boldsymbol{y} \\
&= (\boldsymbol{X}'\boldsymbol{W}\boldsymbol{X})^{-1}\boldsymbol{X}'\boldsymbol{W}\boldsymbol{y}.
\end{aligned}
\tag{3.27}
$$

If we choose $w_i \propto [\text{Var}(y_i)]^{-1}$, then $y_i^* \propto y_i/\sqrt{\text{Var}(y_i)}$, which makes the $\text{Var}(y_i^*)$ constant, thus making them homoscedastic. This is a so-called **variance stabilizing transformation**. These transformations of the response variable are discussed in Section 4.3.1. A common example of $w_i \propto [\text{Var}(y_i)]^{-1}$ occurs when each y_i is an average of n_i i.i.d. repeat observations, y_{ij} ($j = 1, \ldots, n_i$), with a common unknown variance σ^2 so that $\text{Var}(y_i) = \sigma^2/n_i$. For example, the y_{ij} ($j = 1, \ldots, n_i$) may be repeat measurements made under identical experimental conditions, but only their averages are reported. Then the w_i may be taken to be equal to n_i. In that case each term in the WLS criterion (3.25) is weighted by the sample size n_i.

Generally, the $\text{Var}(y_i)$ are completely unknown. Hence, the weights w_i are also unknown. It is often possible to postulate a relationship between $\text{Var}(y_i)$ and $E(y_i) = \mu_i$ based on the residual plots for checking homoscedasticity studied in Section 4.3. For example, if $\text{SD}(y_i) = \sqrt{\text{Var}(y_i)} \propto \mu_i$, then we can use WLS regression with $w_i = 1/\mu_i^2$. However, the μ_i are themselves unknown but can be estimated by $\widehat{y}_i$. The following example illustrates WLS regression using $w_i = 1/\widehat{y}_i^2$. The starting values of $\widehat{y}_i$'s can be obtained using **ordinary least squares (OLS)**. One can then iterate on the new WLS values of $\widehat{y}_i$, which results in a so-called **iteratively reweighted least squares (IRWLS) algorithm** discussed in Section 9.2.2. In the example below, we only give the first step.

Example 3.10 (College GPA and Entrance Test Scores: Weighted Least Squares) In Example 3.3, we fitted an OLS regression model to GPA using Verbal and Math scores as predictors. A residual plot for this model shows that the homoscedasticity assumption is violated, and we may assume $\text{SD}(y_i) = \sqrt{\text{Var}(y_i)} \propto \mu_i$. So we will fit a WLS regression model using $w_i = [\widehat{y}_i^2]^{-1}$ from the OLS regression model. First, we repeat the R output from Example 3.3.

```
> lmfit = lm(GPA ~ Verbal+Math , data = gpa)
> summary(lmfit)

Coefficients:
            Estimate Std. Error t value Pr(>|t|)
(Intercept) -1.570537   0.493749  -3.181  0.00297 **
Verbal       0.025732   0.004024   6.395 1.83e-07 ***
```

```
Math          0.033615   0.004928   6.822 4.90e-08 ***
---
```

```
Residual standard error: 0.4023 on 37 degrees of freedom
Multiple R-squared:  0.6811,    Adjusted R-squared:  0.6638
F-statistic: 39.51 on 2 and 37 DF,  p-value: 6.585e-10
```

Next, we perform WLS regression as described above.

```
> wlsfit=lm(GPA ~ Verbal+Math, weights=1/(lmfit$fitted)^2,data = gpa)
> summary(wlsfit)
```

```
Call: lm(formula = GPA ~ Verbal + Math, data = gpa, weights
  = 1/(lmfit$fitted)^2)
```

```
Coefficients:
             Estimate Std. Error t value Pr(>|t|)
(Intercept) -1.958844   0.410258   -4.775 2.82e-05 ***
Verbal       0.029191   0.003780    7.723 3.16e-09 ***
Math         0.035567   0.004625    7.689 3.49e-09 ***
---
Residual standard error: 0.1495 on 37 degrees of freedom
Multiple R-squared:  0.7658,    Adjusted R-squared:  0.7531
F-statistic: 60.48 on 2 and 37 DF,  p-value: 2.179e-12
```

We see that the WLS model coefficients are qualitatively similar to those from the OLS model, but are much more significant. This is because the residual standard error s equals 0.4023 for the OLS model, whereas it is only 0.1495 for the WLS model. Hence, the t-statistics of the estimated regression coefficients are correspondingly larger.   □

3.4.2   Generalized least squares*

Generalized least squares (GLS) is a generalization of WLS in which W is an arbitrary symmetric positive definite $n \times n$ matrix. This generalization is useful when the y_i's are correlated besides being heteroscedastic. So the covariance structure of the y_i's is nondiagonal. Let us assume that $\text{Cov}(y) = \Sigma = \sigma^2 W^{-1}$, where $\sigma^2 > 0$ is an unknown scalar and W is a known positive definite matrix, which we will refer to as the weight matrix.

The GLS criterion is the same as the WLS criterion (3.26) except that W is now a nondiagonal matrix. For a diagonal matrix we could write $W^{1/2} = \text{diag}\{w_1^{1/2}, \ldots, w_n^{1/2}\}$ as the "square root" of W. When W is nondiagonal, we use the spectral decomposition theorem (see Equation (A.1) from Appendix A), which states that there exists a symmetric positive definite matrix U such that $U'U = W$ and $U W^{-1} U' = I$. In this sense, U can be regarded as the "square root" of $\boldsymbol{W}$. Then the GLS criterion becomes

$$Q = (\boldsymbol{y} - \boldsymbol{X}\boldsymbol{\beta})'\boldsymbol{U}'\boldsymbol{U}(\boldsymbol{y} - \boldsymbol{X}\boldsymbol{\beta})$$
$$= (\boldsymbol{U}\boldsymbol{y} - \boldsymbol{U}\boldsymbol{X}\boldsymbol{\beta})'(\boldsymbol{U}\boldsymbol{y} - \boldsymbol{U}\boldsymbol{X}\boldsymbol{\beta})$$
$$= (\boldsymbol{y}^* - \boldsymbol{X}^*\boldsymbol{\beta})'(\boldsymbol{y}^* - \boldsymbol{X}^*\boldsymbol{\beta}),$$

where $\boldsymbol{y}^* = \boldsymbol{U}\boldsymbol{y}$ and $\boldsymbol{X}^* = \boldsymbol{U}\boldsymbol{X}$. It follows that the **GLS estimator** $\widehat{\boldsymbol{\beta}}_{\mathrm{GLS}}$ of $\boldsymbol{\beta}$ is the same as that given by (3.27). Note that there is no need to find the $\boldsymbol{U}$ matrix explicitly.

In analogy with $w_i \propto [\mathrm{Var}(y_i)]^{-1}$ in the WLS case, note that $\boldsymbol{W} = \boldsymbol{\Sigma}^{-1}$. Then using the sandwich formula (see (A.7)) we get

$$\mathrm{Cov}(\boldsymbol{y}^*) = \boldsymbol{U}\,\mathrm{Cov}(\boldsymbol{y})\,\boldsymbol{U}' = \sigma^2\,\boldsymbol{U}\,\boldsymbol{W}^{-1}\,\boldsymbol{U}' = \sigma^2\boldsymbol{I}.$$

Thus the transformation $\boldsymbol{y}^* = \boldsymbol{U}\boldsymbol{y}$ makes the y_i^* uncorrelated and homoscedastic.

3.4.3 Statistical inference on GLS estimator*

We will consider the GLS estimator since the WLS estimator is a special case of it. Following the same steps as in the derivation of the distribution of the LS estimator $\widehat{\boldsymbol{\beta}}$ given in Section 3.8, it can be shown that $\widehat{\boldsymbol{\beta}}_{\mathrm{GLS}}$ is multivariate normal (MVN) with mean vector and covariance matrix given by

$$E(\widehat{\boldsymbol{\beta}}_{\mathrm{GLS}}) = \boldsymbol{\beta} \quad \text{and} \quad \mathrm{Cov}(\widehat{\boldsymbol{\beta}}_{\mathrm{GLS}}) = (\boldsymbol{X}'\boldsymbol{\Sigma}^{-1}\boldsymbol{X})^{-1} = \sigma^2(\boldsymbol{X}'\boldsymbol{W}\boldsymbol{X})^{-1}. \quad (3.28)$$

An unbiased estimate s^2 of σ^2 with $n - (p+1)$ d.f. can be obtained as

$$s^2 = \frac{(\boldsymbol{y}^* - \boldsymbol{X}^*\widehat{\boldsymbol{\beta}}_{\mathrm{GLS}})'(\boldsymbol{y}^* - \boldsymbol{X}^*\widehat{\boldsymbol{\beta}}_{\mathrm{GLS}})}{n - (p+1)} = \frac{(\boldsymbol{y} - \boldsymbol{X}\widehat{\boldsymbol{\beta}}_{\mathrm{GLS}})'\boldsymbol{W}(\boldsymbol{y} - \boldsymbol{X}\widehat{\boldsymbol{\beta}}_{\mathrm{GLS}})}{n - (p+1)}.$$

Also, $s^2 \sim \chi^2_{n-(p+1)}/[n - (p+1)]$ is independent of $\widehat{\boldsymbol{\beta}}_{\mathrm{GLS}}$. Thus, hypothesis tests and CIs on the components of $\widehat{\boldsymbol{\beta}}_{\mathrm{GLS}}$ can be derived in the same manner as for the LS estimator. The results in the R output in Example 3.10 are calculated using these formulae.

3.5 Partial correlation coefficients

The partial correlation coefficient between two variables measures the residual correlation between them after controlling for the effects of other variables on both of them. The population partial correlation coefficient between y and any predictor x_j is the bivariate correlation coefficient between y and x_j in their conditional joint distribution conditioned on all other predictors x_k for $k \neq j$. Just as there is a direct relation (2.24) between the regression coefficient β_1 and the correlation coefficient ρ between y and x in the context of simple linear regression, there is a corresponding relation between each regression coefficient β_j and the population partial correlation coefficient between y and x_j in the context of multiple linear regression. In fact, the population partial correlation coefficient between y and x_j is a normalized unitless version of the regression coefficient β_j with a range of $[-1, 1]$. An expression for this population partial correlation coefficient can be derived (see Section 3.8.7). However, here we will only consider its sample version.

We will first show how to compute the sample partial correlation coefficient in the case of two predictor variables. Let $r_{yx_2|x_1}$ denote the partial correlation coefficient between y and x_2 conditioned on x_1. The following are alternative ways to compute $r_{yx_2|x_1}$.

- Consider a set of residuals from regression of y on x_1 and another set of residuals from regression of x_2 on x_1. Then $r_{yx_2|x_1}$ is the sample correlation coefficient between these two sets of residuals. Thus, the partial correlation coefficient between y and x_2 is the residual correlation between them after removing the effect of x_1 on both of them by regressing them on x_1.

- Let $\mathrm{SSE}(x_1)$ and $\mathrm{SSE}(x_1, x_2)$ denote the SSEs from regressions of y on x_1 and on x_1, x_2, respectively. Then

$$r_{yx_2|x_1}^2 = \frac{\mathrm{SSE}(x_1) - \mathrm{SSE}(x_1, x_2)}{\mathrm{SSE}(x_1)} = 1 - \frac{\mathrm{SSE}(x_1, x_2)}{\mathrm{SSE}(x_1)} \qquad (3.29)$$

and $r_{yx_2|x_1}$ is the square root of the above, its sign being that of $\widehat{\beta}_2$ in the regression of y on x_1, x_2. Thus, $r_{yx_2|x_1}^2$ is the additional relative contribution to reduction in SSE due to x_2 compared to the SSE with only x_1 in the model.

This formula extends the definition $R^2 = 1 - \mathrm{SSE}/\mathrm{SST}$ by replacing SSE with $\mathrm{SSE}(x_1, x_2)$ and SST with $\mathrm{SSE}(x_1)$. Essentially, R^2 compares the SSE of the full model consisting of p variables with $\mathrm{SSE} = \mathrm{SST}$ of the null model consisting only of the intercept term, which is treated as the partial model. Here $\{x_1, x_2\}$ is the full model and $\{x_1\}$ is the partial model.

- A third method of computing $r_{yx_2|x_1}$ is from the bivariate correlations r_{yx_1}, r_{yx_2}, and $r_{x_1x_2}$ using the formula:

$$r_{yx_2|x_1} = \frac{r_{yx_2} - r_{yx_1}r_{x_1x_2}}{\sqrt{(1 - r_{yx_1}^2)(1 - r_{x_1x_2}^2)}}. \qquad (3.30)$$

This formula is the sample version of the population partial correlation coefficient formula derived in Equation (3.36).

The above formulae can be extended by conditioning on multiple predictors. Denote the partial correlation coefficient between y and x_p conditioned on $x_1, \ldots, x_{p-1}$ by $r_{yx_p|x_1,\ldots,x_{p-1}}$. Equation (3.29) generalizes to

$$r^2_{yx_p|x_1,\ldots,x_{p-1}} = 1 - \frac{\mathrm{SSE}(x_1,\ldots,x_p)}{\mathrm{SSE}(x_1,\ldots,x_{p-1})}. \tag{3.31}$$

Then $r_{yx_p|x_1,\ldots,x_{p-1}}$ is the square root of the above, its sign being that of $\widehat{\beta}_p$ in the regression of y on $x_1, \ldots, x_p$.

3.5.1   Test of significance of partial correlation coefficient

The test of significance on the sample partial correlation coefficient can be derived using the extra SS method of Section 3.3.4. There we saw that the test of significance for entering x_p to the model that already includes $x_1, \ldots, x_{p-1}$ is based on the following F-statistic:

$$F_p = \frac{[\mathrm{SSE}(x_1,\ldots,x_{p-1}) - \mathrm{SSE}(x_1,\ldots,x_p)]/1}{\mathrm{SSE}(x_1,\ldots,x_p)/[n-(p+1)]}. \tag{3.32}$$

Using simple algebra, this F-statistic can be written as

$$F_p = \frac{r^2_{yx_p|x_1,\ldots,x_{p-1}}[n-(p+1)]}{1 - r^2_{yx_p|x_1,\ldots,x_{p-1}}}.$$

The corresponding t-statistic is the square root of F_p:

$$t_p = \sqrt{F_p} = \frac{r_{yx_p|x_1,\ldots,x_{p-1}}\sqrt{n-(p+1)}}{\sqrt{1 - r^2_{yx_p|x_1,\ldots,x_{p-1}}}}. \tag{3.33}$$

We conclude that $r_{yx_p|x_1,\ldots,x_{p-1}}$ is significant at level α if $F_p > f_{1,n-(p+1),\alpha}$ or equivalently if $|t_p| > t_{n-(p+1),\alpha/2}$. By comparing this test with (2.25), we see that it is a generalization of the test on the bivariate correlation coefficient. Furthermore, this t-test is exactly the same as the t-test used to test the significance of $\widehat{\beta}_p$ in the multiple regression of y on $x_1, \ldots, x_p$.

Example 3.11 *(College GPA and Entrance Test Scores: Partial Correlation Coefficients)* Denoting GPA $= y$, Verbal $= x_1$, and Math $= x_2$, we want to calculate $r_{yx_1|x_2}$ and $r_{yx_2|x_1}$. By running separate regressions of GPA on Verbal and on Math, we get $\mathrm{SSE}(x_1) = 13.5186$ and $\mathrm{SSE}(x_2) = 12.6063$. From the ANOVA Table 3.4, we get $\mathrm{SSE}(x_1, x_2) = 5.9876$. Hence,

$$r^2_{yx_1|x_2} = 1 - \frac{5.9876}{12.6063} = 0.5250 \quad \text{and} \quad r^2_{yx_2|x_1} = 1 - \frac{5.9876}{13.5186} = 0.5571.$$

Thus, $r_{yx_1|x_2} = \sqrt{0.5250} = 0.7246$ and $r_{yx_2|x_1} = \sqrt{0.5571} = 0.7464$. The coefficients of both are positive since the corresponding regression coefficients are positive.

The same values can be obtained using the formula (3.30). The bivariate correlations are $r_{yx_1} = 0.5291, r_{yx_2} = 0.5732$, and $r_{x_1x_2} = -0.1069$. By substituting these values in (3.30), we get

$$r_{yx_1|x_2} = \frac{0.5291 + 0.1069 \times 0.5732}{\sqrt{(1 - 0.1069^2)(1 - 0.5732^2)}} = 0.7246$$

and

$$r_{yx_2|x_1} = \frac{0.5732 + 0.1069 \times 0.5291}{\sqrt{(1 - 0.1069^2)(1 - 0.5291^2)}} = 0.7464.$$

The t-statistics to test the significance of $r_{yx_1|x_2}$ and $r_{yx_2|x_1}$ are

$$t_{yx_1|x_2} = \frac{0.7246\sqrt{37}}{\sqrt{1 - 0.7246^2}} = 6.395 \quad \text{and} \quad t_{yx_2|x_1} = \frac{0.7464\sqrt{37}}{\sqrt{1 - 0.7464^2}} = 6.822,$$

which are identical to the ones for the regression coefficients given in the R output in Example 3.3. Both are highly significant. $\square$

3.6   Special topics

3.6.1   Dummy variables

A dummy variable is an indicator variable used to denote the presence or absence of a given categorical attribute. In the case of a binary variable such as treated versus untreated (control) patients in a clinical trial, we need only one dummy variable, e.g. $x_1 = 0$ for control patients and $x_1 = 1$ for treated patients. Suppose the model is $E(y) = \beta_0 + \beta_1 x_1$, where y is some measure of patient response. Then the models for control and treated patients are

$$\text{Control}: \quad E(y) = \beta_0 \quad \text{and} \quad \text{Treated}: \quad E(y) = \beta_0 + \beta_1.$$

Thus, β_1 is the difference between the expected response of the treated patients and of the control patients and is called the **treatment effect**.

Next, consider a categorical variable with $c > 2$ categories, e.g. ethnicity with five groups: White, Black, Hispanic, Asian, and Other. What is wrong with coding them $1, \dots, 5$ and treating them as numerical data? It is wrong since it implies a linear order. Thus, if the coefficient of ethnicity is β, then the coefficient for Whites will be β, for Blacks 2β, etc., which is wrong since not only the coding $1, \dots, 5$ is arbitrary for a nominal variable such as ethnicity, but even if the variable were ordinal, such as a course grade, the grades A, B, C, D, F are not necessarily linearly ordered.

Note that we need only $c - 1$ dummy variables to code c categories. For example, if we use $x_1, \ldots, x_5$ as dummy variables for the five ethnic groups, then since exactly one $x_j = 1$ and other $x_j = 0$ for each person, so we will have $x_1 + \cdots + x_5 = 1$ for all persons. As we will see in Chapter 4, this will cause multicollinearity (linear dependence among the predictor variables). Hence, we need to use only four dummy variables for any four groups leaving the fifth group as the reference for comparison. Any category can be chosen as the reference. The β coefficients for the other categories represent the differences from that reference.

Example 3.12 *(Used Car Prices: Regression Model 1)* In this example, there are six categorical variables, three with two categories each (Cruise, Sound, and Leather) and three with more than two categories (Make, Model, and Type). In fact, Doors is also a categorical variable since it has only two values: 2 and 4. However, it is perfectly correlated with the Type variable since Coupe and Convertible have two doors, while Hatchback, Sedan, and Wagon have four doors. So we drop Doors as a predictor variable. For now, we will also ignore the Model variable.

R chooses the first listed category of any categorical variable as the reference by default. Thus, it chooses Buick as the reference category for Make and Convertible as the reference category for Type. So the regression coefficients for the other categories for Make and Type represent the differences from these reference categories.

For fitting the predictive model, we divided the data equally into a training set and a test set by putting all odd-numbered observations into the training set and all even-numbered observations into the test set. We did this deterministic (instead of random) split to enable readers verify the results given here using their own codes but the same training and test sets.

We log-transformed Price so that it better satisfies the model assumptions such as normality and homoscedasticity. This point will be discussed in more detail in Chapter 4. The coefficients, their standard errors, t-statistics, and P-values are listed in Table 3.6. The R^2 for the model is 95.26% and the F-statistic equals 516.8 on 15 and 386 d.f. which is highly significant.

We see that none of the dummy variables for Cruise, Sound, or Leather is significant at $\alpha = 0.05$. We may drop these variables from the model. Finally, the number of cylinders is also nonsignificant with a P-value of 0.074, the reason being that the engine size (Liter) has a very high correlation of 0.958 with it. So we keep only the engine size in the model. After refitting, the resulting model is

$$\widehat{\log(\text{Price})} = 4.169 - 0.0035 \text{ Mileage} + 0.0976 \text{ Liter} + 0.1948 \text{ Cadillac}$$
$$- 0.0540 \text{ Chevrolet} - 0.0413 \text{ Pontiac} + 0.2463 \text{ SAAB}$$
$$- 0.0463 \text{ Saturn} - 0.1337 \text{ Coupe} - 0.1574 \text{ Hatchback}$$
$$- 0.1412 \text{ Sedan} - 0.0714 \text{ Wagon}.$$

Table 3.6. Statistics for used car prices regression: model 1

Predictor	Coef	SE	t	P	Predictor	Coef	SE	t	P
Constant	4.1883	0.0222	188.352	0.000	Chevrolet	−0.0575	0.0081	−7.129	0.000
Mileage	−0.0035	0.0002	−14.227	0.000	Pontiac	−0.0402	0.0082	−4.905	0.000
Cylinder	−0.0127	0.0071	−1.790	0.074	SAAB	0.2357	0.0103	22.922	0.000
Liter	0.1095	0.0079	13.811	0.000	Saturn	−0.0450	0.0107	−4.221	0.000
Cruise	0.0097	0.0060	1.627	0.105	Coupe	−0.1404	0.0106	−13.183	0.000
Sound	0.0037	0.0046	0.801	0.424	Hatchback	−0.1536	0.0124	−12.375	0.000
Leather	0.0082	0.0048	1.694	0.091	Sedan	−0.1437	0.0092	−15.600	0.000
Cadillac	0.2004	0.0109	18.333	0.000	Wagon	−0.0730	0.0115	−6.372	0.000

All of the variables in this model are highly significant ($P < 0.001$) and the R^2 is 95.10%. When this model is applied to the test set, the resulting R^2 is 95.09%, almost exactly the same. $\square$

Example 3.13 *(Used Car Prices: Regression Model 2)* In this example, we will take the Model variable into account. Note that the model variable is nested under the Make variable, e.g. the Century model is available only for Buick. R uses the first Model as the reference for each make and so omits the dummy variable for that model from the regression equation. Buick has four Models: Century, Lacrosse, Lesabre, and Park Avenue, Cadillac has six models: CST-V, CTS, Deville, STS-V6, STS-V8, and XLR-V8, Chevrolet has eight models: AVEO, Cavalier, Classic, Cobalt, Corvette, Impala, Malibu, and Monte Carlo, Pontiac has seven models: Bonneville, G6, Grand Am, Grand Prix, GTO, Sunfire, and Vibe, SAAB has five models: 9-3, 9-3 HO, 9-5, 9-5 HO, and 9-2X AWD, and Saturn has two models: Ion and L Series.

After eliminating nonsignificant predictors, the regression coefficients for the final model along with their standard errors, t-statistics, and P-values are given in Table 3.7. The R^2 for this model is 97.90%, which is a significant improvement over $R^2 = 95.10\%$ for model 1. When this model is applied to the test set, the resulting R^2 is 97.66%, almost exactly the same. $\square$

3.6.2   Interactions

Often, we want to model the joint effect of two or more variables on the response variable. Such a joint effect is called **interaction**. For example, sale price and discount coupon have separate positive effects on the sales of a product but if both are offered together, then the total effect on the sales can be greater than their sum, in which case we say that there is a positive interaction between sale price and coupon. On the other hand, if the total effect on the sales is less than their sum, then we have a negative interaction. Interaction between two or more predictor variables is typically modeled by their product, i.e. by including the term $x_1 x_2$ as the interaction between x_1 and x_2. If there is no interaction between the given variables, then the model is said to be **additive** in those variables.

Example 3.14 *(Interaction   Between   Two   Categorical   Variables)* Consider the fictitious mean salary data in Table 3.8 and treat the cell entries as the true expected salary values. We want to model the interaction between Gender and Race. Suppose Gender is coded as $x_1 = 0$ for females, $x_1 = 1$ for males and Race is coded as $x_2 = 0$ for non-Whites, $x_2 = 1$ for Whites. Ignoring any other predictors, the model with interaction is $E(y) = \beta_0 + \beta_1 x_1 + \beta_2 x_2 + \beta_3 x_1 x_2$, which can be written as four separate

Table 3.7. Statistics for used car prices regression: model 2

Predictor	Coef	SE	t	P	Predictor	Coef	SE	t	P
Constant	4.1649	0.0131	318.85	0.000	Chevrolet				
Mileage	−0.0034	0.000 17	−19.98	0.000	Cavalier	0.0467	0.0085	5.46	0.000
Liter	0.0556	0.0039	13.44	0.000	Classic	0.0726	0.0136	5.33	0.000
Cadillac	0.3520	0.0129	27.36	0.000	Cobalt	0.0816	0.0087	9.41	0.000
Chevrolet	−0.0727	0.0085	−8.50	0.000	Corvette	0.2711	0.0220	12.34	0.000
Pontiac	0.0870	0.0086	10.15	0.000	Impala	0.1487	0.0129	11.53	0.000
SAAB	0.3151	0.0073	42.99	0.000	Malibu	0.0969	0.0105	9.25	0.000
Coupe	−0.0828	0.0090	−9.20	0.000	Monte Carlo	0.1438	0.0139	10.34	0.000
Hatchback	−0.0786	0.0106	−7.43	0.000	Pontiac				
Sedan	−0.0908	0.0080	−11.34	0.000	Grand Am	−0.0882	0.0106	−8.32	0.000
Wagon	−0.0822	0.0102	−8.08	0.000	Grand Prix	−0.0515	0.0086	−6.00	0.000
Buick					GTO	0.0531	0.0167	3.19	0.002
Lacrosse	0.1176	0.0106	11.46	0.000	Sunfire	−0.1206	0.0147	−8.20	0.000
Lesabre	0.1573	0.0117	13.50	0.000	SAAB				
Park Avenue	−0.0806	0.0125	−6.43	0.000	9-5	0.0264	0.0097	2.74	0.006
Cadillac					9-5 HO	0.1122	0.0211	5.32	0.000
CTS	−0.0398	0.0158	−2.52	0.012	9-2X AWD	−0.0904	0.0211	−4.29	0.000
Deville	−0.0574	0.0098	−5.86	0.000	Saturn				
XLR-V8	0.0876	0.0160	5.48	0.000	L Series	0.0308	0.0134	2.30	0.022

Table 3.8. Average salaries by gender and race

Race

		Non-White	White
Gender	Female	$40K	$50K
	Male	$45K	$65K

models:

$$
E(y) = \begin{cases}
\beta_0 & \text{for Female and Non-Whites} \\
\beta_0 + \beta_1 & \text{for Male and Non-Whites} \\
\beta_0 + \beta_2 & \text{for Female and Whites} \\
\beta_0 + \beta_1 + \beta_2 + \beta_3 & \text{for Male and Whites}
\end{cases}
$$

Solving this system of equations we get

$$
\beta_0 = \$40K, \quad \beta_1 = \$5K, \quad \beta_2 = \$10K, \quad \beta_3 = \$10K.
$$

Observe that $\beta_1 = \$5K$ is the difference in the average salary difference between Males and Females for Non-Whites. A similar difference for Whites equals $65K − $50K = $15K. The difference between these two differences equals $10K, which is the interaction β_3. Similarly, the average salary difference between Whites and Non-Whites is $65K − $45K = $20K for Males and $\beta_2 = $50K − $40K = $10K for Females. The difference between these two differences again equals $10K, which is the interaction β_3. If these two differences were equal, then the interaction would be zero, and the data would satisfy an **additive model**.

Note that $\beta_1 = \$5K$ and $\beta_2 = \$10K$ are not what are commonly referred to as the Gender and Race effects, respectively. The Gender effect is defined as the average of the Males versus Females salary differences averaged over the two races, which equals ($5K + $15K)/2 = $10K. Similarly, the Race effect is defined as the average of the Whites versus Non-Whites salary differences averaged over the two genders, which equals ($10K + $20K)/2 = $15K.   □

Next, let us consider interaction between a dummy variable and a **covariate** (i.e. a continuous predictor variable). In the example from Section 3.6.1 about treated versus untreated (control) patients in a clinical trial with $x_1 = 0$ for control and $x_1 = 1$ for treated patients, suppose we include Age (x_2) as a covariate and postulate the model

$$
E(y) = \beta_0 + \beta_1 x_1 + \beta_2 x_2 + \beta_3 x_1 x_2,
$$

where the x_1x_2 term represents the Treatment $\times$ Age interaction. To understand what this interaction means write this model as two separate models:

Control : $E(y) = \beta_0 + \beta_2x_2$ and   Treated : $E(y) = (\beta_0 + \beta_1) + (\beta_2 + \beta_3)x_2.$

Thus, we get two nonparallel regression lines with different intercepts and different slopes as shown in Figure 3.3. The difference between control and treated patients depends on age, so the treatment effect depends on the patient's age. If $\beta_3 = 0$, then there is no interaction, and we get two parallel regression lines with different intercepts, so the effect of the treatment is the same at all ages and does not vary with the age.

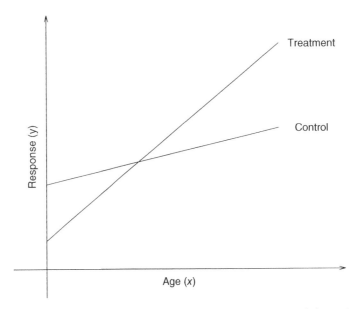

Figure 3.3. *Nonparallel regression lines between response (y) and age (x) for control and treated patients.*

Example 3.15 (*College GPA and Entrance Test Scores: Interaction Effect*) In Example 3.8, we fitted a quadratic model to the GPA data which we found to give a significantly better fit than the linear model. In this example, we will add the interaction term to see whether it further improves the fit of the model significantly. Thus consider the full second-degree model:

$$E(\text{GPA}) = \beta_0 + \beta_1\text{Verbal} + \beta_2\text{Math} + \beta_3\text{Verbal}^2 + \beta_4\text{Math}^2 + \beta_5\text{Verbal} \times \text{Math}.$$

The regression output for this model is given below. We see that the interaction term has a P-value $= 0.103$, which barely fails to be significant at $\alpha = 0.10$.

```
Call:
lm(formula = GPA ~ Verbal * Math + I(Verbal ^ 2) + I(Math ^ 2),
  data = gpa)

Coefficients:
             Estimate Std. Error t value Pr(>|t|)
(Intercept) -9.9167631  1.3544134  -7.322 1.75e-08 ***
Verbal       0.1668098  0.0212447   7.852 3.85e-09 ***
Math         0.1375972  0.0267340   5.147 1.11e-05 ***
I(Verbal ^ 2) -0.0011082  0.0001173 -9.449 4.88e-11 ***
I(Math ^ 2)  -0.0008433  0.0001594 -5.290 7.23e-06 ***
Verbal:Math  0.0002411  0.0001440   1.675   0.103
---

Residual standard error: 0.1871 on 34 degrees of freedom
Multiple R-squared:  0.9366, Adjusted R-squared:  0.9272
F-statistic: 100.4 on 5 and 34 DF,  p-value: < 2.2e-16
```

Table 3.9. ANOVA table for GPA versus entrance test scores: quadratic regression with interaction

Source	SS	d.f.	MS	F	P
Regression	17.5827	5	3.5165	100.41	0.000
Residual error	1.1908	34	0.0350		
Total	18.7735	39			

The ANOVA for this second-degree model with interaction is given in Table 3.9. We get the same result if we use the extra SS test by treating the quadratic model as the reduced model and this model as the full model. Then the F-statistic for $H_0 : \beta_5 = 0$ equals

$$F = \frac{(1.2890 - 1.1908)/1}{1.1908/34} = 2.806,$$

which for $\alpha = 0.10$ can be compared with $f_{1,34,0.10} = 2.859$. Since $F = 2.806 < 2.859$, we fail to reject H_0. The equivalence between the t-test on Verbal×Math interaction in the R output and the F-test calculated using the extra SS method can be verified by checking that $t = \sqrt{2.806} = 1.675$ and $t_{34,0.05} = \sqrt{f_{1,34,0.10}} = \sqrt{2.859} = 1.691$. $\square$

3.6.3   Standardized regression

As mentioned earlier in Section 3.2.2, **standardized regression coefficients** are unitless and hence can be compared with each other. They are also useful for computational purposes as we will explain below.

To perform standardized regression, we first standardize all variables:

$$y_i^* = \frac{y_i - \overline{y}}{s_y} \quad \text{and} \quad x_{ij}^* = \frac{x_{ij} - \overline{x}_j}{s_{x_j}} \quad (i = 1, \ldots, n, j = 1, \ldots, p),$$

where $\overline{y}$ and $\overline{x}_j$'s are the sample means, and s_y and s_{x_j} are the sample standard deviations of the corresponding variables. Denote the estimated standardized regression coefficients by $\widehat{\beta}_j^*$ ($j = 0, \ldots, p$). Since all variables are standardized with 0 means, it follows from (3.8) that $\widehat{\beta}_0^* = 0$. So we can omit the intercept term from the regression equation.

Let $\boldsymbol{y}^*$ denote the n-vector of standardized y_i^*'s, $\boldsymbol{X}^*$ denote the $n \times p$ matrix of standardized x_{ij}^*'s (note that there is no column of all 1's corresponding to the intercept term) and $\widehat{\boldsymbol{\beta}}^* = (\widehat{\beta}_1^*, \ldots, \widehat{\beta}_p^*)'$ denote the p-vector of standardized $\widehat{\beta}_j^*$'s. Then analogous to Equation (3.7), we have

$$\widehat{\boldsymbol{\beta}}^* = (\boldsymbol{X}^{*\prime}\boldsymbol{X}^*)^{-1}\boldsymbol{X}^{*\prime}\boldsymbol{y}^*.$$

The (j, k)th entry of $\boldsymbol{X}^{*\prime}\boldsymbol{X}^*$ equals

$$\sum_{i=1}^n x_{ij}^* x_{ik}^* = \frac{1}{s_{x_j} s_{x_k}} \sum_{i=1}^n (x_{ij} - \overline{x}_j)(x_{ik} - \overline{x}_k) = \frac{(n-1)\mathrm{Cov}(x_j, x_k)}{s_{x_j} s_{x_k}}$$

$$= (n-1) r_{x_j, x_k},$$

where $\mathrm{Cov}(x_j, x_k)$ is the sample covariance between x_j and x_k and r_{x_j, x_k} is the sample correlation coefficient between x_j and x_k. Similarly, the jth entry of $\boldsymbol{X}^{*\prime}\boldsymbol{y}^*$ equals $(n-1) r_{y, x_j}$, where r_{y, x_j} is the sample correlation coefficient between y and x_j.

Let $\boldsymbol{R}$ denote the $p \times p$ sample correlation matrix among $x_1, \ldots, x_p$ with entries r_{x_j, x_k} for $j \neq k$ and let $\boldsymbol{r} = (r_{y, x_1}, \ldots, r_{y, x_p})'$ denote the vector of sample correlations between y and $x_1, \ldots, x_p$. Then from the above it follows that

$$\boldsymbol{X}^{*\prime}\boldsymbol{X}^* = (n-1)\boldsymbol{R} \quad \text{and} \quad \boldsymbol{X}^{*\prime}\boldsymbol{y}^* = (n-1)\boldsymbol{r}.$$

Therefore the formula for $\widehat{\boldsymbol{\beta}}^*$ simplifies to

$$\widehat{\boldsymbol{\beta}}^* = \boldsymbol{R}^{-1}\boldsymbol{r}. \tag{3.34}$$

Thus, to calculate $\widehat{\boldsymbol{\beta}}^*$, we only need to know all the sample correlation coefficients. Then the unstandardized regression coefficients can be calculated using

$$\widehat{\beta}_j = \widehat{\beta}_j^* \left(\frac{s_y}{s_{x_j}} \right) \quad (j = 1, \ldots, p)$$

and $\widehat{\beta}_0$ can be calculated from (3.8).

This method of computation of $\widehat{\beta}_j$ is numerically more stable because all entries of $\boldsymbol{R}$ and $\boldsymbol{r}$ are between -1 and $+1$, whereas entries of $\boldsymbol{X}'\boldsymbol{X}$ and $\boldsymbol{X}'\boldsymbol{y}$ required in the direct computation of $\widehat{\boldsymbol{\beta}}$ using (3.7) can vary over wide ranges and have diverse scales. The formula (3.34) is also useful in that it clearly pinpoints how the multicollinearity problem arises if the x_j's are highly correlated with each other in which case $\boldsymbol{R}$ is close to being singular.

Example 3.16 *(College GPA and Entrance Test Scores: Standardized Regression Coefficients)* The standardized regression coefficients $\widehat{\beta}_1^*$ and $\widehat{\beta}_2^*$ of Verbal and Math can be computed given $\widehat{\beta}_1 = 0.0257$ and $\widehat{\beta}_2 = 0.0336$ from Example 3.3 and the standard deviations

$$s_y = 0.6938, \quad s_{x_1} = 16.1019, \quad s_{x_2} = 13.1481.$$

So we get

$$\widehat{\beta}_1^* = 0.0257 \left(\frac{16.1019}{0.6938} \right) = 0.5964 \quad \text{and} \quad \widehat{\beta}_2^* = 0.0336 \left(\frac{13.1481}{0.6938} \right) = 0.6367.$$

Notice that $\widehat{\beta}_1^*$ and $\widehat{\beta}_2^*$ are close to each other indicating that Verbal and Math have roughly similar effects on GPA. □

3.7 Modern extensions*

In Section 2.4, we briefly reviewed some methods of extending the simple linear regression model using techniques such as regression splines. Those techniques can be extended to $p \geq 2$ predictors by using **generalized additive models (GAMs)**, which have the form $y = \beta_0 + \beta_1 f_1(x_1) + \cdots + \beta_p f_p(x_p) + \varepsilon$, where $f_1(\cdot), \ldots, f_p(\cdot)$ are specified nonlinear functions of the predictors $x_1, \ldots, x_p$, respectively. The main drawback of GAMs is that they are additive, i.e. do not include interactions between predictors. However, it is possible to add interaction terms $f_{jk}(x_j x_k)$ for selected interactions $x_j x_k$. This requires the use of two-dimensional regression splines.

Two other techniques are discussed in more detail below.

3.7.1   Regression trees

A rich class of methods is based on what are called **classification and regression trees (CART)**. They are essentially decision trees that are constructed in a top-down manner by recursive binary splitting on one predictor at a time. The final result is partitioning of the **feature space**, denoted by $\mathcal{X}$, into disjoint rectangular regions, say $R_1, \ldots, R_N$ such that $\cup_{j=1}^{N} R_j = \mathcal{X}$, We will discuss classification trees in Chapter 7. Regression trees use the mean (or the median) of y_i's for all observations $\boldsymbol{x}_i \in R_j$ $(j = 1, \ldots, N)$ as the common fitted value for them. Denote this common fitted value by $\overline{y}_{R_j}$. The partitioning of $\mathcal{X}$ is aimed at minimizing $\text{SSE} = \sum_{j=1}^{N} \sum_{\boldsymbol{x}_i \in R_j} (y_i - \overline{y}_{R_j})^2$.

The recursive binary splitting on one variable at a time proceeds as follows. At the first step, a binary split is made on one of the p predictors, say x_1, using a cutpoint c_1, partitioning $\mathcal{X}$ into regions $R_{1,1} = \{\boldsymbol{x} \in \mathcal{X} | x_1 < c_1\}$ and $R_{1,2} = \{\boldsymbol{x} \in \mathcal{X} | x_1 \geq c_1\}$. These two regions are represented by two **branches** of the decision tree from the primary **node**, which consists of the entire feature space $\mathcal{X}$. The predictor x_1 and its cutpoint c_1 are chosen to minimize

$$\text{SSE} = \sum_{\boldsymbol{x}_i \in R_{1,1}} (y_i - \overline{y}_{R_{1,1}})^2 + \sum_{\boldsymbol{x}_i \in R_{1,2}} (y_i - \overline{y}_{R_{1,2}})^2.$$

Finding the optimum partitioning of $\mathcal{X}$ may appear to be a daunting computational task given that many of the predictors are continuous. However, if a grid of n_j points is specified on each predictor x_j, then a numerical search needs to be made only over $\sum_{j=1}^{p} n_j$ points since we make a binary split on only one variable at a time. So this is a manageable task if the number of predictors is not too large. This process is then repeated to find the best predictor and its cutpoint within each region (branch) $R_{1,1}$ and $R_{1,2}$ to minimize the overall SSE. (Note that the same predictor chosen at an earlier step may be chosen again at any subsequent step, but with a different cutpoint.) The goal at each step is to minimize the SSE summed over the partitioned feature space. This process is continued until some convergence criterion (e.g. decrease in SSE is less than a specified tolerance) is satisfied. The final regions $R_1, \ldots, R_N$ into which the feature space is partitioned correspond to the N terminal nodes, referred to as **leaf nodes**. Oftentimes, the regression tree tends to overgrow (too many leaf nodes) which has the effect of overfitting. It is advisable to prune the tree in such cases by cross-validation. We do not discuss this aspect of building a regression tree.

Example 3.17 *(College GPA and Entrance Test Scores: Regression Tree)* An example of a regression tree for the GPA data is shown in Figure 3.4.

This regression tree can be summarized by the following set of rules.

1. If Verbal < 55.5, then $\widehat{\text{GPA}} = 1.814$.

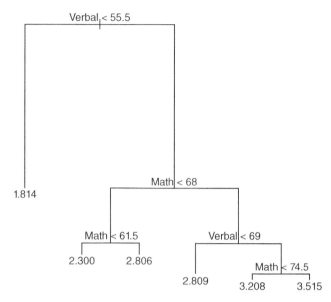

Figure 3.4. *Regression tree for predicting GPA from Verbal and Math scores (left branch at each node corresponds to "yes" and the right branch corresponds to "no").*

   2. If Verbal $\geq$ 55.5 and Math < 61.5, then $\widehat{\text{GPA}}$ = 2.300.

   3. If Verbal $\geq$ 55.5 and 61.5 $\leq$ Math < 68, then $\widehat{\text{GPA}}$ = 2.806.

   4. If 55.5 $\leq$ Verbal < 69 and Math $\geq$ 68, then $\widehat{\text{GPA}}$ = 2.809.

   5. If Verbal $\geq$ 69 and 68 $\leq$ Math < 74.5, then $\widehat{\text{GPA}}$ = 3.208.

   6. If Verbal $\geq$ 69 and Math $\geq$ 74.5, then $\widehat{\text{GPA}}$ = 3.515.

Figure 3.5 gives another way to summarize this regression tree. It shows the six disjoint regions into which the feature space is partitioned and the fitted GPA value for each region. This is referred to as the **partition plot**, which is convenient to use if the feature space is two-dimensional.

The SSE for this regression tree model equals 3.828. Compare this with SSE = 5.988 for the linear model (from the ANOVA in Table 3.4) and SSE = 1.289 for the quadratic model (from the ANOVA in Table 3.5). As seen before, the relationship of GPA with Verbal and Math is quadratic. The regression tree model tries to capture this quadratic relationship by having second splits on both Verbal and Math. However, because the fitted values are restricted to be piecewise constant, the SSE is not as small as achieved by the quadratic model, but is smaller than that achieved by the linear model.   □

Regression trees are easy to interpret and have a nice graphical representation. However, their predictive performance can be improved by further building

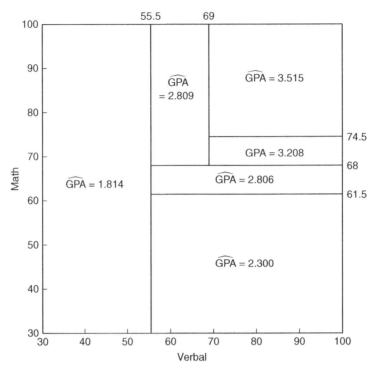

Figure 3.5. *Partitioning of the feature space induced by the regression tree in Figure 3.4.*

on them using the methods of **random forests** and **boosting**. A random forest grows multiple regression trees on multiple randomized data sets and averages them. Randomized data sets are obtained from the training set by resampling observations or variables. Boosting grows successive trees on residuals obtained each time from the previous tree. The final tree is obtained by adding up these successive trees.

3.7.2   Neural nets

Another class of methods is based on **neural nets**, developed in the area of computer science and, as the term indicates, inspired by the architecture of the brain. A neural net consists of nodes (analogous to neurons) connected by links (analogous to minute gaps along which synapses travel across neurons). It consists of an **input layer** and an **output layer** of nodes with intervening layers called **hidden layers**. The nodes in successive layers are connected by links. The number of input nodes corresponds to the number of input variables and the number of output nodes corresponds to the number of output variables. We will restrict to a single output variable y. It is convenient to add what is called a bias node as part of the input layer with a constant input of 1 to account for the intercept term. This node has bypass connections to nodes in all layers since, as we will see below, the intercept term is present in inputs to all nodes.

We will illustrate these concepts with a simple example of a feed-forward neural network with a single hidden layer. Consider $p \geq 2$ input nodes (denoted by $I_1, \ldots, I_p$), corresponding to p input variables, $x_1, \ldots, x_p$, a bias node (denoted by I_0), $q \geq 2$ hidden nodes (denoted by $H_1, \ldots, H_q$) and an output node (denoted by O) for y. The input to each hidden node H_k consists of a weighted linear combination of the outputs from the bias node and the p input nodes:

$$ w_{0k}^{(1)} x_0^{(1)} + \cdots + w_{pk}^{(1)} x_p^{(1)} = w_{0k}^{(1)} + \sum_{j=1}^{p} w_{jk}^{(1)} x_j^{(1)} \quad (k = 1, \ldots, q), $$

where $x_j^{(1)} = x_j \ (1 \leq j \leq p)$, $x_0^{(1)} = x_0 = 1$ and $w_{jk}^{(1)}$ are unknown weights which are estimated to train the network. The hidden node H_k then transforms this input using an **activation function** $g^{(1)}(\cdot)$ to give the output

$$ x_k^{(2)} = g^{(1)} \left(w_{0k}^{(1)} + \sum_{j=1}^{p} w_{jk}^{(1)} x_j^{(1)} \right) \quad (k = 1, \ldots, q). $$

Next, another weighted linear combination of the outputs from the hidden nodes, $w_0^{(2)} + \sum_{k=1}^{q} w_k^{(2)} x_k^{(2)}$, is fed into the terminal output node which transforms it using another activation function $g^{(2)}(\cdot)$ to give

$$ \widehat{y} = g^{(2)} \left(w_0^{(2)} + \sum_{k=1}^{q} w_k^{(2)} x_k^{(2)} \right). $$

A common choice for the activation functions is the sigmoidal function (logistic transform): $g^{(1)}(x) = g^{(2)}(x) = 1/(1 + e^{-x})$. If the activation functions are linear, then the neural net results in a multiple linear regression model. This process is repeated for all n observations.

The weights may be regarded as tuning parameters and are determined to minimize an objective function such as SSE $= \sum_{i=1}^{n} (y_i - \widehat{y}_i)^2$. This is referred to as **training** of a neural net. Figure 3.6 shows a simple feed-forward network with a bias node, two input nodes, one hidden layer with two hidden nodes and an output node. The weights are shown on the links connecting the appropriate nodes. This network is used in the example below.

Example 3.18 (College GPA and Entrance Test Scores: Neural Net) We fitted the neural net model shown in Figure 3.6 to the GPA data using sigmoidal activation functions. This model has nine weights (tuning parameters) that need to be estimated. The predictors were standardized, which is a necessary first step when fitting a neural net model. The resulting

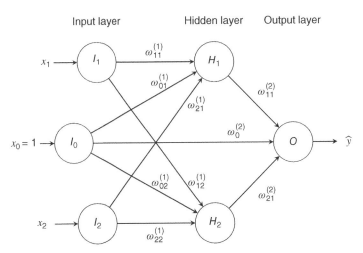

Figure 3.6. *Neural net with two input nodes, two hidden nodes, and one output node.*

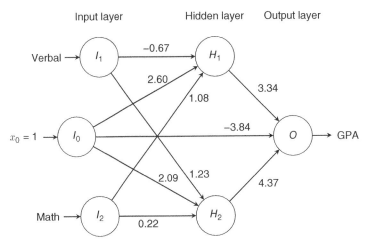

Figure 3.7. *Neural net with two input nodes, two hidden nodes, and one output node for GPA data.*

weights are shown on the links in Figure 3.7. The SSE for this model is 1.439, which is slightly higher than 1.289 obtained with the quadratic model. A model with three hidden nodes (which has 13 tuning parameters) yielded a smaller SSE = 1.162 but seems to be an overfit based on cross-validation SSE. The number of tuning parameters increases exponentially as the number of hidden layers and nodes increase. So complex neural net models tend to overfit unless sufficiently large amount of data is available.   □

Deep learning networks are neural networks with many hidden layers and associated with large number of nodes. They can be applied to supervised learning problems (e.g. predictive analytics) or unsupervised learning problems (e.g. pattern recognition). They are particularly useful in complex problems such as audio/video and speech recognition. The number of tuning parameters can be very large running into millions. Therefore, they require far more data to train them than small neural networks.

3.8 Technical notes*

3.8.1 Derivation of the LS estimators

The LS estimation problem can be written as minimizing the LS criterion Q defined in (3.4), i.e.

$$\min_{\boldsymbol{\beta}} (\boldsymbol{y} - \boldsymbol{X}\boldsymbol{\beta})'(\boldsymbol{y} - \boldsymbol{X}\boldsymbol{\beta}) = \min_{\boldsymbol{\beta}}[\boldsymbol{y}'\boldsymbol{y} - 2\boldsymbol{\beta}'\boldsymbol{X}'\boldsymbol{y} + \boldsymbol{\beta}'\boldsymbol{X}'\boldsymbol{X}\boldsymbol{\beta}]. \qquad (3.35)$$

The LS estimator $\widehat{\boldsymbol{\beta}}$ is found by taking the vector derivative,

$$\frac{dQ}{d\boldsymbol{\beta}} = \left(\frac{\partial Q}{\partial \beta_0}, \frac{\partial Q}{\partial \beta_1}, \dots, \frac{\partial Q}{\partial \beta_p}\right)',$$

and setting it equal to the null vector $\boldsymbol{0}$. The term $\boldsymbol{y}'\boldsymbol{y}$ does not contribute to the derivative since it does not involve $\boldsymbol{\beta}$. The next term $-2\boldsymbol{\beta}'\boldsymbol{X}'\boldsymbol{y}$ is linear in $\boldsymbol{\beta}$ and its derivative can be shown to be $-2\boldsymbol{X}'\boldsymbol{y}$. Finally, the last term $\boldsymbol{\beta}'\boldsymbol{X}'\boldsymbol{X}\boldsymbol{\beta}$ is quadratic in $\boldsymbol{\beta}$ and its derivative can be shown to be $2\boldsymbol{X}'\boldsymbol{X}\boldsymbol{\beta}$. So the equation for $\widehat{\boldsymbol{\beta}}$ is

$$-2\boldsymbol{X}'\boldsymbol{y} + 2\boldsymbol{X}'\boldsymbol{X}\boldsymbol{\beta} = \boldsymbol{0} \quad \Longrightarrow \quad (\boldsymbol{X}'\boldsymbol{X})\boldsymbol{\beta} = \boldsymbol{X}'\boldsymbol{y},$$

which is the normal equation (3.6). If $\boldsymbol{X}'\boldsymbol{X}$ is invertible, then the LS estimator of $\boldsymbol{\beta}$ is given by (3.7).

3.8.2 Distribution of the LS estimators

We use the following result: If $\boldsymbol{u} = (u_1, \dots, u_n)'$ has an n-variate normal distribution with $E(\boldsymbol{u}) = \boldsymbol{\mu}$ and $\text{Cov}(\boldsymbol{u}) = \boldsymbol{\Sigma}$, and $\boldsymbol{A}$ is an $m \times n$ matrix of constants with linearly independent rows with $m \leq n$, then $\boldsymbol{v} = \boldsymbol{A}\boldsymbol{u} = (v_1, \dots, v_m)$ has an m-variate normal distribution with mean vector $\boldsymbol{A}\boldsymbol{\mu}$ and covariance matrix $\boldsymbol{A}\boldsymbol{\Sigma}\boldsymbol{A}'$ using the sandwich formula (A.7) in Appendix A.

Put $\boldsymbol{A} = (\boldsymbol{X}'\boldsymbol{X})^{-1}\boldsymbol{X}'$ which is a $(p+1) \times n$ matrix of constants and $\boldsymbol{u} = \boldsymbol{y}$, which is n-variate normal with $E(\boldsymbol{y}) = \boldsymbol{\mu} = \boldsymbol{X}\boldsymbol{\beta}$ and $\text{Cov}(\boldsymbol{y}) = \sigma^2\boldsymbol{I}$. Then $\widehat{\boldsymbol{\beta}} = \boldsymbol{A}\boldsymbol{u}$ is $(p+1)$-variate normal with mean vector

$$E(\widehat{\boldsymbol{\beta}}) = (\boldsymbol{X}'\boldsymbol{X})^{-1}\boldsymbol{X}'\boldsymbol{\mu} = (\boldsymbol{X}'\boldsymbol{X})^{-1}\boldsymbol{X}'\boldsymbol{X}\boldsymbol{\beta} = \boldsymbol{\beta}$$

and covariance matrix

$$
\begin{aligned}
\mathrm{Cov}(\widehat{\boldsymbol{\beta}}) &= (\boldsymbol{X}'\boldsymbol{X})^{-1}\boldsymbol{X}'\mathrm{Cov}(\boldsymbol{y})\boldsymbol{X}(\boldsymbol{X}'\boldsymbol{X})^{-1} \\
&= \sigma^2(\boldsymbol{X}'\boldsymbol{X})^{-1}\boldsymbol{X}'\boldsymbol{I}\boldsymbol{X}(\boldsymbol{X}'\boldsymbol{X})^{-1} \\
&= \sigma^2(\boldsymbol{X}'\boldsymbol{X})^{-1} = \sigma^2\boldsymbol{V}.
\end{aligned}
$$

3.8.3 Gauss–Markov theorem

Consider the linear model (3.3) and assume homoscedasticity and independence of random errors but normality is not needed. Then among all linear unbiased estimators of any linear parametric function $\theta = \boldsymbol{c}'\boldsymbol{\beta}$, the LS estimator $\widehat{\theta} = \boldsymbol{c}'\widehat{\boldsymbol{\beta}}$ has the smallest variance. This result is usually stated as the LS estimator $\widehat{\boldsymbol{\beta}}$ of $\boldsymbol{\beta}$ is the **best linear unbiased estimator (BLUE)**. Here a linear estimator means that it is a linear function of y_i's.

Proof: Let $\tilde{\boldsymbol{\beta}}$ be any other linear unbiased estimator of $\boldsymbol{\beta}$, i.e. $E(\tilde{\boldsymbol{\beta}}) = \boldsymbol{\beta}$. Write $\tilde{\boldsymbol{\beta}} = [(\boldsymbol{X}'\boldsymbol{X})^{-1}\boldsymbol{X}' + \boldsymbol{A}]\boldsymbol{y} = \widehat{\boldsymbol{\beta}} + \boldsymbol{A}\boldsymbol{y}$, where $\boldsymbol{A}$ is any $(p+1) \times n$ matrix of constants (so that $\tilde{\boldsymbol{\beta}}$ is a linear function of $\boldsymbol{y}$). For $\tilde{\boldsymbol{\beta}}$ to be unbiased, we must have

$$
E(\tilde{\boldsymbol{\beta}}) = E(\widehat{\boldsymbol{\beta}}) + \boldsymbol{A}E(\boldsymbol{y}) = \boldsymbol{\beta} + \boldsymbol{A}\boldsymbol{X}\boldsymbol{\beta} = \boldsymbol{\beta} \quad \text{for all } \boldsymbol{\beta}.
$$

Hence, $\boldsymbol{A}\boldsymbol{X}$ must be equal to a null matrix $\boldsymbol{O}$: $(p+1) \times (p+1)$.

Next, we get

$$
\begin{aligned}
\mathrm{Cov}(\tilde{\boldsymbol{\beta}}) &= \mathrm{Cov}(\widehat{\boldsymbol{\beta}} + \boldsymbol{A}\boldsymbol{y}) \\
&= \mathrm{Cov}(\widehat{\boldsymbol{\beta}}) + \mathrm{Cov}(\boldsymbol{A}\boldsymbol{y}) + 2\mathrm{Cov}(\widehat{\boldsymbol{\beta}}, \boldsymbol{A}\boldsymbol{y}) \\
&= \sigma^2(\boldsymbol{X}'\boldsymbol{X})^{-1} + \boldsymbol{A}\mathrm{Cov}(\boldsymbol{y})\boldsymbol{A}' + 2\mathrm{Cov}((\boldsymbol{X}'\boldsymbol{X})^{-1}\boldsymbol{X}'\boldsymbol{y}, \boldsymbol{A}\boldsymbol{y}) \\
&= \sigma^2(\boldsymbol{X}'\boldsymbol{X})^{-1} + \sigma^2\boldsymbol{A}\boldsymbol{A}' + 2(\boldsymbol{X}'\boldsymbol{X})^{-1}\boldsymbol{X}'\mathrm{Cov}(\boldsymbol{y})\boldsymbol{A}' \\
&= \sigma^2(\boldsymbol{X}'\boldsymbol{X})^{-1} + \sigma^2\boldsymbol{A}\boldsymbol{A}' + 2\sigma^2(\boldsymbol{X}'\boldsymbol{X})^{-1}\boldsymbol{X}'\boldsymbol{A}' \\
&= \sigma^2(\boldsymbol{X}'\boldsymbol{X})^{-1} + \sigma^2\boldsymbol{A}\boldsymbol{A}' \quad (\text{since } \boldsymbol{X}'\boldsymbol{A}' = \boldsymbol{O}') \\
&= \mathrm{Cov}(\widehat{\boldsymbol{\beta}}) + \sigma^2\boldsymbol{A}\boldsymbol{A}'.
\end{aligned}
$$

Hence, $\mathrm{Cov}(\tilde{\boldsymbol{\beta}}) - \mathrm{Cov}(\widehat{\boldsymbol{\beta}}) = \sigma^2\boldsymbol{A}\boldsymbol{A}'$, which is positive semidefinite. Therefore, for any two linear unbiased estimators, $\boldsymbol{c}'\tilde{\boldsymbol{\beta}}$ and $\boldsymbol{c}'\widehat{\boldsymbol{\beta}}$, of a linear parametric function $\boldsymbol{c}'\boldsymbol{\beta}$, we have

$$
\begin{aligned}
\mathrm{Var}(\boldsymbol{c}'\tilde{\boldsymbol{\beta}}) - \mathrm{Var}(\boldsymbol{c}'\widehat{\boldsymbol{\beta}}) &= \boldsymbol{c}'[\mathrm{Cov}(\tilde{\boldsymbol{\beta}}) - \mathrm{Cov}(\widehat{\boldsymbol{\beta}})]\boldsymbol{c} \\
&= \sigma^2\boldsymbol{c}'\boldsymbol{A}\boldsymbol{A}'\boldsymbol{c} \\
&= \sigma^2\boldsymbol{d}'\boldsymbol{d} \geq 0,
\end{aligned}
$$

where $\boldsymbol{d} = \boldsymbol{A}'\boldsymbol{c}$. Hence, $\mathrm{Var}(\boldsymbol{c}'\tilde{\boldsymbol{\beta}}) \geq \mathrm{Var}(\boldsymbol{c}'\widehat{\boldsymbol{\beta}})$. ∎

3.8.4 Properties of fitted values and residuals

We have shown that $\widehat{y}$ and e are orthogonal to each other and e is orthogonal to every column of X; in particular $e'\mathbf{1} = \sum_{i=1}^{n} e_i = 0$, where $\mathbf{1}$ is the n-vector of all 1's, which is the first column vector of X. Next, we derive the distributions of the fitted and residual vectors. Since $\widehat{y} = Hy$ is a linear transform of $y \sim N(X\beta, \sigma^2 I)$, it follows that $\widehat{y}$ has an MVN distribution with mean vector, $E(\widehat{y}) = E(X\widehat{\beta}) = X\beta = \mu$, and covariance matrix (using the sandwich formula (A.7)):

$$\mathrm{Cov}(\widehat{y}) = \mathrm{Cov}(Hy) = H\,\mathrm{Cov}(y)H' = \sigma^2 HIH = \sigma^2 H \quad (\text{since } HH = H).$$

Thus, $\widehat{y}_i \sim N(\mu_i, \sigma^2 h_{ii})$, where h_{ii} is the ith diagonal entry of H $(1 \le i \le n)$. Similarly, $e = (I - H)y$ has an MVN distribution with the mean vector,

$$E(e) = (I - H)E(y)$$
$$= (I - H)X\beta = X\beta - X(X'X)^{-1}X'X\beta$$
$$= X\beta - X\beta = 0$$

and covariance matrix:

$$\mathrm{Cov}(e) = (I - H)\mathrm{Cov}(y)(I - H)' = \sigma^2(I - H)I(I - H) = \sigma^2(I - H).$$

Thus, $e_i \sim N(0, \sigma^2(1 - h_{ii}))$ $(i = 1, \ldots, n)$.

3.8.5 Geometric interpretation of least squares

We have seen above that the fitted vector $\widehat{y}$ and the residual vector e are projections of the observation vector y, and these two vectors are orthogonal to each other. What spaces are these projections made on, and how are they related to the LS estimator $\widehat{\beta}$? To answer these questions geometric interpretation of LS estimation is helpful.

The linear regression model $E(y) = X\beta$ implies that $E(y)$ is a linear combination of the column vectors of X, where the coefficients of the linear combination are the unknown regression coefficients $\beta_0, \beta_1, \ldots, \beta_p$. In other words, $E(y)$ lies in the vector space spanned by the column vectors of X, which is referred to as the **estimation space**. If these column vectors are linearly independent, as assumed in this chapter, the estimation space is of dimension $p + 1$. The LS estimator of $E(y)$ is the fitted vector $\widehat{y}$. LS estimation chooses $\widehat{y}$ to be the closest vector in the estimation space to the observation vector y in terms of the Euclidean distance, i.e. it minimizes the LS criterion $Q = \|y - \widehat{y}\|^2$. Geometrically, $\widehat{y}$ is obtained by projecting y onto the estimation space, defined by the projection matrix $\boldsymbol{H}$, i.e. $\widehat{\boldsymbol{y}} = \boldsymbol{H}\boldsymbol{y}$. The particular linear combination of the column vectors of $\boldsymbol{X}$ corresponding to the projection $\widehat{\boldsymbol{y}}$ is given by the coefficients $\widehat{\beta}_0, \widehat{\beta}_1, \ldots, \widehat{\beta}_p$, which are the LS estimators of $\beta_0, \beta_1, \ldots, \beta_p$, respectively. This linear combination is unique, i.e. $\widehat{\beta}_0, \widehat{\beta}_1, \ldots, \widehat{\beta}_p$, are unique if and only if the estimation space is $(p+1)$-dimensional, which is true if and only if the column vectors of $\boldsymbol{X}$ are linearly independent.

In general, the dimension of the estimation space equals the column rank of $\boldsymbol{X}$. If the column rank is less than $(p+1)$, then the projection $\widehat{\boldsymbol{y}}$ is still unique, but its representation as a linear combination of the columns of $\boldsymbol{X}$ is not unique, i.e. $\widehat{\beta}_0, \widehat{\beta}_1, \ldots, \widehat{\beta}_p$ are not unique.

The space orthogonal to the estimation space consists of all vectors that are orthogonal to every vector in the estimation space. This space is called the **error space**. The residual vector $\boldsymbol{e}$ is the projection of $\boldsymbol{y}$ onto the error space defined by the projection matrix $\boldsymbol{I} - \boldsymbol{H}$, i.e. $\boldsymbol{e} = (\boldsymbol{I} - \boldsymbol{H})\boldsymbol{y} = \boldsymbol{y} - \widehat{\boldsymbol{y}}$. Hence, $\boldsymbol{e}$ is orthogonal to $\widehat{\boldsymbol{y}}$. The dimension of the error space is $n - (p+1)$, which we refer to as the error d.f.

To view these projections pictorially, consider fitting a linear regression model $y = \beta_1 x_1 + \beta_2 x_2 + \varepsilon$ without the intercept term β_0. Assume we have $n = 3$ observations on each variable and let $\boldsymbol{y} = (y_1, y_2, y_3)'$, $\boldsymbol{x}_1 = (x_{11}, x_{21}, x_{31})'$ and $\boldsymbol{x}_2 = (x_{12}, x_{22}, x_{32})'$, where $\boldsymbol{x}_1$ and $\boldsymbol{x}_2$ are the column vectors of $\boldsymbol{X}$. (Note that we have deviated from the previous notation here for convenience in that previously $\boldsymbol{x}_i'$ denoted the ith row vector of $\boldsymbol{X}$, whereas here $\boldsymbol{x}_i$ denotes the ith column vector of $\boldsymbol{X}$.) Then assuming that $\boldsymbol{x}_1$ and $\boldsymbol{x}_2$ are not collinear (i.e. they are not linearly dependent), the estimation space is the two-dimensional plane spanned by $\boldsymbol{x}_1$ and $\boldsymbol{x}_2$ as shown in Figure 3.8. If $\boldsymbol{x}_1$ and $\boldsymbol{x}_2$ are collinear, then the estimation

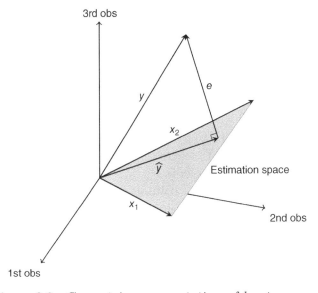

Figure 3.8. *Geometric representation of least squares.*

space is one-dimensional, i.e. it is a straight line. Then the projection $\widehat{\boldsymbol{y}}$ on this line is unique, but its representation $\widehat{\beta}_1\boldsymbol{x}_1 + \widehat{\beta}_2\boldsymbol{x}_2$ is not unique because $\boldsymbol{x}_1$ and $\boldsymbol{x}_2$ are linearly related so one can be expressed in terms of the other.

3.8.6 Confidence ellipsoid for β

As shown above, $\widehat{\boldsymbol{\beta}}$ has an MVN distribution with mean vector $\boldsymbol{\beta}$ and covariance matrix $\sigma^2 \boldsymbol{V}$. Since $\boldsymbol{V} = (\boldsymbol{X}'\boldsymbol{X})^{-1}$ is a positive definite matrix, using the spectral decomposition theorem (see Appendix A), there exists a nonsingular $(p+1) \times (p+1)$ matrix $\boldsymbol{U}$ such that $\boldsymbol{U}'\boldsymbol{U} = \boldsymbol{V}^{-1}$ and $\boldsymbol{U}\boldsymbol{V}\boldsymbol{U}' = \boldsymbol{I}$. Make the transformation $\boldsymbol{z} = \boldsymbol{U}(\widehat{\boldsymbol{\beta}} - \boldsymbol{\beta})$. Then $\boldsymbol{z}$ is MVN with mean vector $\boldsymbol{0}$ and covariance matrix $\sigma^2 \boldsymbol{U}\boldsymbol{V}\boldsymbol{U}' = \sigma^2\boldsymbol{I}$. So the elements z_i $(i = 1, \ldots, p+1)$ of $\boldsymbol{z}$ are i.i.d. $N(0, \sigma^2)$. Hence,

$$\sum_{i=1}^{p+1} z_i^2 = \boldsymbol{z}'\boldsymbol{z} = (\widehat{\boldsymbol{\beta}} - \boldsymbol{\beta})' \boldsymbol{U}'\boldsymbol{U} (\widehat{\boldsymbol{\beta}} - \boldsymbol{\beta}) = (\widehat{\boldsymbol{\beta}} - \boldsymbol{\beta})' \boldsymbol{V}^{-1} (\widehat{\boldsymbol{\beta}} - \boldsymbol{\beta}) \sim \sigma^2 \chi^2_{p+1}.$$

Furthermore, from Section 3.3.2, we also know that $[n - (p+1)]s^2 \sim \sigma^2 \chi^2_{n-(p+1)}$, and these two χ^2 random variables are independent. Hence, their ratio divided by their d.f. is distributed as $F_{p+1, n-(p+1)}$ or

$$\frac{(\widehat{\boldsymbol{\beta}} - \boldsymbol{\beta})' \boldsymbol{X}'\boldsymbol{X} (\widehat{\boldsymbol{\beta}} - \boldsymbol{\beta})}{(p+1)s^2} \sim F_{p+1, n-(p+1)}.$$

The confidence ellipsoid (3.18) for $\boldsymbol{\beta}$ follows from this result.

3.8.7 Population partial correlation coefficient

Consider a general setting of a random vector $\boldsymbol{x}$ of dimension p partitioned into two subvectors $\boldsymbol{x}_1$ and $\boldsymbol{x}_2$ of dimensions p_1 and p_2, respectively, such that $p_1 + p_2 = p$. If $f(\boldsymbol{x}_1, \boldsymbol{x}_2)$ denotes the joint distribution of $(\boldsymbol{x}_1, \boldsymbol{x}_2)$ and $f(\boldsymbol{x}_1)$ denotes the marginal distribution of $\boldsymbol{x}_1$, then the conditional distribution of $\boldsymbol{x}_2$ conditioned on $\boldsymbol{x}_1$ is given by $f(\boldsymbol{x}_2|\boldsymbol{x}_1) = f(\boldsymbol{x}_1, \boldsymbol{x}_2)/f(\boldsymbol{x}_1)$. In particular, suppose that $f(\boldsymbol{x}_1, \boldsymbol{x}_2)$ is an MVN distribution with mean vector $\boldsymbol{\mu}$ and covariance matrix $\boldsymbol{\Sigma}$ partitioned corresponding to $\boldsymbol{x}_1$ and $\boldsymbol{x}_2$ as follows:

$$\boldsymbol{\mu} = \begin{bmatrix} \boldsymbol{\mu}_1 \\ \boldsymbol{\mu}_2 \end{bmatrix} \quad \text{and} \quad \boldsymbol{\Sigma} = \begin{bmatrix} \boldsymbol{\Sigma}_{11} & \boldsymbol{\Sigma}_{12} \\ \boldsymbol{\Sigma}'_{12} & \boldsymbol{\Sigma}_{22} \end{bmatrix}.$$

Then it can be shown that the marginal distribution $f(\boldsymbol{x}_1)$ is MVN with mean vector $\boldsymbol{\mu}_1$ and covariance matrix $\boldsymbol{\Sigma}_{11}$ and the conditional distribution $f(\boldsymbol{x}_2|\boldsymbol{x}_1)$ is also MVN with mean vector and covariance matrix given by

$$\boldsymbol{\mu}_2 + \boldsymbol{\Sigma}'_{12}\boldsymbol{\Sigma}^{-1}_{11}(\boldsymbol{x}_1 - \boldsymbol{\mu}_1) \quad \text{and} \quad \boldsymbol{\Sigma}_{22} - \boldsymbol{\Sigma}'_{12}\boldsymbol{\Sigma}^{-1}_{11}\boldsymbol{\Sigma}_{12}.$$

We now specialize this result to the trivariate case where two of the variables are predictor variables, x_1 and x_2, and the third variable is the response variable y. For convenience, suppose that all three variables are standardized so that they have 0 means, unit variances and their pairwise covariances are pairwise correlation coefficients $\rho_{x_1 x_2}, \rho_{x_1 y}$ and $\rho_{x_2 y}$. Thus,

$$\boldsymbol{\Sigma} = \begin{bmatrix} 1 & \rho_{x_1 x_2} & \rho_{x_1 y} \\ \rho_{x_1 x_2} & 1 & \rho_{x_2 y} \\ \rho_{x_1 y} & \rho_{x_2 y} & 1 \end{bmatrix}.$$

We want to derive a formula for the partial correlation coefficient $\rho_{x_2 y | x_1}$. Identifying $\boldsymbol{x}_1 = x_1$ and $\boldsymbol{x}_2 = (x_2, y)'$, we have $\boldsymbol{\Sigma}_{11} = 1, \boldsymbol{\Sigma}_{12} = [\rho_{x_1 x_2}, \rho_{x_1 y}]$ and

$$\boldsymbol{\Sigma}_{22} = \begin{bmatrix} 1 & \rho_{x_2 y} \\ \rho_{x_2 y} & 1 \end{bmatrix}.$$

Hence, the conditional covariance matrix of (x_2, y) conditioned on x_1 equals

$$\boldsymbol{\Sigma}_{22} - \boldsymbol{\Sigma}'_{12} \boldsymbol{\Sigma}_{11}^{-1} \boldsymbol{\Sigma}_{12} = \begin{bmatrix} 1 & \rho_{x_2 y} \\ \rho_{x_2 y} & 1 \end{bmatrix} - \begin{bmatrix} \rho_{x_1 x_2} \\ \rho_{x_1 y} \end{bmatrix} (1)^{-1} [\rho_{x_1 x_2}, \rho_{x_1 y}]$$

$$= \begin{bmatrix} 1 - \rho_{x_1 x_2}^2 & \rho_{x_2 y} - \rho_{x_1 x_2} \rho_{x_1 y} \\ \rho_{x_2 y} - \rho_{x_1 x_2} \rho_{x_1 y} & 1 - \rho_{x_1 y}^2 \end{bmatrix}.$$

So the partial correlation coefficient is given by

$$\rho_{x_2 y | x_1} = \frac{\rho_{x_2 y} - \rho_{x_1 x_2} \rho_{x_1 y}}{\sqrt{(1 - \rho_{x_1 x_2}^2)(1 - \rho_{x_1 y}^2)}}. \tag{3.36}$$

The sample partial correlation coefficient $r_{x_2 y | x_1}$ given by (3.30) is the sample version of this formula.

Exercises

Theoretical Exercises

3.1 (Regression through origin) Write the model $y_i = \beta x_i + \varepsilon_i$ $(i = 1, \ldots, n)$ in matrix notation. Use the formula (3.7) to derive the LS estimator of β given in Exercise 2.1.

3.2 (Hat matrix for simple linear regression) Show that the elements of the hat matrix $\boldsymbol{H}$ for simple linear regression are given by

$$h_{ij} = \frac{1}{n} + \frac{(x_i - \bar{x})(x_j - \bar{x})}{S_{xx}} \quad \text{and} \quad h_{ii} = \frac{1}{n} + \frac{(x_i - \bar{x})^2}{S_{xx}} \quad (1 \le i \ne j \le n).$$

Further show that $\sum_{j=1}^{n} h_{ij} = 1$ across every row of the hat matrix. So the fitted value $\hat{y}_i = \sum_{j=1}^{n} h_{ij} y_j$ is the weighted average of all the y_j. (*Hint*: Use the equivalent simple linear regression model $y_i = \beta_0 + \beta_1 x_i + \varepsilon_i$ where $x_i \to x_i - \bar{x}$. For this model $\boldsymbol{X}'\boldsymbol{X}$ is a diagonal matrix and so $\boldsymbol{H} = \boldsymbol{X}(\boldsymbol{X}'\boldsymbol{X})^{-1}\boldsymbol{X}'$ is easy to evaluate. Note that the $\boldsymbol{H}$ matrix is the same for the two models since the fitted values obtained by the two models are the same.)

3.3 (Hat matrix for the Anscombe Data Set IV) In this problem, we want to show that for the extreme observation $(x_i = 19, y_i = 12.50)$ in the Anscombe Data Set IV from Table 2.3, we have $h_{ii} = 1$ and $h_{ij} = 0$ for $j \neq i$. So the fitted value of y_i at $x_i = 19$ is always equal to the observed value of y_i. Consider the problem more generally by assuming that the first $n - 1$ observations all have the same $x_i = x'$ $(1 \leq i \leq n - 1)$ and $x_n = x'' \neq x'$.

(a) Show that

$$\bar{x} = \frac{(n-1)x' + x''}{n} \quad \text{and} \quad S_{xx} = \left(\frac{n-1}{n}\right)(x' - x'')^2.$$

(b) Next show that

$$(x_i - \bar{x})(x_n - \bar{x}) = -\frac{(n-1)(x' - x'')^2}{n^2} \quad (1 \leq i \leq n - 1)$$

and

$$(x_n - \bar{x})^2 = \left(\frac{n-1}{n}\right)^2 (x' - x'')^2.$$

(c) Hence, conclude that $h_{in} = 0$ for $1 \leq i \leq n - 1$ and $h_{nn} = 1$.

3.4 (Orthogonal designs) Let the model matrix $\boldsymbol{X} = [\boldsymbol{X}_1 \vdots \boldsymbol{X}_2]$, where $\boldsymbol{X}_1$ is an $n \times p_1$ matrix and $\boldsymbol{X}_2$ is an $n \times p_2$ matrix such that $p_1 + p_2 = p$ (here we assume, for convenience, that there is no constant term in the model). Let the parameter vector $\boldsymbol{\beta}$ be similarly partitioned into two subvectors $\boldsymbol{\beta}_1$: $p_1 \times 1$ and $\boldsymbol{\beta}_2$: $p_2 \times 1$. If every column of $\boldsymbol{X}_1$ is orthogonal to every column of $\boldsymbol{X}_2$, show the following results.

(a) The LS estimators of $\boldsymbol{\beta}_1$ and $\boldsymbol{\beta}_2$ are $\hat{\boldsymbol{\beta}}_1 = (\boldsymbol{X}_1'\boldsymbol{X}_1)^{-1}\boldsymbol{X}_1'\boldsymbol{y}$ and $\hat{\boldsymbol{\beta}}_2 = (\boldsymbol{X}_2'\boldsymbol{X}_2)^{-1}\boldsymbol{X}_2'\boldsymbol{y}$, respectively. Thus, $\hat{\boldsymbol{\beta}}_1$ does not depend on $\boldsymbol{X}_2$ and $\hat{\boldsymbol{\beta}}_2$ does not depend on $\boldsymbol{X}_1$.

(b) The LS estimators $\hat{\boldsymbol{\beta}}_1$ and $\hat{\boldsymbol{\beta}}_2$ are uncorrelated (i.e. all $\mathrm{Corr}(\hat{\beta}_{1j}, \hat{\beta}_{2k}) = 0$) and so under the normality assumption they are independent.

3.5 (Omitted variables) Suppose that the true linear model is

$$\boldsymbol{y} = \boldsymbol{X}_1\boldsymbol{\beta}_1 + \boldsymbol{X}_2\boldsymbol{\beta}_2 + \boldsymbol{\varepsilon},$$

where $\boldsymbol{X}_1 : n \times p_1, \boldsymbol{\beta}_1 : p_1 \times 1, \boldsymbol{X}_2 : n \times p_2, \boldsymbol{\beta}_2 : p_2 \times 1$. However, we mistakenly fit the model $\boldsymbol{y} = \boldsymbol{X}_1\boldsymbol{\beta}_1 + \boldsymbol{\varepsilon}$ and estimate $\boldsymbol{\beta}_1$ by $\widehat{\boldsymbol{\beta}}_1 = (\boldsymbol{X}_1'\boldsymbol{X}_1)^{-1}\boldsymbol{X}_1'\boldsymbol{y}$.

(a) Show that, in general, $\widehat{\boldsymbol{\beta}}_1$ is a biased estimator with

$$\text{Bias}(\widehat{\boldsymbol{\beta}}_1) = (\boldsymbol{X}_1'\boldsymbol{X}_1)^{-1}\boldsymbol{X}_1'\boldsymbol{X}_2\boldsymbol{\beta}_2.$$

Note that this formula generalizes that given in Exercise 2.4.

(b) Under what condition on $\boldsymbol{X}_1$ and $\boldsymbol{X}_2$ is $\widehat{\boldsymbol{\beta}}_1$ unbiased? How is this condition related to that in Exercise 2.4 for $\hat{\beta}_1$ to be unbiased?

3.6 **(Mean and covariance matrix of the GLS estimator)** Derive the formulae (3.28) for the mean and covariance matrix of the GLS estimator.

3.7 **(Derivation of the one-way ANOVA F-test using the extra SS method)** In one-way ANOVA, we have $k \geq 2$ groups (e.g. treatment groups) with n_i observations, $y_{i1}, \ldots, y_{in_i}$, from the ith group $(i = 1, \ldots, k)$. Let $N = \sum n_i$ denote the total sample size. The data are usually modeled as

$$y_{ij} = \mu_i + \varepsilon_{ij} = \mu + \alpha_i + \varepsilon_{ij} \quad (i = 1, \ldots, k),$$

where the ε_{ij} are i.i.d. $N(0, \sigma^2)$ random errors, μ_i is the ith group mean, $\mu = \sum n_i\mu_i/N$ is the overall mean and $\alpha_i = \mu_i - \mu$ is the ith group "effect" subject to the linear restriction $\sum n_i\alpha_i = 0$. We want to test the overall null hypothesis $H_0 : \mu_1 = \cdots = \mu_k$ or equivalently $H_0 : \alpha_1 = \cdots = \alpha_k = 0$. It is easy to show that the LS estimates of the unknown parameters in the above linear model are as follows:

$$\widehat{\mu}_i = \overline{y}_i, \quad \widehat{\mu} = \overline{\overline{y}} = \frac{\sum n_i\overline{y}_i}{N} \quad \text{and} \quad \widehat{\alpha}_i = \overline{y}_i - \overline{\overline{y}}.$$

(Alternatively, the overall mean μ may be defined as the unweighted average of the μ_i's: $\mu = \sum \mu_i/k$, in which case the $\alpha_i = \mu_i - \mu$ satisfy the linear restriction $\sum \alpha_i = 0$. The LS estimates of μ and the α_i will be different for this parameterization, but the estimates of the μ_i will be the same, namely $\widehat{\mu}_i = \overline{y}_i$.)

(a) Show that the ANOVA identity (3.13) can be expressed as

$$\underbrace{\sum_{i=1}^{k}\sum_{j=1}^{n_i}(y_{ij} - \overline{\overline{y}})^2}_{\text{SST}} = \underbrace{\sum_{i=1}^{k}n_i(\overline{y}_i - \overline{\overline{y}})^2}_{\text{SSG}} + \underbrace{\sum_{i=1}^{k}\sum_{j=1}^{n_i}(y_{ij} - \overline{y}_i)^2}_{\text{SSE}},$$

where SST and SSE have their usual meanings and SSG is referred to as the SS between the groups.

(b) Using the above identity, show that the extra SS F-test rejects H_0 at level α if

$$F = \frac{\text{SSG}/(k-1)}{\text{SSE}/(N-k)} > f_{k-1,N-k,\alpha}.$$

Applied Exercises

3.8 (Matrix calculation by hand) This is a hand-calculation exercise to help you get an understanding of the matrix formulae in multiple regression. Consider the following small data set. We want to fit a straight line $y = \beta_0 + \beta_1 x$ to these data.

x	1	2	3	4	5
y	2	6	7	9	10

(a) Write the $\boldsymbol{X}$ matrix and the $\boldsymbol{y}$ vector.

(b) Calculate $\boldsymbol{X'X}$ and its inverse. For a 2×2 matrix, the formula for the inverse is simple:

$$\begin{bmatrix} a & c \\ d & b \end{bmatrix}^{-1} = \frac{1}{ab - cd} \begin{bmatrix} b & -c \\ -d & a \end{bmatrix}.$$

Check that the product of the original matrix and its inverse equals the identity matrix.

(c) Calculate the $\boldsymbol{X'y}$ vector.

(d) Finally, calculate the LS estimates $\widehat{\beta}_0$ and $\widehat{\beta}_1$ using the formula (3.7).

3.9 (Alternate coding of categorical variables) Refer to Example 3.14 and the data in Table 3.8. Suppose that the Gender is coded as $x_1 = -1$ for females and $x_1 = +1$ for males. Similarly, Race is coded as $x_2 = -1$ for non-Whites and $x_2 = +1$ for Whites. What are the new values of $\beta_0, \beta_1, \beta_2$, and β_3? Interpret them.

3.10 (Cobb–Douglas production function) Data on 569 European companies on their capital (K) measured as the total fixed assets (in millions of euros) at the end of 1995, labor (L) measured as the number of workers and output (O) measured as the value added (in millions of euros) are available in file `cobbdouglas.csv`. The companies in this data set are from different industry sectors in which different production functions may apply since their capital and labor requirements are different, but we ignore this problem.

(a) Fit the Cobb–Douglas production function $y = \gamma K^{\alpha} L^{\beta} \varepsilon$ by making the log-transformation, so that the model becomes $y = \beta_0 + \beta_1 x_1 + \beta_2 x_2 + \ln \varepsilon$, where $y = \ln O, x_1 = \ln K, x_2 = \ln L, \beta_0 = \ln \gamma, \beta_1 = \alpha$, and $\beta_2 = \beta$.

(b) If $\alpha + \beta = \beta_1 + \beta_2 = 1$, then it is easy to check that if the capital and labor are changed by a common scaling factor then the output is changed by the same factor. In economics this is known as the **constant returns to scale**. Test the null hypothesis of the constant returns to scale for these data by testing $H_0 : \beta_1 + \beta_2 = 1$ using the t-statistic

$$t = \frac{\widehat{\beta}_1 + \widehat{\beta}_2 - 1}{\sqrt{\left(\widehat{\text{Var}}(\widehat{\beta}_1) + \widehat{\text{Var}}(\widehat{\beta}_2) + 2\widehat{\text{Cov}}(\widehat{\beta}_1, \widehat{\beta}_2)\right)}},$$

where the estimates of $\text{Var}(\widehat{\beta}_1), \text{Var}(\widehat{\beta}_2)$, and $\text{Cov}(\widehat{\beta}_1, \widehat{\beta}_2)$ can be obtained from the `vcov` function in R.

(c) An alternative way to do the above test is to use $y - x_2$ as the response variable and fit the model $y - x_2 = \beta_0 + \beta_1(x_1 - x_2) + (\beta_1 + \beta_2 - 1)x_2 + \ln \varepsilon$, where $x_1 - x_2$ and x_2 are the new predictor variables. Test the significance of the coefficient of x_2 in this regression, i.e. test $H_0 : \beta_1 + \beta_2 - 1 = 0$ using a t-test. Is the result the same as that obtained in Part (b)?

3.11 **(Research expenditures data)** Research expenditures is an important factor in the algorithm used by *US News & World* to rank graduate engineering programs. It carries 25% weight (15% for total research expenditures and 10% for research expenditures per faculty). The file `Research.csv` gives data on research expenditures in millions of dollars (Research), number of faculty (Faculty), and number of PhD students (PhD) in top 30 US Universities according to *US News & World* 2017 rankings. The data are taken from *ASEE Profiles*. We want to build a predictive model for research expenditures as a function of the number of faculty and the number of PhD students.

(a) Make a matrix scatter plot and compute the correlation matrix of all three variables. Comment on the relationships between the variables.

(b) Fit a regression model for Research versus Faculty and PhD. From this model note that PhD is a significant predictor of Research but Faculty is not. Why can't research expenditure be increased simply by increasing the number of PhD students? Given that faculty with more grants fund more PhD students (i.e. the causal arrow is Faculty $\rightarrow$ PhD) explain the apparently anomalous result obtained.

(c) Calculate the partial correlation coefficients between Research and each predictor controlling for the other predictor and their t-statistics. Check that these t-statistics are the same as those given by the regression analysis.

3.12 (Standardized regression) For a two-variable regression of y on x_1, x_2, we have the following correlation matrix and vector:

$$\boldsymbol{R} = \begin{bmatrix} 1 & 0.5 \\ 0.5 & 1 \end{bmatrix}, \boldsymbol{r} = \begin{bmatrix} 0.4 \\ 0.8 \end{bmatrix}.$$

Calculate the standardized regression coefficients $\widehat{\beta}_1^*$ and $\widehat{\beta}_2^*$. Hence, find the unstandardized LS regression equation if $\bar{y} = 10, \bar{x}_1 = 3, \bar{x}_2 = 5, s_y = 5, s_{x_1} = 2, s_{x_2} = 4$.

3.13 (Sales data) Consider the following data on sales (y) of a company in 10 sales territories. The predictors are the number of salesmen (x_1) and the amount of sales expenditures in millions of dollars (x_2). The data are in file `Sales Data.csv`.

No.	x_1	x_2	y	No.	x_1	x_2	y
1	31	1.85	4.20	6	49	2.80	7.42
2	46	2.80	7.28	7	31	1.85	3.36
3	40	2.20	5.60	8	38	2.30	5.88
4	49	2.85	8.12	9	33	1.60	4.62
5	38	1.80	5.46	10	42	2.15	5.88

Source: Tamhane and Dunlop (2000), Example 11.7.

(a) Calculate the correlation matrix $\boldsymbol{R}$ between x_1 and x_2 and the correlation vector $\boldsymbol{r}$ between y and x_1, x_2. From these bivariate correlations, calculate the partial correlations $r_{yx_1|x_2}$ and $r_{yx_2|x_1}$.

(b) Calculate the standardized regression coefficients $\widehat{\beta}_1^*$ and $\widehat{\beta}_2^*$ from $\boldsymbol{R}$ and $\boldsymbol{r}$ for the model. How do they compare with the partial correlation coefficients $r_{yx_1|x_2}$ and $r_{yx_2|x_1}$?

(c) Check that you get the same values for $\widehat{\beta}_1^*$ and $\widehat{\beta}_2^*$ from the unstandardized LS estimates $\widehat{\beta}_1$ and $\widehat{\beta}_2$ by scaling them appropriately.

(d) Compare $(\widehat{\beta}_1, \widehat{\beta}_2)$ with $(\widehat{\beta}_1^*, \widehat{\beta}_2^*)$. Which variable is a better predictor of sales and why?

Table 3.10. Salary data variables

Variable	Explanation
Salary	Annual salary in US$
YrsEm	No. of years employed with the company
PriorYr	No. of years of prior experience
Educ	No. of years of education after high school
Super	No. of people supervised
Gender	M = Male, F = Female
Dept	Advertising, Engineering, Purchase, Sales

Source: McKenzie and Goldman (1999, Temco Data Set).

3.14 **(Salary data)** File `salaries.csv` contains data on annual salaries of 46 employees of a company and possible predictors. The variable definitions are given in Table 3.10. Use $\log_{10}(\text{Salary})$ as the response variable.

(a) Fit a prediction model using the given data. Using Male and Purchase as the reference categories for Gender and Department categorical variables, check that the fitted equation is

$$\widehat{\log_{10}(\text{Salary})} = 4.429 + 0.0075 \text{ YrsEm} + 0.0017 \text{ PriorYr} + 0.0170 \text{ Educ}$$
$$+ 0.0004 \text{ Super} + 0.0231 \text{ Female} - 0.0388 \text{ Adver}$$
$$- 00057 \text{ Engg} - 0.0938 \text{ Sales}.$$

(b) If we use Female and Sales as the reference categories, what will be the new coefficients for Male and for the other three departments?

(c) Predict a person's salary along with the associated prediction interval for the following values of predictor variables: YrsEm = 8, PriorYr = 10, Education = 12, Gender = Male, Dept = Sales, Super = 5.

(d) The coefficient of Engg is highly nonsignificant with a P-value = 0.774 in the above regression. But if Sales is used as the reference category, the coefficient of Engg is highly significant with a P-value < 0.001. Interpret this result.

(e) In the above model, the coefficient of Female is nonsignificant with $P = 0.115$, so there is not a significant difference between the salaries of Males and Females, controlling for other variables. Should we drop Gender as a predictor variable? For Dept, the question is more complicated since the coefficients of Advert and Engg are nonsignificant, but the coefficient of Sales is highly significant with $P = 0.0002$. Should the nonsignificant categories, Advert and Engg, be pooled with the reference category, Purchase, or should the categories be left unpooled or should the Dept be dropped as a predictor variable?

Chapter 4

Multiple linear regression: model diagnostics

This chapter focuses on detecting violations of standard multiple linear regression assumptions such as normality, homoscedasticity, linearity, and independence. We describe methods to address these violations through data transformations and other methods. In addition to testing these model assumptions, we also discuss other data problems that adversely affect regression results, namely, outliers, influential observations and multicollinearity. We discuss both graphical and formal statistical methods based on residuals.

4.1 Model assumptions and distribution of residuals

The following assumptions underlie the multiple regression model, in particular about the random errors ε_i.

1. *Normality*: The ε_i are normally distributed.

2. *Homoscedasticity*: The ε_i have a constant variance σ^2.

3. *Independence*: The ε_i are statistically independent.

4. *No outliers*: No observations deviate significantly from the specified model.

5. The model is correctly specified.

These assumptions imply that the y_i are independent $N(\mu_i, \sigma^2)$ r.v.s., where $\mu_i = \beta_0 + \beta_1 x_{i1} + \cdots + \beta_p x_{ip}$ for $i = 1, \ldots, n$.

Predictive Analytics: Parametric Models for Regression and Classification Using R,
First Edition. Ajit C. Tamhane.
© 2021 John Wiley & Sons, Inc. Published 2021 by John Wiley & Sons, Inc.
Companion website: www.wiley.com/go/tamhane/predictiveanalytics

Under these assumptions, the distribution of the residuals has been shown in Section 3.8 to be multivariate normal with a null mean vector and variance–covariance matrix $\sigma^2(\boldsymbol{I} - \boldsymbol{H})$, where $\boldsymbol{H} = \{h_{ij}\}$ is the hat matrix defined in (3.10). Thus, each $e_i \sim N(0, \sigma^2(1 - h_{ii}))$. We will use this distributional result to test the assumptions listed above.

Note that $\mathrm{Var}(e_i) = \sigma^2(1 - h_{ii})$ is not constant even if $\mathrm{Var}(\varepsilon_i) = \sigma^2$ is constant. Similarly, $\mathrm{Cov}(e_i, e_j) = \sigma^2 h_{ij}$, where the $h_{ij} \neq 0$ in general even if $\mathrm{Cov}(\varepsilon_i, \varepsilon_j) = 0$. Thus, the e_i's are not independent even if the ε_i's are. This is clear from, as noted in Section 3.2.3, that the n residuals are subject to $p + 1$ linear constraints and so they are linearly dependent. However, if n is large relative to p, then the e_i's imitate the distribution of the ε_i's closely, and we can use them to test the homoscedasticity and independence assumptions.

4.2   Checking normality

To test the normality assumption on the ε_i's, the recommended method is the normal quantile–quantile plot (called the **normal Q–Q plot** or simply the **normal plot**) of the residuals. A histogram is not particularly useful for checking normality as its closeness to a normal bell-shaped curve is difficult to judge by eye, and there are other distributions that are also bell-shaped, e.g. the t-distribution and the Cauchy distribution, The normal plot is the plot of the quantiles of the residuals versus theoretical standard normal distribution quantiles. Denote the ordered residuals by $e_{(1)} \leq e_{(2)} \leq \cdots \leq e_{(n)}$ and define the ith ordered residual $e_{(i)}$ as the $[i/(n + 1)]$th quantile of the residuals. For example, if $n = 25$, then $e_{(1)}$ is the $1/26 = 0.0385$th quantile (or 3.85th percentile) and the corresponding standard normal distribution quantile is -1.769. Other definitions of a sample quantile may also be used, e.g. many authors define $e_{(i)}$ as the $[(i - 0.5)/n]$th sample quantile.

The following points should be noted about normal plots:

- Standardized residuals defined in (4.5) are preferred for making a normal plot because they have a common variance equal to 1, whereas raw residuals have slightly unequal variances.

- Normality of the ε's is tested by making a normal plot of the residuals and not of the y's. The reason is that the residuals have a constant zero mean, while the means of the y's are not constant since they depend on the associated x's.

- Some typical normal plots are shown in Figure 4.1 for simulated data. Panel (a) shows a plot for normal data, (b) shows a plot for a right-skewed distribution such as the lognormal distribution, (c) shows a plot for a symmetric long-tailed distribution (with longer tails than the normal distribution such as the t-distribution with a small number of degrees of freedom) and (d) shows a plot for a symmetric short-tailed distribution (with shorter tails than the normal distribution such as the uniform distribution which has no tails).

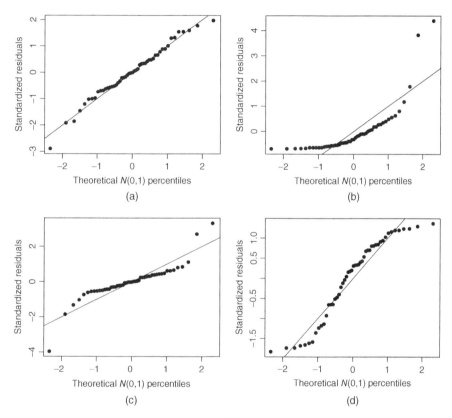

Figure 4.1. *Some typical normal plots: (a) normally distributed data, (b) right-skewed data, (c) long-tailed data, (d) short-tailed data.*

There are several formal normality tests, e.g. the Anderson–Darling test and the Shapiro–Wilk test. For most common applications, a normal plot is usually sufficient.

It should be noted that nonnormality of the data is not as serious a problem as is often perceived because the LS estimators are still approximately normally distributed for large n, thanks to the central limit theorem. Normality is a crucial assumption mainly in case of small samples in order for the normal theory inferences on the regression coefficients to be valid; LS estimation itself does not require normality.

Example 4.1 *(Used Car Prices: Checking Normality)* Figure 4.2 shows the normal plots of the residuals obtained using Price and log(Price) as the response variables (with the same predictors as in Example 3.12). These plots are obtained using the following R code:

```
> carprices = read.csv("c:/data/usedcarprices.csv")
> fit1 = lm(Price ~ Mileage + Liter + factor(Make)
>    + factor(Type), carprices)
> fit2 = lm(log(Price) ~ Mileage + Liter + factor(Make)
> +factor(Type), carprices)
> plot(fit1,which=2)
> plot(fit2,which=2)
```

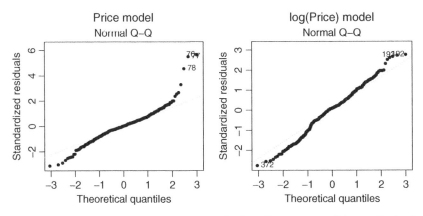

Figure 4.2. *Normal plots of residuals with (a) Price and (b) log(Price) as the response variables.*

We see that the normal plot for Price shows more extreme departures in the upper tail than does the normal plot for log(Price). If the outliers in the upper tail are excluded, then the normal plot for Price also becomes fairly linear. Thus, the departures from normality are caused by outliers; the log transformation helps to mitigate these outliers.   □

4.3   Checking homoscedasticity

Violation of the homoscedasticity assumption can lead to invalid inferences on the regression coefficients since the test statistics incorrectly use a common pooled MSE as an estimate of the error variance. If the homoscedasticity assumption does not hold, then typically $\sigma_i = \text{SD}(y_i)$ is a function of $\mu_i = E(y_i)$, which is estimated by the fitted value $\widehat{y}_i$. Therefore, to test homoscedasticity, we make the plot of the raw residuals e_i against the fitted values $\widehat{y}_i$. This is called the **fitted values plot**. Remember that the spread of the residuals is proportional to $\text{SD}(e_i)$. If the residuals spread out evenly forming a roughly parallel band around the zero line, then it indicates that the σ_i is roughly constant supporting the homoscedasticity assumption.

Note that μ_i is different for each observation and hence if σ_i is a function of μ_i, then it is different for each observation thus resulting in heteroscedasticity. For example, the variability in household incomes increases with the mean income levels. If the incomes follow a lognormal distribution, then σ_i is proportional to μ_i. In that case, the residuals fan out approximately linearly in a funnel shape. In general, the shape of the plot tells us something about the relationship between σ and μ. Some typical fitted values plots are shown in Figure 4.3. The transformations of y suggested by these relations to stabilize their variances are discussed in the next section.

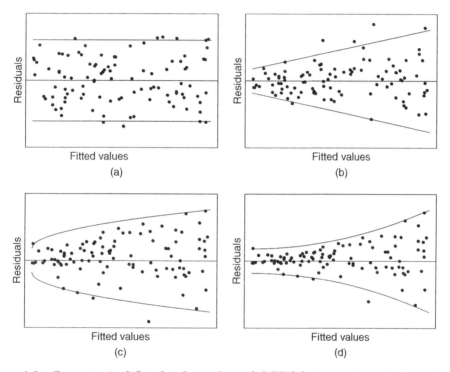

Figure 4.3. *Some typical fitted values plots: (a)* $\mathrm{SD}(y)$ *is constant independent of* μ, *(b)* $\mathrm{SD}(y) \propto \mu$, *(c)* $\mathrm{SD}(y) \propto \sqrt{\mu}$, *(d)* $\mathrm{SD}(y) \propto \mu^2$.

4.3.1 Variance stabilizing transformations

Suppose that $\sigma = \mathrm{SD}(y)$ is a known smooth function of $\mu = E(y)$, say $\sigma = g(\mu)$, and we want to find a transformation $f(y)$ such that $\mathrm{SD}[f(y)]$ is approximately constant. Consider the first-order Taylor series expansion of $f(y)$ around μ, namely $f(y) \approx f(\mu) + (y - \mu)f'(\mu)$, where $f'(\mu)$ is the first derivative of $f(\mu)$. Then $\mathrm{Var}[f(y)]$ can be approximated as

$$\mathrm{Var}[f(y)] \approx \mathrm{Var}(y)[f'(\mu)]^2. \qquad (4.1)$$

This is called the **delta method** for approximating the variance of a nonlinear function of y. We want $\mathrm{Var}[f(y)]$ to be constant, so we set it equal to 1 (or any other positive constant). Then putting $\mathrm{Var}(y) = g^2(\mu)$, we get the equation $g^2(\mu)[f'(\mu)]^2 = 1$, which upon solving for $f(\mu)$ and changing the variable of integration from μ to y gives

$$f(y) = \int \frac{dy}{g(y)}. \qquad (4.2)$$

This is the desired **variance stabilizing transformation** $f(y)$.

Consider the example of lognormal data, where $\mathrm{SD}(y)$ is proportional to μ, i.e. $g(\mu) = c\mu$, where $c > 0$. The corresponding fitted values plot is shown in Figure 4.3b. Ignoring the constant of proportionality c, we get

$$f(y) = \int \frac{dy}{y} = \ln y,$$

which is the **logarithmic transformation**. If y has a lognormal distribution, then by definition $\ln y$ has a normal distribution with a constant variance.

For another example, count data (e.g. the number of calls received at a call center in one hour) are often modeled by a Poisson distribution for which we have $E(y) = \mu$ and $\mathrm{SD}(y) = \sqrt{\mu}$. The corresponding fitted values plot is shown in Figure 4.3c. Then

$$f(y) = \int \frac{dy}{\sqrt{y}} = 2\sqrt{y},$$

which is the **square root transformation**.

Still another example is the **inverse transformation**, $f(y) = y^{-1}$, which arises when $\sigma = g(\mu) \propto \mu^2$. The corresponding fitted values plot is shown in Figure 4.3d.

Example 4.2 *(Used Car Prices: Checking Homoscedasticity)* Continuing with Example 4.1, we give the fitted values plots for Price and log(Price) as the response variables in Figure 4.4. These plots are produced using the same R code as in Example 4.1 except using `which=1` in the `plot` function. Notice that the plot for Price is funnel-shaped indicating variance increasing with mean, while the plot for log(Price) exhibits a random parallel band. Thus log transformation stabilizes the variance. □

4.3.2   Box–Cox transformation*

The logarithmic, square root, and inverse transformations are special cases of a general family of power transformations:

$$f_\lambda(y) = \begin{cases} \dfrac{y^\lambda - 1}{\lambda} & \text{if } \lambda \neq 0 \\ \ln y & \text{if } \lambda = 0. \end{cases} \qquad (4.3)$$

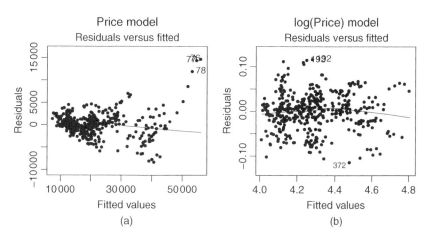

Figure 4.4. *Fitted values plots of residuals with Price and log(Price) as the response variables.*

Here λ is a parameter to be determined. Note that the log transformation is obtained as the limit of $f_\lambda(y)$ as $\lambda \to 0$ by using l'Hospital's rule.

How to choose λ? One can fit models using different transformations $f_\lambda(y)$ for a selected grid of λ-values, calculate SSE for each model and then choose λ that minimizes SSE. However, the SSEs for different λ are not comparable because they are not scale-free. To address this issue, Box and Cox (1964) suggested the following modification of (4.3):

$$f^*_\lambda(y) = \begin{cases} \dfrac{y^\lambda - 1}{\lambda \tilde{y}^{\lambda-1}} & \text{if } \lambda \neq 0 \\[2mm] \tilde{y}\ln y & \text{if } \lambda = 0, \end{cases} \tag{4.4}$$

where $\tilde{y} = \left(\prod_{i=1}^{n} y_i\right)^{1/n}$ is the geometric mean of the y_i's (assumed to be all positive). With this modification, the SSEs for different λ can be compared with each other.

Usually, λ-values in the range $[-1, +1]$ are of interest. Note that for given λ, $f^*_\lambda(y)$ is just a multiple of $f_\lambda(y)$, and the multiplication factor $1/(\tilde{y})^{\lambda-1}$ is common to all observations. Therefore, the two transformations are equivalent as response variables. So regression can be performed using $f_\lambda(y)$ as the response variable instead of $f^*_\lambda(y)$ after the minimizing value of λ has been determined.

Example 4.3 *(Textile Experiment Data: Finding the Best λ)* Box and Cox (1964) gave data from a 3^3 factorial experiment to study the effects of three factors on the cycles to failure (y) of worsted yarn. The three factors and their experimental levels were

x_1: length of test specimen (250 mm, 300 mm, 350 mm)

x_2: amplitude of loading cycle (8 mm, 9 mm, 10 mm)

x_3: load (40 g, 45 g, 50 g).

For convenience, we have coded the levels of all the three factors as $-1, 0, +1$ in Table 4.1.

Table 4.1. Cycles of failure of worsted yarn from a 3^3 factorial experiment

x_1	x_2	x_3	y	x_1	x_2	x_3	y	x_1	x_2	x_3	y
-1	-1	-1	674	0	-1	-1	1414	$+1$	-1	-1	3636
-1	-1	0	370	0	-1	0	1198	$+1$	-1	0	3184
-1	-1	$+1$	292	0	-1	$+1$	634	$+1$	-1	$+1$	2000
-1	0	-1	338	0	0	-1	1022	$+1$	0	-1	1568
-1	0	0	266	0	0	0	620	$+1$	0	0	1070
-1	0	$+1$	210	0	0	$+1$	438	$+1$	0	$+1$	566
-1	$+1$	-1	170	0	$+1$	-1	442	$+1$	$+1$	-1	1140
-1	$+1$	0	118	0	$+1$	0	332	$+1$	$+1$	0	884
-1	$+1$	$+1$	90	0	$+1$	$+1$	220	$+1$	$+1$	$+1$	360

Source: Box and Cox (1964).

As recommended by Box and Cox, the following model was fitted to the data:

$$\widehat{\log y} = 1.6784 + 4.9504 \log x_1 - 5.6537 \log x_2 - 3.5030 \log x_3;$$

here x_1, x_2, x_3 are in their original units. All the regression coefficients are highly significant. The fitted values and the normal plots are shown in Figure 4.5.

The plots look satisfactory, so we decide to stay with this model. It may be of interest to check whether the log transformation ($\lambda = 0$) of the response is in fact the best choice. For this purpose, we computed the transformation (4.4) for $\lambda = -1$ to $\lambda = +1$ in steps of 0.1 and regressed them on $\log x_1, \log x_2$, and $\log x_3$. The resulting SSEs are plotted in Figure 4.6. We see that SSE is minimized at $\lambda = 0$ which is indeed the best choice. $\square$

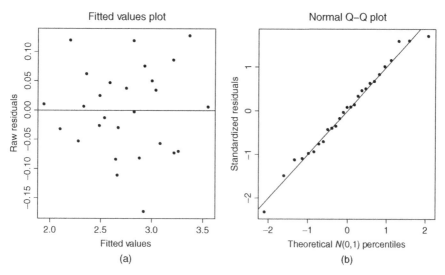

Figure 4.5. *(a) Fitted values and (b) normal plots of residuals from the fitted model for textile data.*

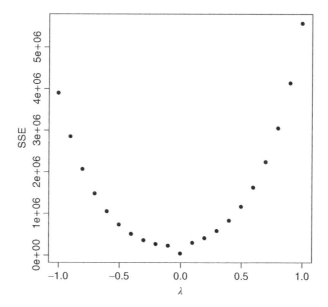

Figure 4.6. *Plot of SSE versus λ for textile data.*

4.4 Detecting outliers

Outliers are observations that deviate significantly from the fitted model, e.g. by more than two or three standard deviations. From the properties of the residuals given in Section 4.1, it follows that $\mathrm{SE}(e_i) = s\sqrt{1 - h_{ii}}$, where $s = \sqrt{\mathrm{MSE}}$.

The standardized residual

$$e_i^* = \frac{e_i}{s\sqrt{1 - h_{ii}}} \quad (i = 1, \ldots, n) \tag{4.5}$$

is used to test if the ith observation is an outlier by checking if e_i^* exceeds a specified critical constant, e.g. 3, in absolute value.

Outliers can significantly affect the fit of the model. So it is not correct to use an observation to fit the model and also test its residual for outlierness. A better way is to fit the model by deleting that observation and then testing its residual. Denote by $\widehat{y}_{i(i)}$ the fitted value of y_i after omitting it when fitting the model. Then the **deleted residual** is defined as $e_{(i)} = y_i - \widehat{y}_{i(i)}$ $(i = 1, \ldots, n)$.

In fact, it is not necessary to fit n separate models to compute n deleted residuals by omitting one observation at a time. They can be computed from the regular residuals by using the following formula:

$$e_{(i)} = \frac{e_i}{1 - h_{ii}} \quad (i = 1, \ldots, n). \tag{4.6}$$

Since $e_{(i)}$ is a scaled multiple of e_i, it follows that $\mathrm{SE}(e_{(i)}) = \mathrm{SE}(e_i)/(1 - h_{ii})$ and so the standardized deleted residual $e_{(i)}^*$ is the same as the standardized regular residual e_i^*.

In (4.5), s is the usual RMSE computed from all residuals including e_i. Therefore, if the ith observation is an outlier, then e_i will be large and will inflate s^2. This will deflate the standardized residual e_i^* making it less likely to be detected as an outlier. To obviate this difficulty, it has been suggested that s^2 should also be computed by deleting the ith observation. We denote the resulting s^2 by $s_{(i)}^2$. Thus, instead of using e_i^* given by (4.5), we use

$$r_i^* = \frac{e_i}{s_{(i)}\sqrt{1 - h_{ii}}}. \tag{4.7}$$

The e_i^* given by (4.5) are called **internally studentized residuals**, while the r_i^* given by (4.7) are called **externally studentized residuals**. The two are related by

$$r_i^* = e_i^* \sqrt{\frac{n - (p + 2)}{n - (p + 1) - e_i^{*2}}}. \tag{4.8}$$

Once again, it is not necessary to fit n separate models to compute the n separate estimates $s_{(i)}^2$. They can be computed from s^2 for the single model estimated from all observations by using the following formula:

$$s_{(i)}^2 = s^2 \left[\frac{(n - (p + 1)) - e_i^{*2}}{n - (p + 2)} \right] \quad (i = 1, \ldots, n). \tag{4.9}$$

Graphical plots are also useful for identifying outliers as we have seen from Figure 4.2, where departures of a few points from a linear normal plot indicate outliers. A plot of observed y_i's versus fitted $\widehat{y}_i$'s is also useful for this purpose.

Outliers (and influential observations discussed later in Section 4.7) should not be deleted automatically without additional inspection. First, they must be checked for validity and should be deleted only if they are erroneous. If they are valid observations, then they may indicate model misspecification. For example, we may be fitting a straight line to data that actually follow a quadratic or an exponential model. Thus, an outlier may be useful for revealing a misspecified model, which is the topic of discussion in Section 4.5.

Example 4.4 (College GPA and Entrance Test Scores: Checking Outliers) Refer to Example 3.3 where we fitted the model

$$\widehat{\text{GPA}} = -1.5705 + 0.0257\text{Verbal} + 0.0336\text{Math}.$$

Figure 4.7 gives the sequence plot of the residuals. This plot is obtained by specifying `which=4` in the `plot` function. We see that observation ♯4 appears to be an outlier. The same observation appears to be an outlier in the normal plot of the residuals. Let us check this by calculating the corresponding standardized residual. The fitted value for this observation is

$$\widehat{y}_4 = -1.5705 + 0.0257 \times 100 + 0.0336 \times 49 = 2.650,$$

so the residual is

$$e_4 = 1.54 - 2.650 = -1.110.$$

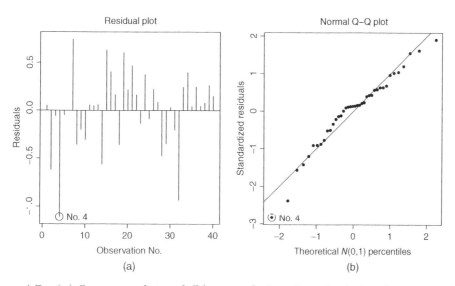

Figure 4.7. (a) Sequence plot and (b) normal plot of residuals for the college GPA data.

Now, $s^2 = 0.1618$ from the ANOVA Table 3.4 and the leverage $h_{44} = 0.1784$ from Table 4.2. Therefore, the internally studentized residual equals

$$e_4^* = \frac{-1.110}{\sqrt{0.1618(1 - 0.1784)}} = -3.044,$$

which is clearly significant.

Next, we will calculate the externally studentized residual. We first calculate $s_{(4)}^2$ using (4.9):

$$s_{(4)}^2 = 0.1618 \left[\frac{37 - (-3.044)^2}{36} \right] = 0.1246.$$

Hence, $r_4^* = -1.110/\sqrt{0.1246(1 - 0.1784)} = -3.469$, which is even more significant than the internally studentized residual, as expected. $\qquad\square$

4.5   Checking model misspecification*

Models can be misspecified in any number of ways. So it is not possible to recommend a single diagnostic that will detect any type of misspecification. The most common type of misspecification is linearity when the true relationship is nonlinear. One can plot residuals versus each x_j to see if they are randomly distributed or exhibit some pattern indicating departure from the assumed linearity in x_j. However, this plot could be influenced by possible nonlinearities in other predictors. Therefore better plots are needed.

The **added variables (AV) plot** (also known as the **partial regression plot**) removes the effects of other predictors by computing two sets of residuals. The first set of residuals, called the **y-residuals**, is the set of usual residuals, obtained by regressing y on all other predictors except x_j. The second set of residuals, called the **x_j-residuals**, is obtained by regressing x_j on all other predictors. Then the two sets of residuals are plotted against each other. If the relationship between y and x_j is linear then this plot will be roughly linear. In fact, $\widehat{\beta}_j$ is the slope coefficient of the LS fit to this plot and the correlation coefficient between the two sets of residuals is the partial correlation coefficient between y and x_j conditioned on all other predictors. If the relationship is nonlinear then the corresponding pattern is reflected in the plot. This plot can also be used to determine whether a variable not currently in the model should be added to the model. If the partial regression plot for that variable is random, then that variable should not be added. On the other hand, if the plot shows a clear trend, e.g. linear, then that variable should be added.

A variation on the added variables plot is the **component plus residuals (CR) plot** (also known as the **partial residual plot**) where the idea is to plot the **partial residuals**, defined as $e_i + \widehat{\beta}_j x_{ij}$, versus x_{ij} for $i = 1, \dots, n$. Here the e_i are the regular residuals from the full model with all predictors and $\widehat{\beta}_j$ is the regression coefficient of x_j from the same model. Note that the partial residuals are the regular residuals with the effect of the predictor x_j removed as can be seen from the following:

$$
\begin{aligned}
e_i + \widehat{\beta}_j x_{ij} &= y_i - \widehat{y}_i + \widehat{\beta}_j x_{ij} \\
&= y_i - [\widehat{\beta}_0 + \widehat{\beta}_1 x_{i1} + \cdots + \widehat{\beta}_p x_{ip}] + \widehat{\beta}_j x_{ij} \\
&= y_i - [\widehat{\beta}_0 + \widehat{\beta}_1 x_{i1} + \cdots + \widehat{\beta}_{j-1} x_{i,j-1} + \widehat{\beta}_{j+1} x_{i,j+1} + \cdots + \widehat{\beta}_p x_{ip}].
\end{aligned}
$$

If the plot is linear, then no nonlinear terms in x_j should be added to the model. The interpretation of the CR plot is similar to that of the AV plot, but it is more effective in displaying the relationship of y with each x_j since the plot is made against x_j.

Example 4.5 *(College GPA and Entrance Test Scores: AV and CR Plots)* Consider the linear model of GPA versus Verbal and Math scores fitted in Example 3.3. We want to check whether linear terms in Verbal and Math are adequate or should quadratic terms be added to the model. We assess this graphically using the AV plots and the CR plots shown in Figures 4.8 and 4.9 for the two variables. The following R code produced these plots.

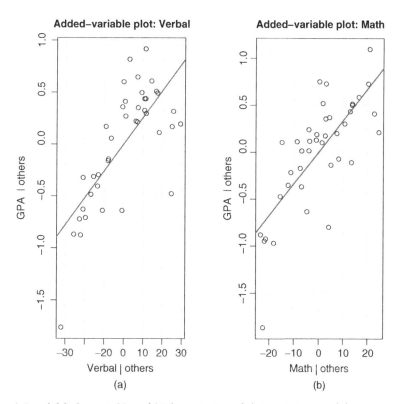

Figure 4.8. *Added variables (AV) plots for (a) Verbal and (b) Math scores.*

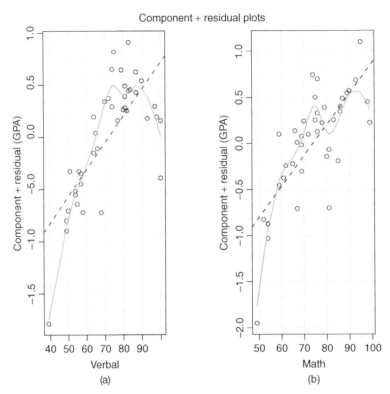

Figure 4.9. *Component plus residuals (CR) plots for (a) Verbal and (b) Math scores.*

```
> library("car")
> gpa = read.csv("c:/data/GPA.csv")
> fit = lm(GPA ~ Verbal + Math, data = gpa)
> crPlots(fit,"Verbal")
> crPlots(fit,"Math")
> avPlot(fit,"Verbal")
> avPlot(fit,"Math")
```

Both these plots show quadratic patterns, so quadratic terms in Math and Verbal should be added to the linear model. □

4.6   Checking independence

Lack of independence among the ε_i's typically occurs in **time-series data**. Such data are said to be autocorrelated or serially correlated. To denote that the observations are taken over time, we use t as the index of the observations instead of i ($t = 1, 2, \ldots, n$). For a stationary process, the first-order **autocorrelation coefficient** is defined as $\phi = \mathrm{Corr}(y_{t-1}, y_t) = \mathrm{Corr}(\varepsilon_{t-1}, \varepsilon_t)$. If $\phi > 0$, then the time-series is said to be positively autocorrelated while if $\phi < 0$, then it is said to be negatively autocorrelated. In time-series literature, the ε_t's are called **disturbances**.

The LS estimators $\widehat{\beta}_j$ are unbiased under autocorrelation, but their variances are different from those under independence. Under positive autocorrelation, the true variances of the $\widehat{\beta}_j$ are higher. Also, the MSE underestimates σ^2. Therefore the estimated variances of $\widehat{\beta}_j$ are underbiased, and so the $\widehat{\beta}_j$ appear more significant than they actually are. The situation is reversed for negatively autocorrelated data.

4.6.1 Runs test

Independence can be assessed by making a **run chart** of the residuals, i.e. by plotting e_t versus t. The number of runs of positive and negative residuals can be used to test independence. A run is defined as a sequence of like-signed residuals. For example, the sequence $+, +, -, -, -, +, +$ has two runs of $+$s and one run of $-$s for a total of three runs. If $\phi > 0$, then there will be less than expected number of runs under independence resulting in relatively few long cycles since each residual will tend to be followed by a like-signed residual. On the other hand, if $\phi < 0$, then there will be more than expected number of runs under independence runs resulting in a zigzag pattern since each residual will tend to be followed by an opposite signed residual.

How do we know that the number of runs is significantly less than or more than the expected number of runs under independence? To answer this question we can perform the runs test. Denote the total number of runs by R, the number of $+$ signs by n_1 and the number of $-$ signs by n_2, where $n_1 + n_2 = n$. Then under the null hypothesis of independence and conditioned on n_1 and n_2, it can be shown that

$$E(R) = \frac{2n_1 n_2}{n} + 1, \quad \text{and} \quad \text{Var}(R) = \frac{2n_1 n_2 (2n_1 n_2 - n)}{n^2 (n - 1)}. \qquad (4.10)$$

For large n_1, n_2, the statistic

$$z = \frac{R - E(R)}{\sqrt{\text{Var}(R)}}$$

can be used as a standard normal test statistic. So we can reject H_0 in favor of the alternative hypothesis $H_1 : \phi > 0$ at level α if $z < -z_\alpha$. Similarly, we can reject H_0 in favor of the alternative hypothesis $\phi < 0$ at level α if $z > z_\alpha$.

4.6.2 Durbin–Watson test*

This test assumes that the ε_t follow the so-called **first-order autoregressive (AR(1)) model**:

$$\varepsilon_t = \phi \varepsilon_{t-1} + \eta_t \quad (t = 2, \dots, n), \qquad (4.11)$$

where the η_t are i.i.d. $N(0, \sigma_0^2)$ r.v.s. Then it follows that $\mathrm{Corr}(\varepsilon_{t-1}, \varepsilon_t) = \phi$. Further it is easy to show that the ε_t are $N(0, \sigma^2)$ r.v.s with $\sigma^2 = \sigma_0^2/(1 - \phi^2)$.

For $n \gg p$, an approximate sample estimate of ϕ is given by

$$\widehat{\phi} \approx \frac{\sum_{t=2}^n e_t e_{t-1}}{\sum_{t=1}^n e_t^2} \tag{4.12}$$

The Durbin–Watson statistic equals

$$
\begin{aligned}
d &= \frac{\sum_{t=2}^n (e_t - e_{t-1})^2}{\sum_{t=1}^n e_t^2} \\
&= \frac{\sum_{t=2}^n e_t^2 + \sum_{t=2}^n e_{t-1}^2 - 2\sum_{t=2}^n e_t e_{t-1}}{\sum_{t=1}^n e_t^2} \\
&\approx 2(1 - \widehat{\phi}). \tag{4.13}
\end{aligned}
$$

Note that $0 \le d \le 4$ with $d < 2$ if $\widehat{\phi} > 0$, $d > 2$ if $\widehat{\phi} < 0$ and $d = 2$ if $\widehat{\phi} = 0$.

The Durbin–Watson test uses two critical constants, d_L and d_U, which depend on p, n, and α. These constants are given in Table D.5. If the alternative hypothesis is $\phi > 0$, the test operates as follows:

1. If $d < d_L$, then reject H_0 and conclude that $\phi > 0$.

2. If $d > d_U$, then do not reject H_0 and conclude that there is not sufficient evidence to conclude that $\phi > 0$.

3. If $d_L \le d \le d_U$, then the test is inconclusive.

If the alternative hypothesis is $\phi < 0$, then simply transform $d \to 4 - d$ and apply the same test. Note that this test has three possible decisions unlike the usual hypothesis tests that involve only two decisions (reject H_0 or do not reject H_0).

We conclude this section with two brief remarks.

- In many applications, response at time t is influenced by predictor values at previous times, e.g. $t - 1$, $t - 2$. In such cases, **lagged variables**, x_{t-1}, x_{t-2}, etc. must be used as predictors.

- Seasonality and cyclical variations are other common features of time series data. Such features can be modeled using a combination of sinusoidal functions. Often, the period of cyclical variation is known, e.g. monthly or quarterly, which can be incorporated in the sinusoidal functions.

4.7   Checking influential observations

The idea of fitting a model is to capture the overall pattern of variation in the response variable as a function of the predictor variables. So the fit of the model should be determined by the majority of the data points and not by a few so-called **influential observations** (also called **high leverage observations**). An extreme example of an influential observation is provided by the Anscombe Data Set IV given in Table 2.3 and plotted in Figure 2.4. It is obvious that $(x = 19, y = 12.50)$ is an influential observation since it alone determines the slope of the LS line as the line must pass through it and the midpoint of the remaining 10 observations, all at $x = 8$.

4.7.1   Leverage

How can we define undue influence and detect influential observations? Recall that the fitted or predicted vector is given by $\widehat{\boldsymbol{y}} = \boldsymbol{H}\boldsymbol{y}$, where $\boldsymbol{H}$ is the hat matrix. So the ith fitted value is given by

$$\widehat{y}_i = h_{i1}y_1 + \cdots + h_{ii}y_i + \cdots + h_{in}y_n,$$

where h_{ij} is the (i, j)th entry of $\boldsymbol{H}$. If the constant term is included in the model, it can be shown that the h_{ij}'s sum to 1 across each row. Thus $\widehat{y}_i$ is a weighted average of all observations y_j and h_{ii} is the weight on y_i in its own fitted value. If h_{ii} is too large, then y_i can be said to have too much influence on its own fitted value. We refer to h_{ii} as the **leverage** of the observation. For the observation $(x = 19, y = 12.50)$ in the Anscombe data set IV, it can be shown that $h_{ii} = 1$ and all other h_{ij}'s are zero (see Exercise 3.3). So the fitted $\widehat{y}_i$ is identically equal to the observed y_i.

How large must the leverage be in order for the observation to be regarded influential? To answer this question, h_{ii} is compared to the average of the diagonal elements of $\boldsymbol{H}$. The trace of $\boldsymbol{H}$ can be shown to be $p + 1$. So the average of the diagonal elements of $\boldsymbol{H}$ is $(p + 1)/n$. A rule of thumb is to declare the ith observation as influential if h_{ii} exceeds twice this average, i.e. if

$$h_{ii} > \frac{2(p + 1)}{n}. \tag{4.14}$$

Note that if $2(p + 1)/n > 1$, then this rule is not applicable since all $h_{ii} \leq 1$. Usually, $n \gg 2(p + 1)$ and so this threshold is too small resulting in too many observations flagged as influential.

4.7.2   Cook's distance

In general, leverage identifies those observations as influential that are outliers in the x-space. But an observation can be influential also because it is an outlier in the y-space. Cook's distance takes the effects of both these outliers into account.

Cook's distance for the ith observation measures the effect of deleting that observation on the fitted values of all observations. Let $\widehat{y}_{j(i)}$ denote the fitted value of y_j based on the regression model when the ith observation is omitted. Then Cook's distance for the ith observation is defined as

$$D_i = \frac{\sum_{j=1}^{n} (\widehat{y}_j - \widehat{y}_{j(i)})^2}{s^2(p + 1)}. \tag{4.15}$$

D_i measures the total amount by which all fitted values change when the ith observation is deleted. The denominator $s^2(p+1)$ is just a scaling factor.

It can be shown that

$$D_i = \left(\frac{e_i^*}{\sqrt{p+1}} \right)^2 \left(\frac{h_{ii}}{1 - h_{ii}} \right). \tag{4.16}$$

We see that D_i combines the outlierness of the ith observation in the y-space through the first term, which involves the standardized residual e_i^*, and in the x-space through the second term, which is an increasing function of the leverage h_{ii}.

Another interpretation of Cook's distance can be obtained as follows. Denote by $\widehat{\boldsymbol{\beta}}_{(i)}$ the LS estimator of $\boldsymbol{\beta}$ when the ith observation is deleted and let $\widehat{\boldsymbol{y}}_{(i)} = \boldsymbol{X}\widehat{\boldsymbol{\beta}}_{(i)}$, the corresponding fitted vector. Then the quantity $\sum_{j=1}^{n}(\widehat{y}_j - \widehat{y}_{j(i)})^2$ in the numerator of (4.15) equals

$$(\widehat{\boldsymbol{y}} - \widehat{\boldsymbol{y}}_{(i)})'(\widehat{\boldsymbol{y}} - \widehat{\boldsymbol{y}}_{(i)}) = (\boldsymbol{X}\widehat{\boldsymbol{\beta}} - \boldsymbol{X}\widehat{\boldsymbol{\beta}}_{(i)})'(\boldsymbol{X}\widehat{\boldsymbol{\beta}} - \boldsymbol{X}\widehat{\boldsymbol{\beta}}_{(i)})$$
$$= (\widehat{\boldsymbol{\beta}} - \widehat{\boldsymbol{\beta}}_{(i)})'\boldsymbol{X}'\boldsymbol{X}(\widehat{\boldsymbol{\beta}} - \widehat{\boldsymbol{\beta}}_{(i)}).$$

So another formula for Cook's distance is

$$D_i = \frac{(\widehat{\boldsymbol{\beta}} - \widehat{\boldsymbol{\beta}}_{(i)})'\boldsymbol{X}'\boldsymbol{X}(\widehat{\boldsymbol{\beta}} - \widehat{\boldsymbol{\beta}}_{(i)})}{(p+1)s^2}. \tag{4.17}$$

Comparing this formula with that for the $100(1-\alpha)\%$ confidence ellipsoid for the parameter vector $\boldsymbol{\beta}$ given by Equation (3.18), a possible decision rule for deciding the ith observation as influential is if $\widehat{\boldsymbol{\beta}}_{(i)}$ falls outside this ellipsoid for some chosen confidence level $1-\alpha$ and hence can be regarded as a significant change in the LS estimate of $\boldsymbol{\beta}$. This is equivalent to $D_i > f_{p+1,n-(p+1),\alpha}$. Montgomery et al. (2012, p. 216) have suggested using a confidence level of 10–20%, which may seem very low but keep in mind that the goal here is to identify whether an observation is influential – not to estimate $\boldsymbol{\beta}$ with high confidence.

Example 4.6 (College GPA and Entrance Test Scores: Detecting Influential Observations)   Again, refer to Example 3.3. We now check for influential observations by computing the leverage and Cook's distance values for the 40 observations which are given in Table 4.2. The threshold for the leverage is $2(p+1)/n = 6/40 = 0.15$. We see that observations 4 and 31 exceed this threshold; observation 32 falls slightly short. For Cook's distance, using $1-\alpha = 0.1$ (i.e. a 10% confidence ellipsoid) we have $f_{3,37,0.9} = 0.1937$, which is exceeded by observations 4 and 32; observation 31 falls significantly short. Thus, we may regard observations 4 and 32 as influential.   □

Table 4.2. Influence statistics for regression of GPA on Verbal and Math

No.	h_{ii}	D_i	No.	h_{ii}	D_i	No.	h_{ii}	D_i
1	0.0613	0.0004	15	0.0381	0.0337	29	0.0963	0.0289
2	0.1178	0.1187	16	0.0429	0.0158	30	0.0679	0.0002
3	0.0647	0.0005	17	0.0369	0.0022	31	0.1505	0.0181
4	0.1784	0.6703	18	0.1178	0.0399	32	0.1359	0.3283
5	0.0657	0.0004	19	0.0584	0.0493	33	0.0490	0.0066
6	0.0595	0.0000	20	0.0691	0.0077	34	0.0905	0.0353
7	0.0258	0.0310	21	0.0301	0.0143	35	0.0756	0.0003
8	0.1224	0.0415	22	0.0517	0.0031	36	0.0324	0.0042
9	0.1232	0.0132	23	0.0805	0.0036	37	0.1088	0.0006
10	0.0815	0.0183	24	0.0274	0.0084	38	0.0650	0.0010
11	0.0317	0.0002	25	0.1408	0.0030	39	0.0340	0.0051
12	0.0649	0.0004	26	0.0329	0.0036	40	0.0414	0.0021
13	0.0539	0.0005	27	0.0516	0.0009			
14	0.1291	0.1099	28	0.0947	0.0533			

The next example illustrates the use of these two statistics in graphical formats.

Example 4.7 *(Used Car Prices: Checking Influential Observations)*
Consider the used car prices data and the regression model fitted to $\log_{10}(\text{Price})$ based on the training data set in Example 3.12. Figure 4.10a shows the sequence plot of Cook's distances in which large D_i's show up as spikes. These observations (nos. 76, 77, and 78) can be regarded as influential.

Figure 4.10b shows the plot of D_i versus $h_{ii}/(1 - h_{ii})$ (although the label on the x-axis says "Leverage h_{ii}"). The same three observations are clustered in the north-east corner because they have both high D_i and high h_{ii}. Hence, they are identified as influential. These plots are produced using the same R code as in Example 4.1 except using which=c(4,6) in the plot function. In this example, $p = 10$ and $n = 402$, so the threshold for the leverage is $2(p + 1)/n = 22/402 = 0.0547$, which is exceeded by quite a few observations as can be seen from Figure 4.10b. Using $1 - \alpha = 0.1$, the threshold for Cook's distance is $f_{11,391,0.9} = 0.5049$, which is not exceeded by any of the observations

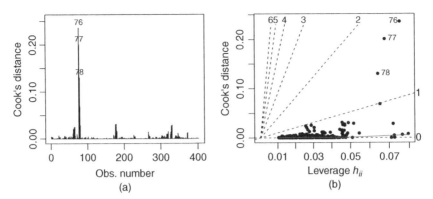

Figure 4.10. *(a) Sequence plot of Cook's distance and (b) plot of Cook's distance versus leverage for the regression model fitted in Example 3.12.*

including nos. 76, 77, and 78. Another threshold for Cook's distance used in the literature is $4/[n - (p + 1)]$, which equals 0.0102 in the present example, but it is exceeded by many observations as can be seen from Figure 4.10a. So why only these three observations are marked as influential? A possible reason is that they have D_i values that are many times larger than other D_i values, which is the reason they show up as tall spikes in the sequence plot. The dotted straight lines in Figure 4.10b are drawn at constant values of slope $D_i/[h_{ii}/(1 - h_{ii})] = (e_i^*)^2/(p + 1) = k$ or $e_i^* = \sqrt{k(p + 1)}$ for $k = 1, \ldots, 6$. Thus, large values of this ratio indicate outlier observations having large $|e_i^*|$ values. □

There are several other deletion diagnostics similar to Cook's distance. For example, DFFITS is a standardized measure of the change in the fitted value of the ith observation due to deleting the same observation. Similarly, DFBETA is a standardized measure of the change in $\widehat{\beta}_j$ due to deleting the ith observation (thus there is a matrix of $n \times p$ DFBETA values). We do not discuss them here. Interested reader is referred to the books by Montgomery et al. (2012) and Chatterjee and Hadi (2012) for more details.

4.8 Checking multicollinearity

4.8.1 Multicollinearity: causes and consequences

As seen in Chapter 3, mathematically multicollinearity is the result of the columns of $\boldsymbol{X}$ being approximately or exactly linearly dependent. This causes $\boldsymbol{X}'\boldsymbol{X}$ to be nearly singular making it difficult or impossible to invert, which in turn makes the computation of $\widehat{\boldsymbol{\beta}}$ difficult and subject to numerical errors. Furthermore, $\mathrm{Cov}(\widehat{\boldsymbol{\beta}}) = \sigma^2(\boldsymbol{X}'\boldsymbol{X})^{-1}$ has large entries, which imply large variances of the $\widehat{\beta}_j$'s.

The following are some of the common causes of multicollinearity:

- The most obvious cause is structural linear relationships among the predictors. One example is the percentages of ingradients in a chemical product that add up to 100%. A second example is a categorical variable for which the dummy variables for all of its categories are included in the model. A third example was mentioned in Chapter 3 concerning the data on income, expenditure, and saving.

- In any big data set with a large number of predictors, there are often approximate linear dependencies among subsets of predictors which can go unnoticed.

- Predictors with spurious correlations because of omitted lurking variables.

To see how the $\widehat{\beta}_j$'s deviate from the true β_j's in case of multicollinearity, consider the eigenvalues λ_j ($1 \leq j \leq p$) of $\boldsymbol{R}$, the correlation matrix of the x_j's. Note that the $\lambda_j \geq 0$ since $\boldsymbol{R}$ is a positive semidefinite matrix ($\lambda_j > 0$ if $\boldsymbol{R}$ is positive definite). Furthermore, since the $\widehat{\beta}_j$'s are standardized regression coefficients, their covariance matrix equals $\boldsymbol{R}^{-1}$. Therefore,

$$\sum_{j=1}^{p} \text{Var}(\widehat{\beta}_j) = \text{tr}(\boldsymbol{R}^{-1}) = \sum_{j=1}^{p} (1/\lambda_j). \tag{4.18}$$

So if some of the $\lambda_j \to 0$, then $\sum_{j=1}^{p} \text{Var}(\widehat{\beta}_j) \to \infty$. Note that

$$\sum_{j=1}^{p} \text{Var}(\widehat{\beta}_j) = E(\widehat{\boldsymbol{\beta}} - \boldsymbol{\beta})'(\widehat{\boldsymbol{\beta}} - \boldsymbol{\beta}) = E(\widehat{\boldsymbol{\beta}}'\widehat{\boldsymbol{\beta}}) - \boldsymbol{\beta}'\boldsymbol{\beta}. \tag{4.19}$$

Thus, if $\sum_{j=1}^{p} \text{Var}(\widehat{\beta}_j)$ becomes extremely large, then the squared norm (length) of the $\widehat{\boldsymbol{\beta}}$ vector tends to be much greater than that of the true $\boldsymbol{\beta}$ vector. Therefore, the individual $\widehat{\beta}_j$ coefficients tend to be too large in absolute value. Ridge and lasso regressions discussed in Chapter 5 address this problem by putting a constraint on the length of the $\widehat{\boldsymbol{\beta}}$ vector.

4.8.2 Multicollinearity diagnostics

Pairwise correlations

The simplest multicollinearity diagnostic is the pairwise correlations among the x's. If some or all of them are large, e.g. greater than 0.8 in absolute value, then they indicate multicollinearity. Note that multicollinearity refers to linear dependencies among the x's, not between the x's and y.

Example 4.8 *(Hald Cement Data: Correlation Matrix)* Hald (1952) gave the data shown in Table 4.3 on the heat evolved in calories (y) during hardening of cement per gram for 13 samples of cement and the percentages of their four ingradients: tricalcium aluminate (x_1), tricalcium silicate (x_2), tetracalcium alumino ferrite (x_3), and dicalcium silicate (x_4). Obviously, the four percentages add up to 100%, except for rounding errors and any impurities. So these data are structurally multicollinear.

Table 4.3. Hald cement data

No.	x_1	x_2	x_3	x_4	y	No.	x_1	x_2	x_3	x_4	y
1	7	26	6	60	78.5	8	1	31	22	44	72.5
2	1	29	15	52	74.3	9	2	54	18	22	93.1
3	11	56	8	20	104.3	10	21	47	4	26	115.9
4	11	31	8	47	87.6	11	1	40	23	34	83.8
5	7	52	6	33	95.9	12	11	66	9	12	113.3
6	11	55	9	22	109.2	13	10	68	8	12	109.4
7	3	71	17	6	102.7						

Source: Hald (1952).

The correlation matrix for these data is shown below.

$$
\begin{array}{c}
\\
x_1 \\
x_2 \\
x_3 \\
x_4
\end{array}
\begin{array}{cccc}
x_1 & x_2 & x_3 & x_4 \\
\left[\begin{array}{cccc}
1 & 0.229 & -0.824 & -0.245 \\
0.229 & 1 & -0.139 & -0.973 \\
-0.824 & -0.139 & 1 & 0.030 \\
-0.245 & -0.973 & 0.030 & 1
\end{array}\right]
\end{array}
$$

We notice that two correlations, $\mathrm{Corr}(x_1, x_3) = -0.824$ and $\mathrm{Corr}(x_2, x_4) = -0.973$, are large and negative. Examining the data closely we see that $x_1 + x_3$ is roughly constant equal to 20% and $x_2 + x_4$ is also roughly constant equal to 80%, so there are two approximate linear dependencies among the x's resulting in these two large negative correlations. □

Although large pairwise correlations between the x's are indicative of multicollinearity, the converse is not necessarily true. If a linear dependency exists among multiple x's (referred to as **multivariate linear dependency**), then none of the pairwise correlations may be large; see Exercise 4.10 for an example. In the following, we introduce diagnostic measures that take into account such multivariate linear dependencies.

t- and F-Statistics

Because multicollinearity results in large variances of the $\widehat{\beta}_j$'s, most of them turn out to be statistically nonsignificant. On the other hand, the ANOVA F-statistic can be highly significant. Thus, the overall fit is significant but none of the individual $\widehat{\beta}_j$'s is significant. The Hald cement data from Example 4.8 illustrates this phenomenon.

Example 4.9 *(Hald Cement Data: Regression)*   The regression output for the cement data is shown in the R output below. We see that all four regression coefficients are nonsignificant, whereas the overall F-statistic is highly significant with $P = 4.756 \times 10^{-7}$ pointing to a multicollinearity problem.

```
> lsfit=lm(y~x1+x2+x3+x4,data=cement)
> summary(lsfit)

Coefficients:
            Estimate Std. Error t value Pr(>|t|)
(Intercept)  62.4054    70.0710   0.891   0.3991
x1            1.5511     0.7448   2.083   0.0708 .
x2            0.5102     0.7238   0.705   0.5009
x3            0.1019     0.7547   0.135   0.8959
x4           -0.1441     0.7091  -0.203   0.8441
---

Residual standard error: 2.446 on 8 degrees of freedom
Multiple R-squared:  0.9824,    Adjusted R-squared:  0.9736
F-statistic: 111.5 on 4 and 8 DF,  p-value: 4.756e-07
```                                                                    □

Variance inflation factors

The most widely used and a reliable statistical measure of multicollinearity is the **variance inflation factor (VIF)**. To understand the idea behind the VIF, again refer to Equation (3.34) for $\widehat{\boldsymbol{\beta}}^*$. It follows from this equation that $\text{Cov}(\widehat{\boldsymbol{\beta}}^*) = \boldsymbol{R}^{-1}$. So $\text{Var}(\widehat{\beta}_j^*)$ is the jth diagonal entry of $\boldsymbol{R}^{-1}$. The variances of the unstandardized regression coefficients, $\text{Var}(\widehat{\beta}_j)$ are proportional to the $\text{Var}(\widehat{\beta}_j^*)$ since $\widehat{\beta}_j = (s_y/s_{x_j})\widehat{\beta}_j^*$ $(1 \leq j \leq p)$. In the ideal case of uncorrelated x's, $\boldsymbol{R}$ is an identity matrix and so $\text{Var}(\widehat{\beta}_j^*) = 1$ for all j. If $\boldsymbol{R}$ is not an identity matrix, then the $\text{Var}(\widehat{\beta}_j^*)$ are greater than 1 and represent the factors by which the $\text{Var}(\widehat{\beta}_j)$ are inflated because of the correlations among the x's. Therefore, the jth diagonal entry of $\boldsymbol{R}^{-1}$ is defined as the VIF for x_j and is denoted by VIF_j.

As an example, consider a multiple regression problem with two predictors, x_1 and x_2. Let r denote the sample correlation coefficient between x_1 and x_2. So

$$\boldsymbol{R} = \begin{bmatrix} 1 & r \\ r & 1 \end{bmatrix} \quad \text{and} \quad \boldsymbol{R}^{-1} = \frac{1}{1 - r^2} \begin{bmatrix} 1 & -r \\ -r & 1 \end{bmatrix}.$$

Hence, $\text{VIF}_1 = \text{VIF}_2 = 1/(1 - r^2)$. If $r = 0$, then there is no variance inflation but as $|r| \to 1$, $\text{VIF}_1 = \text{VIF}_2 \to \infty$.

This example is a special case of an alternative definition of the VIFs for multiple predictors. Let R_j^2 denote the R^2 from the regression of x_j on all other predictors. Then VIF_j is given by

$$\text{VIF}_j = \frac{1}{1 - R_j^2} \quad (j = 1, \ldots, p). \tag{4.20}$$

If x_j is approximately linearly dependent on other predictors, then R_j^2 will be close to 1 and so VIF_j will be large.

A common rule of thumb is to decide that there is a multicollinearity problem if several of the VIF_j are > 10. This means that their $R_j^2 > 0.90$, i.e. more than 90% of the variation in x_j is accounted for by a linear least squares fit with respect to other predictors.

Example 4.10 *(**Hald Cement Data: Variance Inflation Factors**)* We calculate the VIFs for the Hald cement data using two different methods described above. In the first method, we calculate the inverse of the correlation matrix between $x_1, \ldots, x_4$ given in Example 4.8:

$$
\boldsymbol{R}^{-1} =
\begin{array}{c}
\\ x_1 \\ x_2 \\ x_3 \\ x_4
\end{array}
\begin{array}{cccc}
x_1 & x_2 & x_3 & x_4 \\
\left[\begin{array}{cccc}
38.496 & 94.120 & 41.884 & 99.786 \\
94.120 & 254.423 & 105.091 & 267.539 \\
41.884 & 105.091 & 46.868 & 111.145 \\
99.786 & 267.539 & 111.145 & 282.513
\end{array}\right]
\end{array}.
$$

The diagonal entries of this matrix are the four VIFs, which can be seen to be very large.

In the second method, we regress each x_j on the other three predictors. The resulting R^2's are as follows:

$$R_1^2 = 0.9740, \quad R_2^2 = 0.9961, \quad R_3^2 = 0.9785, \quad R_4^2 = 0.9965.$$

So the corresponding VIFs are

$$\text{VIF}_1 = \frac{1}{1 - 0.9740} = 38.476, \quad \text{VIF}_2 = \frac{1}{1 - 0.9961} = 254.453,$$

$$\text{VIF}_3 = \frac{1}{1 - 0.9785} = 46.577, \quad \text{VIF}_4 = \frac{1}{1 - 0.9965} = 282.486,$$

which agree with the values obtained by the first method except for round-off errors. All four VIFs exceed the threshold of 10 by large margins indicating a serious multicollinearity problem. □

Exercises

Theoretical Exercises

4.1 (Internally and externally studentized residuals) Prove the relationship (4.8) between internally and externally studentized residuals by using the formula (4.9) for $s_{(i)}^2$.

4.2 (Leverage for simple linear regression) Refer to Exercise 3.2 which gives the following expression for leverage in case of simple linear regression:

$$h_{ii} = \frac{1}{n} + \frac{(x_i - \bar{x})^2}{S_{xx}} \quad (1 \le i \le n).$$

Show that the ith observation is flagged as influential using the rule of thumb (4.14) if

$$(x_i - \bar{x})^2 > \frac{3S_{xx}}{n}.$$

Interpret this result in terms of outlierness of x_i.

4.3 (Relation between the $\mathrm{Var}(\widehat{\beta}_j)$ and the VIF_j) Show that

$$\sum_{j=1}^{p} \mathrm{Var}(\widehat{\beta}_j) = \sum_{j=1}^{p} \mathrm{VIF}_j,$$

where the $\widehat{\beta}_j$ are standardized regression coefficients.

Applied Exercises

4.4 (College GPA and entrance test scores: Checking normality and homoscedasticity) Refer to Example 3.15 in which we fitted the model

$$\mathrm{GPA} = \beta_0 + \beta_1 \mathrm{Verbal} + \beta_2 \mathrm{Math} + \beta_3 \mathrm{Verbal}^2 + \beta_4 \mathrm{Math}^2$$
$$+ \beta_5 \mathrm{Verbal} \times \mathrm{Math} + \varepsilon.$$

(a) Make the normal and fitted values plots of residuals. Comment on why the normality and especially the homoscedasticity assumption seem to be violated. Does the fitted values plot suggest the log transformation of GPA?

(b) Fit the same model using log(GPA) as the response variable. Make the normal and fitted values plots of residuals. Are the normality and homoscedasticity assumptions satisfied now?

4.5 (Research expenditures data) Refer to Exercise 3.11 on modeling research expenditures of the top 30 engineering schools using the number of faculty and the number of PhD students as predictor variables. The two scatter plots are shown in Figures 4.11 and 4.12 with each data point labeled by the abbreviated name of the university. Identify the outliers and influential observations in the data using appropriate diagnostic statistics. Provide plausible explanations for why these universities are flagged.

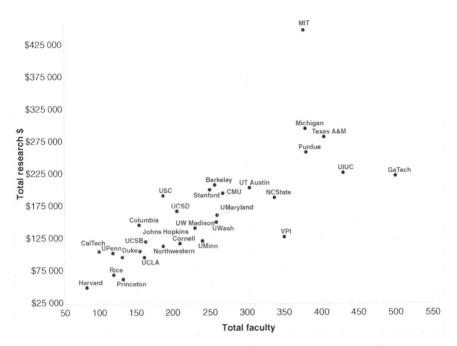

Figure 4.11. *Plot of research expenditures (in millions of dollars) versus number of faculty for the top 30 graduate engineering programs.*

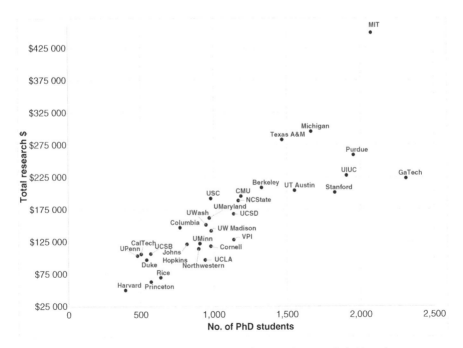

Figure 4.12. *Plot of research expenditures (in millions of dollars) versus number of PhD students for the top 30 graduate engineering programs*

4.6 (Employee salaries: Checking normality and homoscedasticity)
Refer to Exercise 3.14. Using the same predictors that were found significant in Part (a) of that exercise fit two regressions, one using Salary as the response variable and the other using log(Salary) as the response variable.

(a) Make normal plots for residuals from both regressions. Has the log transformation of Salary improved normality?

(b) Make fitted values plots for residuals from both regressions. Has the log transformation of Salary improved homoscedasticity?

4.7 (Soft drink sales: Testing independence) Data on sales of a soft drink, and advertising expenditures are available for 20 successive years. The sequence plot of residuals from the simple linear regression of sales on advertising expenditures is shown below (open circles represent negative residuals, dark circles represent positive residuals).

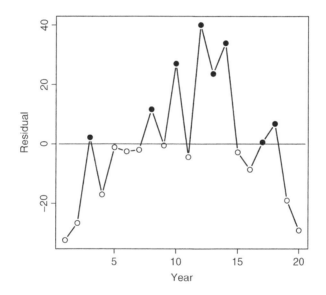

(a) Do the runs test to check if the autocorrelation coefficient is significantly greater than 0.

(b) The sample autocorrelation coefficient for this data is 0.34. Do the Durbin–Watson test to check if it is significantly greater than 0?

4.8 (Woodbeam data: Influential observations) Table 4.4 gives data on the specific gravity (x_1), moisture content (x_2) and strength (y) of wood beams.

(a) Make a scatter plot of the two predictors. Which observation appears to be influential?

(b) Is that observation influential using the leverage rule $h_{ii} > 2(p+1)/n$ and Cook's distance rule $D_i > f_{p+1,n-(p+1),\alpha}$ for $1 - \alpha = 0.10$ (i.e. 10% confidence ellipsoid).

Table 4.4. Woodbeam strength data

Obser-vation No.	Specific gravity	Moisture content	Strength	Obser-vation No.	Specific gravity	Moisture content	Strength
1	0.499	11.1	11.14	6	0.528	9.9	12.60
2	0.558	8.9	12.74	7	0.418	10.7	11.13
3	0.604	8.8	13.13	8	0.480	10.5	11.70
4	0.441	8.9	11.51	9	0.406	10.5	11.02
5	0.550	8.8	12.38	10	0.467	10.7	11.41

Source: Draper and Stoneman (1966), Table 1.

(c) Fit the equation $y = \beta_0 + \beta_1 x_1 + \beta_2 x_2 + \varepsilon$ from all data, and compare it with the fit obtained after omitting the influential observation. Does the fit change much?

4.9 (Anscombe data: Cook's distance) We have seen in Section 4.7.1 that the observation $(x = 19, y = 12.50)$ in the Anscombe Data Set IV is influential and that its leverage equals 1. Is the Cook's distance defined for this observation? Why or why not?

4.10 (Multivariate linear dependency) Webster et al. (1974) gave the data shown in Table 4.5 on four predictor variables. The data are constructed such that x_1, x_2, x_3, x_4 add up to 10 in all 12 observations except the first one where they add up to 11. Thus, there is an almost exact linear dependence among the four observations.

Table 4.5. Data illustrating multivariate linear dependence

No.	x_1	x_2	x_3	x_4	No.	x_1	x_2	x_3	x_4
1	8	1	1	1	7	2	7	0	1
2	8	1	1	0	8	2	7	0	1
3	8	1	1	0	9	2	7	0	1
4	0	0	9	1	10	0	0	0	10
5	0	0	9	1	11	0	0	0	10
6	0	0	9	1	12	0	0	0	10

Source: Webster et al. (1974). Reproduced with permission of Taylor and Francis.

(a) Calculate the correlation matrix among the four predictors. Check that no correlation exceeds 0.5 in absolute value. Thus, correlations do not provide an indication of multicollinearity.

(b) Calculate the four VIFs and check that they all exceed 150 with the maximum equal to 289.23. Thus, VIFs indicate serious multicollinearity.

4.11 (Gas mileages of cars: Multicollinearity and variance stabilizing transformation) The file `mpg.csv` contains data on gas mileages (`mpg`) of 392 cars and their number of cylinders, piston displacement, horsepower, weight, and acceleration. We want to build a predictive model for `mpg` based on these five predictors.

(a) Calculate the correlation matrix between the five predictors. Does it indicate presence of multicollinearity in the data?

(b) Fit a full regression model with all five predictors. How is multicollinearity reflected in this fit?

(c) To improve the model suppose we drop `displacement` from the above model since it is least significant. In what ways does the resulting model improve upon the full model?

(d) Make the normal Q–Q plot and the fitted values plot for the above fit. Note that the spread of the residuals in the fitted values plot is roughly proportional to the square of the fitted values, which suggests the inverse transformation. Use the inverse transformation `gp100m` = `100/mpg` (the gallons of fuel per 100 miles) as the dependent variable and rerun the regression, and make the normal Q–Q and the fitted values plots. Has this transformation helped to remove the flaws of the previous model? Does this transformation change VIFs?

(e) Calculate the estimated mpg of a car with 6 cylinders, 105 HP, 3000 lbs weight, and 15 miles per second^2 acceleration using the above fitted model.

4.12 (Acetylene data: Multicollinearity statistics) Table 4.6 gives data from Marquardt and Snee (1975) on conversion of n-heptane to acetylene (y) as a function of three reaction conditions: reactor temperature (x_1), ratio of H_2 to n-heptane (x_2) and contact time (x_3).

The following full second-degree model is to be fitted to the data:

$$y = \beta_0 + \beta_1 x_1 + \beta_2 x_2 + \beta_3 x_3 + \beta_{12} x_1 x_2 + \beta_{13} x_1 x_3 + \beta_{23} x_2 x_3$$
$$+ \beta_{11} x_1^2 + \beta_{22} x_2^2 + \beta_{33} x_3^2 + \varepsilon.$$

(a) Plot the three predictor variables against each other. Also calculate the correlation coefficients between them. Do you see any indications of multicollinearity?

Table 4.6. Acetylene data

x_1 Reactor temperature (°C)	x_2 Ratio of H$_2$ to n-heptane (mole ratio)	x_3 Contact time (s)	y Conversion of n-heptane to acetylene (%)
1300	7.5	0.0120	49.0
1300	9.0	0.0120	50.2
1300	11.0	0.0115	50.5
1300	13.5	0.0130	48.5
1300	17.0	0.0135	47.5
1300	23.0	0.0120	44.5
1200	5.3	0.0400	28.0
1200	7.5	0.0380	31.5
1200	11.0	0.0320	34.5
1200	13.5	0.0260	35.0
1200	17.0	0.0340	38.0
1200	23.0	0.0410	38.5
1100	5.3	0.0840	15.0
1100	7.5	0.0980	17.0
1100	11.0	0.0920	20.5
1100	17.0	0.0860	29.5

Source: Marquardt and Snee (1975). Reproduced with permission of Taylor and Francis.

(b) Calculate the VIFs for all the terms in the above model. Comment on your results.

(c) Center x_1, x_2, x_3 by subtracting the mean of each predictor from its values. Compute the remaining terms (pairwise products and squares) from these centered values. Now calculate the VIFs for all the terms. Compare the results with those from (b). Has centering made the multicollinearity problem less severe?

Chapter 5

Multiple linear regression: shrinkage and dimension reduction methods

As we have seen in the previous chapter, multicollinearity can result in serious difficulties for LS regression. Multicollinearity is often caused by having too many predictors. In some practical examples, we even have $p \gg n$, in which case standard regression methods break down. This occurs when the samples are expensive, but collecting data on many variables is inexpensive. For example, spectroscopic methods are often used in chemical analysis of specimens. The number of specimens is generally small, but spectroscopic intensity measurements can be made at hundreds of frequencies inexpensively. The same is the case with genomic data where the number of blood samples is small but the number of genes for which the measurements are made run into thousands.

LS estimators have the nice property of unbiasedness. However, when multicollinearity is present, this property is not particularly relevant since LS estimators have very large variances and become unstable if the data are perturbed. One can trade bias for variance, which can lead to a smaller mean square error (MSE) (defined as the expected value of the squared difference between an estimator and the true parameter it is estimating) since MSE equals $\text{Bias}^2 + \text{Variance}$. In this chapter, we discuss some alternative estimation methods that improve upon LS estimation by the reducing variance substantially at the expense of small bias, thus reducing overall MSE.

The point of departure of ridge and lasso regressions is the problem noted in (4.18) that under multicollinearity, the LS estimator vector $\widehat{\boldsymbol{\beta}}$ tends to be too long in comparison to the true $\boldsymbol{\beta}$ vector. So these methods put a

Predictive Analytics: Parametric Models for Regression and Classification Using R,
First Edition. Ajit C. Tamhane.

constraint on the length of $\widehat{\boldsymbol{\beta}}$. In ridge regression, the length is defined by the $\boldsymbol{L_2}$-**norm**: $\sqrt{\sum_{j=1}^{p} \beta_j^2} = \sqrt{\boldsymbol{\beta}'\boldsymbol{\beta}}$, while in lasso regression the length is defined by the $\boldsymbol{L_1}$-**norm**: $\sum_{j=1}^{p} |\beta_j|$. The specified constraint is incorporated by adding a penalty term to the LS criterion, which is then minimized. The resulting regression is known as **penalized regression** or **regularized regression**. Putting such a constraint causes the estimators of the regression coefficients to shrink toward zero. Therefore, these two methods of regression belong to a class of so-called **shrinkage methods**.

For both ridge and lasso regressions, all the variables must be standardized so that the β_j coefficients are unitless. Otherwise, the lengths of the $\boldsymbol{\beta}$ vectors can be artificially inflated or deflated depending on their units. In Sections 5.1 and 5.2, we will assume the standardized regression setting with no intercept term; for convenience, we will not use the asterisk notation.

Another class of regression methods considered in this chapter involves reducing the dimensionality of the predictor space by replacing the original predictors by a few key linear combinations of them that capture most of the variation among the former, which are then used to perform regression. If $p > n$, then this is necessitated by the fact that the rank of $\boldsymbol{X}'\boldsymbol{X}$ is less than p, and so $(\boldsymbol{X}'\boldsymbol{X})^{-1}$ is not invertible. Even if $p < n$, $\boldsymbol{X}'\boldsymbol{X}$ may be less than full rank because of multicollinearity. **Principal components regression (PCR)** and **partial least squares (PLS)** are two methods in this class. In PCR, a few key linear combinations of the predictors (called **principal components (PCs)**) are chosen to capture most of the variation among the original predictors; these PCs are then used as predictors. PLS follows a similar approach but uses the criterion of maximizing the covariance between the linear combinations of the original predictors and the response variable. These two methods of regression belong to the class of the so-called **dimension reduction methods**.

5.1   Ridge regression

The method of ridge regression was originally proposed by Hoerl and Kennard (1970). There have been many extensions and improvements, but here we will only present the basic method.

5.1.1   Ridge problem

Ridge estimator of $\boldsymbol{\beta}$ is defined as the estimator that minimizes the LS criterion subject to a constraint on its squared length:

$$\min_{\boldsymbol{\beta}} (\boldsymbol{y} - \boldsymbol{X}\boldsymbol{\beta})'(\boldsymbol{y} - \boldsymbol{X}\boldsymbol{\beta}) \quad \text{subject to} \quad \boldsymbol{\beta}'\boldsymbol{\beta} \le d^2 \qquad (5.1)$$

for some specified $d > 0$. This is a convex optimization problem, which can be reformulated using the **Lagrangian multiplier method** as

$$\min_{\boldsymbol{\beta}}[(\boldsymbol{y} - \boldsymbol{X}\boldsymbol{\beta})'(\boldsymbol{y} - \boldsymbol{X}\boldsymbol{\beta}) + \lambda\boldsymbol{\beta}'\boldsymbol{\beta}], \tag{5.2}$$

where $\boldsymbol{\beta}'\boldsymbol{\beta}$ is the penalty term and λ is the Lagrangian multiplier. Note that λ is a decreasing function of d^2 with $\lambda = 0$ when $d^2 = \infty$ (the unconstrained LS minimization problem). The ridge estimator of $\boldsymbol{\beta}$ is then given by

$$\widehat{\boldsymbol{\beta}}^R(\lambda) = (\boldsymbol{X}'\boldsymbol{X} + \lambda\boldsymbol{I})^{-1}\boldsymbol{X}'\boldsymbol{y}, \tag{5.3}$$

where $\boldsymbol{I}$ is a $p \times p$ identity matrix. Note that $\widehat{\boldsymbol{\beta}}^R(0)$ is the usual LS estimator.

As we saw in the previous chapter, multicollinearity is caused by the $\boldsymbol{X}'\boldsymbol{X}$ matrix being close to singular, which results in its minimum eigenvalue to be close to zero. The ridge estimator $\widehat{\boldsymbol{\beta}}^R(\lambda)$ can be viewed as ameliorating this singularity by adding a positive constant λ to each diagonal term of $\boldsymbol{X}'\boldsymbol{X}$ and consequently to each of its eigenvalues.

Geometrically, the ridge estimator can be interpreted as follows. The contours of the constant values of the LS criterion $(\boldsymbol{y} - \boldsymbol{X}\boldsymbol{\beta})'(\boldsymbol{y} - \boldsymbol{X}\boldsymbol{\beta})$ can be plotted in the $\boldsymbol{\beta}$ space. These contours are ellipsoids. In two dimensions, they are ellipses as shown in Figure 5.3. The center of the ellipse is where the LS criterion is minimum and corresponds to the LS estimator $\widehat{\boldsymbol{\beta}}$. As one moves away from the center of the ellipse, the value of the LS criterion increases. The ridge estimator is the shortest $\boldsymbol{\beta}$ vector that touches the contour with the circle of radius equal to d as shown in Figure 5.3. As d^2 gets smaller (or as λ gets bigger), the ridge estimator shrinks, approaching the null vector in the limit. Thus, its variance becomes zero, but its bias equals $\boldsymbol{\beta}$.

5.1.2   Choice of λ

Choice of the tuning parameter λ is a difficult problem. In the **ridge trace** method, the individual estimates $\widehat{\beta}_j^R(\lambda)$ $(1 \leq j \leq p)$ are plotted against λ. As λ increases, the magnitudes of the estimates shrink approaching to zero in the limit, where the Bias^2 term approaches $\boldsymbol{\beta}'\boldsymbol{\beta}$. One could choose λ, where the ridge traces for most coefficients stabilize. However, such a value of λ may not exist or may be difficult to identify, particularly when there are many β's. Also, this choice is subjective; so this method is no longer used.

The `glmnet` function in R chooses the optimal λ via cross-validation. It selects a grid of λ-values. For each value of λ in the grid, it performs m-**fold cross-validation**, which divides the data randomly into $m \geq 2$ approximately equal size subsets. One of these subsets is set aside as the test set and the model is fitted on the remaining $m - 1$ subsets treated together as the training set. The MSE of prediction is then computed for the test set. This process is repeated

m times, each time leaving a different subset out as the test set and using the remaining $m-1$ subsets as the training set. The MSE is averaged over these m replicates which gives one realization of MSE. To minimize the sampling error associated with dividing the data into m random subsets, this process is repeated multiple times each time randomly drawing different subsets. The MSE is averaged over all these subset divisions and also their standard deviation is computed. The associated $\widehat{\boldsymbol{\beta}}^{R}(\lambda)$ and MSE-values are calculated for each value of λ, resulting in a range of MSE values. The means of these MSE values along with one standard deviation bars around them are plotted as shown in Figure 5.2. From this plot one can find not just the optimal λ which minimizes the estimated MSE, but a range of acceptable λ-values which give MSE-values within the limits of sampling error from the estimated minimum MSE.

Ridge regression shrinks all coefficient estimates simultaneously. Some estimates are driven to zero faster than others, but generally no estimates are set exactly equal to zero. Therefore, no predictors are dropped from the model. As a result, ridge regression cannot be used as a variable selection method as can lasso regression discussed in Section 5.2.

Both ridge and lasso regressions can be performed using `glmnet`. This package uses standardized data but reports unstandardized regression coefficients. It minimizes a generalized criterion:

$$\frac{1}{2n}(\boldsymbol{y}-\boldsymbol{X}\boldsymbol{\beta})'(\boldsymbol{y}-\boldsymbol{X}\boldsymbol{\beta}) + \lambda\left[\frac{(1-\alpha)}{2}\boldsymbol{\beta}'\boldsymbol{\beta} + \alpha\sum_{j=1}^{p}|\beta_j|\right],$$

where the quantity in square brackets in the second term is a generalized penalty. For ridge regression, we set $\alpha=0$, so the penalty term is $(1/2)\boldsymbol{\beta}'\boldsymbol{\beta} = (1/2)\sum_{j=1}^{p}\beta_j^2$. For lasso regression, we set $\alpha=1$, so the penalty term is $\sum_{j=1}^{p}|\beta_j|$. For general α $(0<\alpha<1)$, the method is known as **elastic net regression**.

Example 5.1 *(Hald Cement Data: Ridge Regression)* Ridge regression was performed on the Hald cement data using `glmnet`. The R script given below was used to generate the results including the ridge trace plot shown in Figure 5.1. The plot of the MSE minimized using cross-validation is shown in Figure 5.2. For each given value of λ in the grid, the black dots show the averages (over all cross-validations) of the MSE with one standard deviation error bars around them. You can choose an initial seed for generating random subsets needed in cross-validation. Here we have chosen 123456789 to be the initial seed. Different initial seeds give different results.

From the R output below, we get the optimum $\lambda = 0.563$. Denote this optimum λ by $\lambda_{\min}$. Let $\lambda_{\max}$ denote the maximum acceptable value of optimum λ that takes into account the sampling variation in the cross-validation process. In

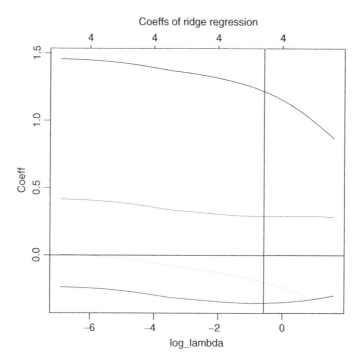

Figure 5.1. *Ridge trace plot for Hald cement data.*

Figure 5.2, a vertical dotted line is drawn at $\ln(\lambda_{\min}) = \ln(0.563) = -0.574$. A second vertical line is drawn at $\ln(\lambda_{\max}) \approx 0.9$, so $\lambda_{\max} \approx 2.46$. The fitted ridge regression model corresponding to the optimum $\lambda = 0.563$ is

$$\widehat{y} = 85.4819 + 1.2176x_1 + 0.2887x_2 - 0.2013x_3 - 0.3559x_4.$$

Compare this model with the LS fitted model

$$\widehat{y} = 62.4063 + 1.5511x_1 + 0.5102x_2 + 0.1019x_3 - 0.1441x_4.$$

```
> library(glmnet)
> cement=read.csv("c:/data/cement.csv")
> y=cement$y
> x=model.matrix(y~ .,cement)
> ridgefit=glmnet(x, y, alpha=0,lambda=seq(0,5,0.0001))
> set.seed(123456789)
> ridgecv=cv.glmnet(x, y, alpha=0,lambda=seq(0,5,0.001),nfold=3)
> lambdaridge=ridgecv$lambda.min
> print(lambdaridge)
[1] 0.563
> plot(ridgecv)
> plot(ridgefit,xvar="lambda", main="Coeffs of Ridge Regression",
+       type="l", xlab=expression("log_lambda"), ylab="Coeff")
```

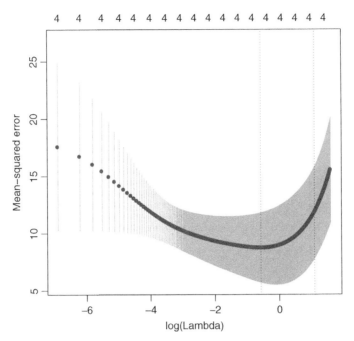

Figure 5.2. *Mean-squared error plot for ridge regression of Hald cement data.*

```
> abline(h=0); abline(v=log(ridgecv$lambda.min))
> small.lambda.index <- which(ridgecv$lambda == ridgecv$lambda.min)
> small.lambda.betas <- coef(ridgecv$glmnet.fit)[,small.lambda.index]
> print(small.lambda.betas)
(Intercept) (Intercept)          x1          x2          x3          x4
 85.4819340   0.0000000   1.2175599   0.2886873  -0.2012739  -0.3558764
```

5.2   Lasso regression

The method of lasso regression was proposed by Tibshirani (1996). This method has become extremely popular in recent years, and there have been many extensions of it summarized in the book by Hastie et al. (2015). As in the case of ridge regression, we will present only the basic method.

5.2.1   Lasso problem

Analogous to the ridge estimation problem (5.1), the lasso estimator solves the following constrained minimization problem:

$$\min_{\boldsymbol{\beta}} (\boldsymbol{y} - \boldsymbol{X}\boldsymbol{\beta})'(\boldsymbol{y} - \boldsymbol{X}\boldsymbol{\beta}) \quad \text{subject to} \quad \sum_{j=1}^{p} |\beta_j| \le d, \qquad (5.4)$$

where $d > 0$ is a specified constant. This is also a convex optimization problem, which can be reformulated using the Lagrangian multiplier method as

$$\min_{\boldsymbol{\beta}} \left[(\boldsymbol{y} - \boldsymbol{X}\boldsymbol{\beta})'(\boldsymbol{y} - \boldsymbol{X}\boldsymbol{\beta}) + \lambda \sum_{j=1}^{p} |\beta_j| \right], \tag{5.5}$$

This minimization problem does not have a closed form solution as in the case of ridge regression. We denote the lasso estimator by $\widehat{\boldsymbol{\beta}}^{L}(\lambda)$. The optimal λ can be chosen by the same m-fold cross-validation method given for ridge regression.

The advantage of lasso regression is that the solution to the above minimization problem generally falls at one of the corners or the edges or the faces of the polyhedron $\sum_{j=1}^{p} |\beta_j| \leq d$ setting some of the β_j's equal to zero; see Figure 5.3. Hence, it effectively performs variable selection.

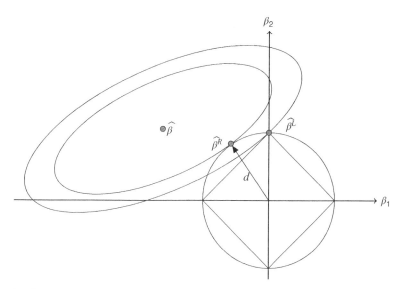

Figure 5.3. *Least squares contours with ridge and lasso regression feasible regions for $p = 2$.*

The feasible region of the lasso problem in two dimensions is a rotated square as shown in Figure 5.3, and in three dimensions, it is an octahedron as shown in Figure 5.4. On the other hand, for ridge regression, the feasible region for the $\boldsymbol{\beta}$ vector, namely $\sum_{j=1}^{p} \beta_j^2 \leq d^2$, is a smooth hypersphere with no sharp corners and so, in general, it does not set any β_j's equal to zero. For this reason, lasso regression is well-suited when there are only a few signals among many predictors, most being noise (nonsignficant). This problem is referred to as **sparse regression**.

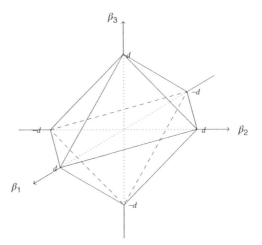

Figure 5.4. *Octahedron in the* $(\beta_1, \beta_2, \beta_3)$ *space satisfying* $|\beta_1| + |\beta_2| + |\beta_3| \leq d.$

Example 5.2 *(Hald Cement Data: Lasso Regression)* Lasso regression was performed on the Hald cement data using `glmnet`. The only change needed in the R code given in Example 5.1 was to change `alpha = 0` to `alpha = 1`. The optimum $\lambda = 0.475$ from the R output (not shown here). The lasso trace plot is shown in Figure 5.5. Note that the coefficient of one of the variables (x_3) is driven to 0 for very small λ. The plot of the objective function (MSE) minimized using cross-validation is shown in Figure 5.6. The definitions of $\lambda_{\min}$

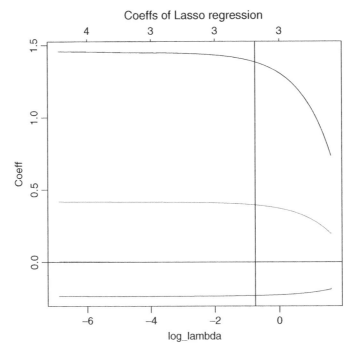

Figure 5.5. *Lasso trace plot for Hald data.*

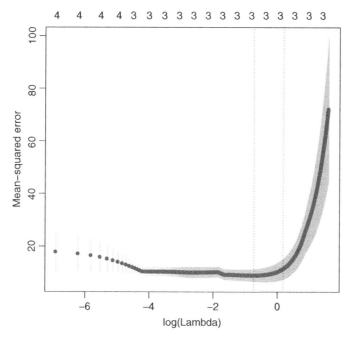

Figure 5.6. *Mean-squared error plot for lasso regression of Hald data.*

(optimum λ) and $\lambda_{\max}$ are the same as those given for ridge regression. Note that $\ln \lambda_{\min} = \ln(0.475) = -0.744$. From Figure 5.6, we see that $\ln(\lambda_{\max}) \approx 0.1$, which corresponds to $\lambda_{\max} \approx 1.105$. The fitted lasso regression model corresponding to optimum $\lambda = 0.475$ is

$$\widehat{y} = 73.0230 + 1.3839x_1 + 0.3952x_2 - 0.2318x_4.$$

Note that the coefficient of x_3 is zero and so x_3 is dropped from the model. □

5.3 Principal components analysis and regression*

5.3.1 Principal components analysis (PCA)

Consider n i.i.d. observations $\boldsymbol{x}_i$ on a random data vector $\boldsymbol{x} = (x_1, \ldots, x_p)'$. The objective of principal components analysis (PCA) is to find linear combinations (called the **principal components** or **PCs**):

$$z_j = u_{1j}x_1 + \cdots + u_{pj}x_p \quad (j = 1, \ldots, p), \tag{5.6}$$

which are applied to all n observations, such that $\mathrm{Var}(z_1) \geq \cdots \geq \mathrm{Var}(z_p)$ and $\mathrm{Corr}(z_j, z_k) = 0$ for $j \neq k$. The vectors $\boldsymbol{u}_j = (u_{1j}, \ldots, u_{pj})'$ are of unit length and are mutually orthogonal. The idea here is that z_1 captures the most variation among the x_j's, z_2 captures the second most variation and is uncorrelated with

z_1, and so on. We refer to z_1 as the first PC (PC1), z_2 as the second PC (PC2), and so on. Often, the first few PCs capture most of the variation among the x's and so can be used to summarize the x's without much loss of information, thus reducing the dimensionality of the data. The vectors $\boldsymbol{u}_j$ are called the **loading vectors** of the PCs. Another term used for PCs is **factors**. This term comes from **factor analysis**, which is a closely related technique.

PCA involves finding the best orthogonal directions on which the x's are be projected in order to capture most variation among them. Along the first PC direction, the x's vary the most, and along the last PC direction, they vary the least. Thus, the most information among the x's is captured by projecting them on the first PC direction, and the least information is captured by projecting them on the last PC direction. If the x's do not vary along some direction, then projecting them on that direction results in a constant with no useful information.

Geometrically, the PCs correspond to the orthogonal principal axes of the ellipsoidal scatter of the x's. This is the case if the x's have a multivariate normal distribution. When $p = 2$, PC1 corresponds to the major axis of the elliptical scatter and PC2 corresponds to the minor axis. Figure 5.7 shows this for a narrow elongated elliptical scatter. The projection on the major axis contains most of the information (most of the variation) among the x's and the projection on the minor axis contains very little information (very little variation) among the x's.

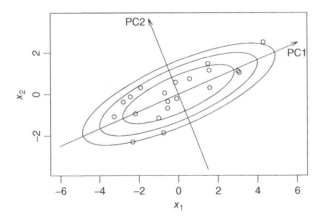

Figure 5.7. *PC directions for an elliptically contoured bivariate distribution.*

The vector $\boldsymbol{z} = (z_1, \ldots, z_p)'$ of **PC scores** is a nonsingular orthogonal linear transformation of the data vector $\boldsymbol{x} = (x_1, \ldots, x_p)'$ of predictor variables. This transformation is defined by a $p \times p$ matrix $\boldsymbol{U} = \{u_{ij}\}$ with column vectors $\boldsymbol{u}_j$ $(j = 1, \ldots, p)$ so that $\boldsymbol{z} = \boldsymbol{U}'\boldsymbol{x}$. We show below that the loading vectors $\boldsymbol{u}_j = (u_{1j}, \ldots, u_{pj})'$ are the eigenvectors of the covariance matrix $\boldsymbol{\Sigma}$ of $\boldsymbol{x}$ corresponding to its eigenvalues $\lambda_1 \geq \cdots \geq \lambda_p \geq 0$. Then $\boldsymbol{U}'\boldsymbol{U} = \boldsymbol{U}\boldsymbol{U}' = \boldsymbol{I}$, and from the spectral decomposition theorem (see (A.1)), we know that $U'\Sigma U = \Lambda = \text{diag}\{\lambda_1, \ldots, \lambda_p\}$. However, from the sandwich formula (A.7), we also know that $\text{Cov}(z) = U'\Sigma U$. Therefore, $\text{Cov}(z) = \text{diag}\{\text{Var}(z_1), \ldots, \text{Var}(z_p)\} = \text{diag}\{\lambda_1, \ldots, \lambda_p\}$. Thus, the z_j are uncorrelated, and their variances are the ordered eigenvalues of Σ. Next, we give a computational algorithm to determine them.

First, we want to find the vector u_1 that maximizes $u_1'\Sigma u_1 = \text{Var}(z_1)$. However, if we multiply u_1 by any constant, then $\text{Var}(z_1)$ gets multiplied by that constant squared. To avoid such an artificial inflation of $\text{Var}(z_1)$, we constrain u_1 to be of unit length, i.e. $\|u_1\|^2 = u_1'u_1 = 1$. Thus, we solve the constrained maximization problem:

$$\max_{u_1} u_1'\Sigma u_1 \quad \text{subject to} \quad u_1'u_1 = 1. \tag{5.7}$$

Using the Lagrangian multiplier method, we get

$$\max_{u_1}[u_1'\Sigma u_1 - \lambda_1(u_1'u_1 - 1)]. \tag{5.8}$$

In Section 5.5.2, we show that the solution to this Lagrangian problem is $\max_{u_1}(u_1'\Sigma u_1) = \lambda_1$, which is the largest eigenvalue of Σ, and the maximizing vector u_1 is the associated eigenvector. (To be notationally correct, (5.7) should be written as $\max_u u'\Sigma u$ subject to $u'u = 1$ and the solution to this constrained maximization problem should be denoted by u_1. To simplify the notation, we have denoted the variable of maximization and the solution to the constrained maximization problem both by u_1. The same convention is followed below.)

The loading vector u_2 of the second PC is found by the same method, but with an additional constraint that u_1 and u_2 are orthogonal, i.e. $u_1'u_2 = 0$. We get $\max_{u_2} u'_2\Sigma u_2 = \text{Var}(z_2) = \lambda_2$, the second largest eigenvalue of Σ, and u_2 is the eigenvector associated with λ_2. This process can be continued until all p PCs are computed. Usually, the process is stopped if the first $r < p$ PCs account for a high fraction (say, 90%) of the total variance among the x's defined to be $\text{tr}(\Sigma) = \sum_{j=1}^{p} \lambda_j$.

In practice, the covariance matrix Σ is unknown, so the sample covariance matrix S estimated from the observations x_i ($i = 1 \ldots, n$) is used instead. We will continue to denote the eigenvalues of S by $\lambda_1, \ldots, \lambda_p$, and the associated eigenvectors by $u_1, \ldots, u_p$. Let Λ and U be as defined before. Then by the spectral decomposition theorem, we have

$$USU' = \Lambda \quad \text{and} \quad U'\Lambda U = S.$$

The PC scores matrix Z is obtained from the data matrix $X : n \times p$ by the orthogonal rotation:

$$Z = XU. \tag{5.9}$$

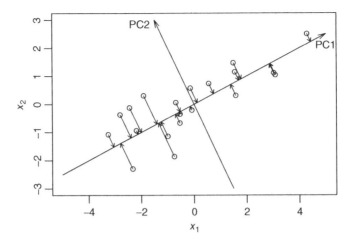

Figure 5.8. *Principal component direction minimizes the sum of squared projections of all data points on it.*

The PCs in the sample space can also be found geometrically as follows. To determine the first PC direction, find the straight line that minimizes the sum of the squared projections on the line from all the data points x_i. The second PC direction is found in the same way subject to the restriction that it is perpendicular to the first PC direction. Figure 5.8 illustrates this for $p = 2$.

An important question to answer before performing PCA is whether the x's should be standardized or not. If the x's are not standardized, then we will be computing linear combinations of variables with different units, e.g. age, weight, and blood pressure of patients. Furthermore, inherent variations in these variables may artificially influence the PCs. For example, weights of patients vary much more than their ages. Because PCA aims at finding linear combinations of the original variables that maximize the variance, it puts higher loadings on variables with higher variances (in this example on the weights). Standardization makes the variables unitless and scale-free. However, if the variables are standardized, then they all have a unit variance. So it is not clear what is meant by maximizing the variation among the x's. General consensus is that if the variables are commensurate, i.e. if they are measured in the same units and have roughly the same scales, then they should not be standardized; otherwise, they should be standardized. Even in case of commensurate variables, standardization may be necessary if some variables have much higher inherent variability than others.

Next, we give two examples of PCA using R. The "stats" module in R has two functions to perform PCA: `prcomp` and `princomp`. The variable names used are different for the two functions. For example, factor loadings are named `rotation` in `prcomp` and `loadings` in `princomp`, while factor scores are stored in a matrix labeled `x` in `prcomp` and in `scores` in `princomp`.

Example 5.3 *(GPA Data: Principal Components Analysis)* In Chapter 3, we used the GPA data in many examples. In Example 3.3, we found that the two predictors, Verbal and Math, are roughly uncorrelated and have about equal linear effects on GPA. Here we use the same data to illustrate PCA of Verbal and Math, both in unstandardized and standardized forms. The R code for performing PCA on unstandardized data and the associated results are shown below:

```
> gpa = read.csv("c:/data/GPA.csv")
> cov(gpa[,1:2])
        Verbal      Math
Verbal 259.2718 -22.6410
Math   -22.6410 172.8718
> fit1=prcomp(gpa[,1:2], scale=FALSE)
> summary(fit1)
Importance of components:
                          PC1     PC2
Standard deviation     16.2741 12.9344
Proportion of Variance  0.6129  0.3871
Cumulative Proportion   0.6129  1.0000
> fit1$sdev ^2 # eigenvalues
[1] 264.8453 167.2983
> fit1$rotation
           PC1    PC2
Verbal -0.971 -0.239
Math    0.239 -0.971
```

The PCs are extracted from the covariance matrix between Verbal and Math shown in the above output. PC1 corresponds to $\lambda_1 = 264.8453$ and PC2 corresponds to $\lambda_2 = 167.2983$. Thus, PC1 accounts for 61.29% of the total variation and PC2 accounts for the remaining 38.71%. For PC1, the loadings on Verbal and Math are -0.971 and 0.239, respectively, and for PC2, the loadings are -0.239 and -0.971, respectively. Thus, PC1 puts four times as much weight on Verbal compared to Math, and PC2 reverses these weights. PC1 and PC2 are orthogonal to each other and are normalized to have unit lengths as can be verified from $0.971^2 + 0.239^2 = 1$. The PC1 and PC2 directions are shown on the scatterplot of Verbal and Math in Figure 5.9.

We can do PCA on standardized data by specifying `scale = TRUE` in the `prcomp` function. This extracts PCs from the correlation matrix. The results are as seen below:

```
> cor(gpa[,1:2])
        Verbal     Math
Verbal  1.0000 -0.1069
Math   -0.1069  1.0000

> fit2=prcomp(gpa[,1:2],scale=TRUE)
```

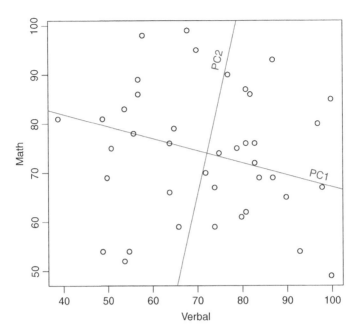

Figure 5.9. *Principal component directions shown on the Verbal-Math scatter-plot.*

```
> summary(fit2)
Importance of components:
                        PC1     PC2
Standard deviation     1.0521  0.9450
Proportion of Variance 0.5535  0.4465
Cumulative Proportion  0.5535  1.0000
> fit1$sdev^2 # eigenvalues
[1] 1.1069 0.8931
> fit2$rotation
          PC1     PC2
Verbal -0.7071 -0.7071
Math    0.7071 -0.7071
```

Notice that the PC1 and PC2 loadings on both the predictors are equal in magnitude, namely $0.7071 = 1/\sqrt{2}$. This is not very useful since the loadings will be always equal to $\pm 1/\sqrt{2}$ for standardized data regardless of how different the variances of the two variables are, and this fact generalizes to $p > 2$ for equicorrelated data; see Exercise 5.4. In the present example, Verbal has higher variance than Math and so should have a higher loading as found for unstandardized data. This puts into context our earlier comment that for standardized data, it is not clear what is meant by maximizing the variance among the x's.   □

Next, we illustrate PCA on the Hald cement data.

Example 5.4 (*Hald Cement Data: Principal Components Analysis*) We use essentially the same R script used in the previous example; in addition we print the PC scores which are stored in a matrix labeled "x."

```
> fit1=prcomp(cement[,1:4])
> summary(fit1)
Importance of components:
                            PC1     PC2      PC3     PC4
Standard deviation       22.755  8.2156  3.52213  0.4870
Proportion of variance   0.866   0.1129  0.02075  0.0004
Cumulative proportion    0.866   0.9789  0.99960  1.0000
> fit1$sdev^2 # eigenvalues
[1] 517.7969  67.4964  12.4054   0.2372
> fit1$rotation
        PC1      PC2      PC3     PC4
x1   0.0678   0.6460  -0.5673 0.5062
x2   0.6785   0.0200   0.5440 0.4933
x3  -0.0290  -0.7553  -0.4036 0.5156
x4  -0.7309   0.1085   0.4684 0.4844
> fit1$x
            PC1       PC2      PC3      PC4
 [1,] -36.8218    6.8709   4.5909   0.3967
 [2,] -29.6073   -4.6109   2.2476  -0.3958
 [3,]  12.9818    4.2049  -0.9022  -1.1261
 [4,] -23.7147    6.6341  -1.8547  -0.3786
 [5,]   0.5532    4.4617   6.0874   0.1424
 [6,]  10.8125    3.6466  -0.9130  -0.1350
 [7,]  32.5882   -8.9798   1.6063   0.0818
 [8,] -22.6064  -10.7259  -3.2365   0.3243
 [9,]   9.2626   -8.9854   0.0169  -0.5437
[10,]   3.2840   14.1573  -7.0465   0.3405
[11,]  -9.2200  -12.3861  -3.4283   0.4352
[12,]  25.5849    2.7817   0.3867   0.4468
[13,]  26.9032    2.9310   2.4455   0.4116
```

The first two PCs account for nearly 98% of the total variation among x_1, x_2, x_3, x_4. PC1 is roughly proportional to the difference $x_2 - x_4$, PC2 to the difference $x_1 - x_3$, PC3 to the difference in the average of x_2 and x_3 and that of x_1 and x_4, and PC4 to the average of all four x's. The opposite signs on the factor loadings on x_2 and x_4 in PC1 and on x_1 and x_3 in PC2 correspond to the large negative correlations between these pairs of variables as shown in Example 4.8. □

5.3.2   Principal components regression (PCR)

To compare the multiple linear regression (MLR) and PCR models, it will be convenient to center both the y_i's and the x_{ij}'s; thus the columns of $\boldsymbol{X}$ sum to 0. Since $\boldsymbol{Z} = \boldsymbol{X}\,\boldsymbol{U}$, it follows that the columns of $\boldsymbol{Z}$ also sum to 0, i.e. the z_{ij} are also centered. Thus, both MLR and PCR models do not have the intercept terms, β_0 and γ_0, respectively. The LS estimates of β_0 and γ_0 for the uncentered data can be readily computed from $\widehat{\beta}_0 = \overline{y} - \sum_{j=1}^{p} \widehat{\beta}_j \overline{x}_j$ and $\widehat{\gamma}_0 = \overline{y}$.

Denoting the parameter vectors for the MLR model by $\boldsymbol{\beta} = (\beta_1, \dots, \beta_p)'$ and that for the PCR model by $\boldsymbol{\gamma} = (\gamma_1, \dots, \gamma_p)'$, the two models are $\boldsymbol{y} = \boldsymbol{X}\boldsymbol{\beta} + \varepsilon$ and $\boldsymbol{y} = \boldsymbol{Z}\boldsymbol{\gamma} + \varepsilon$. Since $\boldsymbol{U}\,\boldsymbol{U}' = \boldsymbol{I}$, it follows that

$$\boldsymbol{\beta} = \boldsymbol{U}\boldsymbol{\gamma} \quad \text{and} \quad \boldsymbol{\gamma} = \boldsymbol{U}'\boldsymbol{\beta}. \tag{5.10}$$

If the full model with all p predictors is the true model, then the PCR model with $r < p$ PCs is biased. But the bias should be small if the first r PCs capture most of the variation among the predictor vectors (the column vectors of $\boldsymbol{X}$).

Denote the submatrix of $\boldsymbol{Z}$ with only the first $r < p$ column vectors by $\boldsymbol{Z}^{(r)}$ and the corresponding subvector of $\boldsymbol{\gamma}$ by $\boldsymbol{\gamma}^{(r)} = (\gamma_1, \dots, \gamma_r)'$. Note that $\boldsymbol{Z}^{(r)} = \boldsymbol{X}\,\boldsymbol{U}^{(r)}$, where $\boldsymbol{U}^{(r)}$ is the submatrix of $\boldsymbol{U}$ consisting of the first r loading vectors, $\boldsymbol{u}_1, \dots, \boldsymbol{u}_r$. We have $\boldsymbol{Z}^{(r)'}\boldsymbol{Z}^{(r)} = \operatorname{diag}\{\lambda_1, \dots, \lambda_r\}$. Hence,

$$\widehat{\boldsymbol{\gamma}}^{(r)} = (\boldsymbol{Z}^{(r)'}\boldsymbol{Z}^{(r)})^{-1}\boldsymbol{Z}^{(r)'}\boldsymbol{y} = \operatorname{diag}\{1/\lambda_1, \dots, 1/\lambda_r\}\boldsymbol{Z}^{(r)'}\boldsymbol{y}. \tag{5.11}$$

Thus, $\widehat{\gamma}_j^{(r)} = (1/\lambda_j)(\boldsymbol{Z}^{(r)'}\boldsymbol{y})_j$ $(1 \leq j \leq r)$. Furthermore, since

$$\operatorname{Cov}(\widehat{\boldsymbol{\gamma}}^{(r)}) = \sigma^2(\boldsymbol{Z}^{(r)'}\boldsymbol{Z}^{(r)})^{-1} = \sigma^2 \operatorname{diag}\{1/\lambda_1, \dots, 1/\lambda_r\},$$

the estimates $\widehat{\gamma}_j^{(r)}$ are uncorrelated. If some PCs z_k are added to or deleted from the model, then the estimates $\widehat{\gamma}_j^{(r)}$ for $j \neq k$ are unchanged. Hence, a stepwise method is not needed to build a regression model. Thus, it is much easier to fit the PCR model than to fit the MLR model.

One drawback of PCR is that the PCs are not always easily interpretable, being linear combinations of the original predictors x_j. However, the final model in original variables can always be recovered using the relationship

$$\widehat{\boldsymbol{\beta}}^{(r)} = \boldsymbol{U}^{(r)}\widehat{\boldsymbol{\gamma}}^{(r)} = \boldsymbol{U}^{(r)}(\boldsymbol{Z}^{(r)'}\boldsymbol{Z}^{(r)})^{-1}\boldsymbol{Z}^{(r)'}\boldsymbol{y}. \tag{5.12}$$

Note that although $\widehat{\boldsymbol{\gamma}}^{(r)}$ has r components, in general, $\widehat{\boldsymbol{\beta}}^{(r)}$ has all p components, i.e. the model includes all p original predictors. But it is not a full model because it is not based on all PCs.

Example 5.5 *(Hald Cement Data: Principal Components Regression)* First, we fit a full PCR model using all four PCs, where PCs are extracted using `prcomp` as in Example 5.4 and the scores are stored in matrix x.

```
> fullfit=lm(cement$y ~ fit1$x)
> summary(fullfit)

Call:
lm(formula = cement$y ~ fit1$x)

Coefficients:
            Estimate Std. Error t value Pr(>|t|)
(Intercept) 95.42308    0.67840 140.659 7.30e-15 ***
fit1$xPC1    0.55365    0.03103  17.842 9.97e-08 ***
fit1$xPC2    0.91964    0.08595  10.700 5.11e-06 ***
fit1$xPC3   -0.71105    0.20048  -3.547 0.00754 **
fit1$xPC4    1.01954    1.44995   0.703 0.50190
---

Residual standard error: 2.446 on 8 degrees of freedom
Multiple R-squared:  0.9824,    Adjusted R-squared:  0.9736
F-statistic: 111.5 on 4 and 8 DF,  p-value: 4.756e-07
```

The coefficients $\widehat{\gamma}_j$ of the first three PCs are highly significant and the intercept $\widehat{\gamma}_0 = 95.4231 = \overline{y}$.

The $\widehat{\beta}_j$ coefficients can be computed from the $\widehat{\gamma}_j$ coefficients as follows:

$$\widehat{\boldsymbol{\beta}} = \boldsymbol{U}\widehat{\boldsymbol{\gamma}} = \begin{bmatrix} 0.068 & 0.640 & -0.567 & 0.506 \\ 0.679 & 0.020 & 0.544 & 0.493 \\ -0.029 & -0.755 & -0.404 & 0.516 \\ -0.731 & 0.108 & 0.468 & 0.484 \end{bmatrix} \begin{bmatrix} 0.554 \\ 0.920 \\ -0.711 \\ 1.020 \end{bmatrix} = \begin{bmatrix} 1.551 \\ 0.510 \\ 0.102 \\ -0.144 \end{bmatrix}.$$

This matrix multiplication can be done in R using the command

```
fit1$rotation[,1:4] %*% matrix(fullfit$coefficients[2:5]).
```

Finally, using $\overline{x}_1 = 7.462, \overline{x}_2 = 48.150, \overline{x}_3 = 11.770, \overline{x}_4 = 30.000$ we can calculate

$$\widehat{\beta}_0 = \overline{y} - \sum_{j=1}^{4} \widehat{\beta}_j \overline{x}_j$$

$$= 95.423 - [(1.551)(7.462) + (0.510)(48.150) + (0.102)(11.770)$$

$$+ (-0.144)(30.000)]$$

$$= 62.412.$$

Thus, the final model is

$$\hat{y} = 62.412 + 1.551x_1 + 0.510x_2 + 0.102x_3 - 0.144x_4,$$

which is the same model (except for round-off errors) that we obtained in Example 4.9. We conclude that the full MLR and PCR models are the same.

Since only the first three PCs have significant coefficients and those three PCs account for almost 100% of the variation among the x's, we next fit a partial PCR model using PC1, PC2, and PC3 as predictors resulting in the following output:

```
> partialfit=lm(cement$y~PC1+PC2+PC3, cbind(cement, fit1$x))
> summary(partialfit)

Call:
lm(formula = cement$y ~ PC1 + PC2 + PC3, data = cbind(cement,
fit1$x))

Coefficients:
            Estimate Std. Error t value Pr(>|t|)
(Intercept) 95.42308    0.65907 144.784  < 2e-16 ***
PC1          0.55365    0.03015  18.366 1.92e-08 ***
PC2          0.91964    0.08350  11.014 1.59e-06 ***
PC3         -0.71105    0.19476  -3.651 0.00531 **
---
Residual standard error: 2.376 on 9 degrees of freedom
Multiple R-squared:  0.9813,    Adjusted R-squared:  0.975
F-statistic: 157.3 on 3 and 9 DF,  p-value: 4.307e-08
```

Notice that the estimated regression coefficients are unchanged from the full model because of the orthogonality of the PCs. Their standard errors have changed, of course, because the MSE of the partial model has changed, both because of the increase in SSE and in error d.f. In this example, the MSE and hence the standard errors have actually decreased. Also R^2 has decreased only slightly from 98.24% to 98.13%.

The coefficients of the regression model in terms of the x's based on three PCs can be obtained using matrix multiplication as mentioned before.

```
> fit1$rotation[,1:3] %*% matrix(partialfit$coefficients[2:4])
x1  1.035031573
x2  0.007260283
x3 -0.423732982
x4 -0.637943480
```

Next, we calculate

$$\widehat{\beta}_0 = \overline{y} - \sum_{j=1}^{4} \widehat{\beta}_j \overline{x}_j$$

$$= 95.423 - [(1.035)(7.462) + (0.007)(48.150) + (-0.424)(11.770)$$
$$+ (-0.638)(30.000)]$$
$$= 111.49.$$

Thus, the final model is

$$\widehat{y} = 111.49 + 1.035x_1 + 0.007x_2 - 0.424x_3 - 0.638x_4,$$

which is quite different from the model obtained using all four PCs, but it gives essentially the same fit.

The functions `prcomp` and `princomp` perform PCA. To obtain the PCR model in terms of the original predictor variables, two further steps are required. First, we fit a regression model with the PCs as the predictors. Next, we transform the regression coefficients of the PCs to those of the x's by doing matrix multiplication as illustrated above. The function `pcr` in the `pls` library (Mevik and Wehrens 2007) can be used to perform PCR directly. Another advantage of using `pcr` is that it allows choosing the number of PCs via cross-validation instead of fixing them *a priori*. We illustrate the use of this function below. We use the option `scale=FALSE` to fit a model with unstandardized data so that the results are comparable to those obtained above.

```
> library(pls)
> cement=read.csv("c:/data/cement.csv")
> fit3=pcr(y~., data=cement, scale=FALSE, validation="CV")
> summary(fit3)
Data:   X dimension: 13 4
        Y dimension: 13 1
Fit method: svdpc
Number of components considered: 4

VALIDATION: RMSEP
Cross-validated using 10 random segments.
       (Intercept)  1 comps  2 comps  3 comps  4 comps
CV           15.66    9.286    4.466    2.785    3.006
adjCV        15.66    9.214    4.425    2.736    2.942

TRAINING: % variance explained
    1 comps  2 comps  3 comps  4 comps
X     86.60    97.89    99.96   100.00
```

```
y     70.13     95.36      98.13     98.24
> fit3$loadings
```

```
Loadings:
   Comp 1 Comp 2 Comp 3 Comp 4
x1          0.646 -0.567  0.506
x2  0.679         0.544  0.493
x3         -0.755 -0.404  0.516
x4 -0.731  0.108  0.468  0.484
```

The root mean square error of prediction (RMSEP) is calculated using ordinary cross-validation (CV) and bias-corrected cross-validation (called adjCV). We see that RMSEP using both CV and adjCV is minimized with three PCs, which also explain almost 100% of the variance among the x's and more than 98% of the variance of y. The loading matrix is the same as that obtained with `prcomp` except that very small loadings (< 0.1) are zeroed out. The coefficients for the regression model using three PCs are shown below:

```
> fit3$coefficients[,,3]
          x1             x2             x3             x4
 1.035031573   0.007260283  -0.423732982  -0.637943480
```

They match exactly with those obtained previously using prcomp.   □

5.4   Partial least squares (PLS)*

The PLS methodology was proposed by Wold (1966) as an econometric technique. It is now applied in a wide variety of disciplines, especially where the predictors are **high dimensional** ($p > n$) and highly correlated. PLS can also deal with multivariate responses (**multivariate regression**). Such applications arise in chemometrics, genomics, proteotomics, and social sciences. For example, in chemometrics PLS is used to build a regression model to estimate or predict the percentages of various ingradients in chemical products from spectroscopic measurements at hundreds of frequencies. Thus the response is multivariate (percentages of ingradients) and predictors (spectroscopic measurements) are very large in number (p) typically exceeding the number of samples (n) used for calibration. In the following, initially we assume the multivariate response setting, but then specialize to the univariate response setting that is the focus of the present book.

Suppose that there are p predictors, $x_1, \ldots, x_p$, and q responses, $y_1, \ldots, y_q$. For convenience, assume that all variables are mean-centered. Denote the data matrix of the x's by $\boldsymbol{X} : n \times p$ and that of the y's by $\boldsymbol{Y} : n \times q$. It is desired to fit a regression model for $\boldsymbol{Y}$ on $\boldsymbol{X}$. Because of the high dimensionality and resulting multicollinearity among the x's, this is a very difficult if not an impossible problem for LS estimation. The goal of PLS is to find lower dimensional representations, $\boldsymbol{Z}$ and $\boldsymbol{W}$ of $\boldsymbol{X}$ and $\boldsymbol{Y}$, respectively, such that $\boldsymbol{Z}$ and $\boldsymbol{W}$ capture most of the covariance between $\boldsymbol{X}$ and $\boldsymbol{Y}$, and then run a regression of $\boldsymbol{W}$ on $\boldsymbol{Z}$. An iterative algorithm is needed to find these matrices. The PLS1 algorithm is used for univariate response ($q = 1$), while the PLS2 algorithm is used for multivariate response ($q > 1$). Here we consider only PLS1.

5.4.1 PLS1 algorithm

Since the response is univariate, the matrix $\boldsymbol{Y}$ is actually a vector $\boldsymbol{y}$; thus, its dimension cannot be reduced. The idea of the algorithm is to extract loading vectors $\boldsymbol{u}_j$ for the x's such that the resulting score vectors $\boldsymbol{z}_j$ are good predictors of $\boldsymbol{y}$. This is done by sequentially extracting a single loading vector $\boldsymbol{u}_j$ at each step j by regressing the residuals of $\boldsymbol{y}$ on the residuals of $\boldsymbol{X}$ from the previous step. The final result is the reduced rank scores matrix $\boldsymbol{Z} = [\boldsymbol{z}_1, \ldots, \boldsymbol{z}_r]$. The column vectors of $\boldsymbol{Z}$, which are analogous to PCs, are called **latent variables (LVs)**. In the examples in the sequel, they are labeled as components in the R outputs, but to distinguish them from PCs, we refer to them as LVs.

The PLS1 algorithm operates as follows:

Step 0: Let $\boldsymbol{X}_0 = \boldsymbol{X}$ and $\boldsymbol{y}_0 = \boldsymbol{y}$. In subsequent iterations, $\boldsymbol{X}_j$ and $\boldsymbol{y}_j$ consist of residuals of $\boldsymbol{X}_{j-1}$ and $\boldsymbol{y}_{j-1}$. Set $j = 1$ and go to the next step.

Step j:

1. Calculate the loading vector by

$$\boldsymbol{u}_j = \frac{\boldsymbol{X}'_{j-1}\boldsymbol{y}_{j-1}}{||\boldsymbol{X}'_{j-1}\boldsymbol{y}_{j-1}||}.$$

2. Compute the score vector

$$\boldsymbol{z}_j = \boldsymbol{X}_{j-1}\boldsymbol{u}_j.$$

Note that the columns in $\boldsymbol{X}_{j-1}$, which are highly correlated with $\boldsymbol{y}_{j-1}$ receive high loadings and vice versa.

3. Regress the predictor and response variable residuals on the scores obtained in the previous step. Denote the regression coefficient vector for $\boldsymbol{X}_{j-1}$ by $\boldsymbol{a}_j$ and the regression coefficient for $\boldsymbol{y}_{j-1}$ by b_j, where

$$\boldsymbol{a}_j = \frac{\boldsymbol{X}'_{j-1}\boldsymbol{z}_j}{\boldsymbol{z}'_j\boldsymbol{z}_j} \quad \text{and} \quad b_j = \frac{\boldsymbol{y}'_{j-1}\boldsymbol{z}_j}{\boldsymbol{z}'_j\boldsymbol{z}_j}.$$

4. Compute the residuals of $\boldsymbol{X}_{j-1}$ and $\boldsymbol{y}_{j-1}$:

$$\boldsymbol{X}_j = \boldsymbol{X}_{j-1} - \boldsymbol{z}_j \boldsymbol{a}'_j \quad \text{and} \quad \boldsymbol{y}_j = \boldsymbol{y}_{j-1} - \boldsymbol{z}_j b_j.$$

Note that $\boldsymbol{X}'_j \boldsymbol{z}_j = \boldsymbol{0}$ (where $\boldsymbol{0}$ is a p-dimensional null vector) and $\boldsymbol{y}'_j \boldsymbol{z}_j = 0$.

5. Increment $j \to j+1$ and return to the beginning of this step until $\boldsymbol{X}_j$ and $\boldsymbol{y}_j$ become sufficiently small. Denote by r the value of j at the last step (the reduced rank of $\boldsymbol{X}$) and let $\boldsymbol{Z}^{(r)} = [\boldsymbol{z}_1, \ldots, \boldsymbol{z}_r]$, $\boldsymbol{U}^{(r)} = [\boldsymbol{u}_1, \ldots, \boldsymbol{u}_r]$ and $\boldsymbol{A}^{(r)} = [\boldsymbol{a}_1, \ldots, \boldsymbol{a}_r]$.

The PLS model is obtained by regressing $\boldsymbol{y}$ on $\boldsymbol{Z}^{(r)}$. For practical use, this model needs to be expressed in terms of the original predictor matrix $\boldsymbol{X}$. This relationship is not easy to derive since $\boldsymbol{Z}^{(r)} \neq \boldsymbol{X}\boldsymbol{U}^{(r)}$ because the column vectors $\boldsymbol{u}_j$ of $\boldsymbol{U}^{(r)}$ are not obtained from $\boldsymbol{X}$ itself but from its residuals at successive steps. Also, the $\boldsymbol{u}_j$ are not orthonormal vectors, so $\boldsymbol{U}^{(r)} \boldsymbol{U}^{(r)'} \neq \boldsymbol{I}$ as in the case of PCA. Using some algebra, it can be shown that $\boldsymbol{Z}^{(r)} = \boldsymbol{X}\boldsymbol{R}$, where $\boldsymbol{R} = \boldsymbol{U}^{(r)}(\boldsymbol{A}^{(r)'}\boldsymbol{U}^{(r)})^{-1}$. Thus, $\widehat{\boldsymbol{\beta}}$ can be obtained by regressing $\boldsymbol{y}$ on $\boldsymbol{Z}^{(r)}$ and then premultiplying the resulting LS estimator by $\boldsymbol{R}$, which gives

$$\widehat{\boldsymbol{\beta}} = \boldsymbol{R}(\boldsymbol{Z}^{(r)'}\boldsymbol{Z}^{(r)})^{-1}\boldsymbol{Z}^{(r)'}\boldsymbol{y}.$$

Then the PLS model can be written as $\widehat{\boldsymbol{y}} = \boldsymbol{X}\widehat{\boldsymbol{\beta}}$.

To illustrate the PLS methodology, we will give two examples with small data sets: the GPA data and the Hald cement data, and one example with high-dimensional ($p > n$) data.

Example 5.6 (GPA Data: Partial Least Squares) The R script and the resulting output are as follows:

```
> library(pls)
> gpa = read.csv("c:/data/GPA.csv")
> plsfit=plsr(GPA~.,data=gpa, scale=FALSE)
> summary(plsfit)
Data:   X dimension: 40 2
        Y dimension: 40 1
Fit method: kernelpls
Number of components considered: 2
TRAINING: % variance explained
      1 comps   2 comps
X       48.44    100.00
GPA     65.05     68.11
> plsfit$loadings
```

```
Loadings:
       Comp 1 Comp 2
Verbal  0.901 -0.663
Math    0.491  0.749

              Comp 1 Comp 2
SS loadings    1.053  1.000
Proportion Var 0.526  0.500
Cumulative Var 0.526  1.026
> plsfit$coefficients
, , 1 comps

              GPA
Verbal 0.02972051
Math   0.02629076

, , 2 comps

              GPA
Verbal 0.02573212
Math   0.03361487
```

We see that the loadings on the two components are quite different from those for PCA. Here LV1 loadings on Verbal and Math are 0.901 and 0.491, respectively; both are positive unlike those for PCA. Thus, LV1 is positively sloping, while PC1 was negatively sloping in the predictor space. This is because the PLS loadings take into account the correlations of Verbal and Math with GPA, and both Verbal and Math are positively correlated with GPA. Also notice that the proportion of variance explained by the two LVs exceeds 100%. This is because the LVs are not orthogonal. Finally note that in the full model (with two LVs), coefficients of Verbal and Math are the same as those for the MLR model from Example 3.3. □

Next, we apply PLS to the Hald cement data.

Example 5.7 *(Hald Cement Data: Partial Least Squares)*

```
> library(pls)
> cement=read.csv("c:/data/cement.csv")
> plsfit=plsr(y~.,data=cement, scale=FALSE)
> summary(plsfit)
Data:   X dimension: 13 4
        Y dimension: 13 1
Fit method: kernelpls Number of components considered: 4
TRAINING: % variance explained
   1 comps  2 comps  3 comps  4 comps
X    86.14    97.84    99.96   100.00
```

```
y     76.53      96.38      98.13      98.24
Loadings:
   Comp 1 Comp 2 Comp 3 Comp 4
x1           0.638  0.459  0.487
x2   0.690 -0.141 -0.538  0.512
x3          -0.726  0.518  0.501
x4  -0.741  0.250 -0.482  0.500


                 Comp 1 Comp 2 Comp 3 Comp 4
SS loadings        1.036  1.016  1.001  1.000
Proportion Var     0.259  0.254  0.250  0.250
Cumulative Var     0.259  0.513  0.763  1.013
```

Note that the LV1 loadings on x_2 and x_4 are almost identical to the PC1 loadings obtained in Example 5.4, and they are roughly proportional to $\mathrm{Corr}(x_2, y) = 0.816$ and $\mathrm{Corr}(x_4, y) = -0.821$, respectively. We fit the model using three LVs since they capture almost all of the variance of the x's as well as of y. The regression coefficients of the x's corresponding to the three LVs are as follows:

```
> plsfit$coefficients[,,3]
         x1           x2           x3           x4
 1.04728616  0.01848865  -0.41135995  -0.62687397
```

Next, $\widehat{\beta}_0$ can be calculated as before:

$$\widehat{\beta}_0 = \overline{y}_0 - \sum_{j=1}^{4} \widehat{\beta}_j \overline{x}_j$$

$$= 95.423 - [(1.047)(7.462) + (0.018)(48.150) + (-0.411)(11.770)$$

$$+ (-0.627)(30.000)]$$

$$= 110.39.$$

Thus, the final model is

$$\widehat{y} = 110.39 + 1.047x_1 + 0.018x_2 - 0.411x_3 - 0.627x_4.$$

This model is almost the same as the model obtained using PCR with three PCs. □

Next, we give an example that illustrates the use of PLS for high dimensional data $(p > n)$.

Example 5.8 (Soybean Data Set: Partial Least Squares) The data set (soybeandata.csv) is taken from Minitab: (https://support.minitab.com/en-us/datasets/regression-data-sets/soybean-flour-data/). It consists of spectroscopic measurements on $n = 54$ soybean flour samples at $p = 88$ frequencies.

The goal is to build a predictive model for the response variable Fat (fat content of soybean sample) from the spectroscopic measurements. The maximum number of LVs that can be extracted from these data equals $\min(n-1, p) = 53$. Obviously, we do not want so many LVs. Here we give RMSEP results using fivefold cross-validation for the first 13 LVs out of the first 20 LVs considered (results for the additional seven LVs are suppressed to save space).

```
> library(pls)
> soybean=read.csv("c:/data/soybeandata.csv")
> plsfit=plsr(Fat~.,data=soybean, scale=FALSE, validation="CV")
> fit1=plsr(Fat~.,data=soybeantrain, ncomp=20, scale=FALSE,
  validation="CV",segments=5)
> summary(fit1)
Data:    X dimension: 54 88
         Y dimension: 54 1
Fit method: kernelpls
Number of components considered: 20

VALIDATION: RMSEP
Cross-validated using five random segments.
        (Intercept)  1 comps  2 comps  3 comps  4 comps  5 comps  6 comps
CV          2.391     2.527    1.721    1.407    1.243    1.181    1.129
adjCV       2.391     2.500    1.699    1.428    1.238    1.163    1.131
         7 comps  8 comps  9 comps  10 comps  11 comps  12 comps  13 comps
CV        1.214    1.257    1.248    1.265     1.299     1.369     1.403
adjCV     1.194    1.201    1.191    1.207     1.236     1.289     1.308

TRAINING: % variance explained
       1 comps  2 comps  3 comps  4 comps  5 comps  6 comps  7 comps  8 comps
X       99.547   99.73    99.85    99.94    99.98    99.99    100.00   100.00
Fat      4.427   56.02    69.28    77.27    80.62    81.70    83.87     88.26
       9 comps  10 comps  11 comps  12 comps  13 comps
X       100.00   100.00    100.00    100.00    100.00
Fat     89.02    89.59     89.95     91.12     92.05
```

From the above output, we observe that RMSEP using both CV and adjCV (which are quite close) is minimized when the number of LVs is six. The percentage of the variance of the x's accounted for by six LVs is 99.99% and that of y is 81.70%. To see how RMSEP behaves as a function of the number of LVs, it is useful to plot it, which is done in Figure 5.10. We see that RMSEP drops steeply until the number of LVs equal to six and then it increases steadily.

A plot of predicted Fat values versus measured Fat values is shown in Figure 5.11. It shows that the relationship between predicted and measured values is linear with no outliers or systematic departures.

The final model consists of 88 predictors, which are the spectral measurements at 88 frequencies. It is not convenient to give the regression coefficients

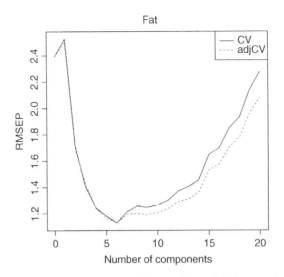

Figure 5.10. *RMSEP plot for the PLS model for soybean data.*

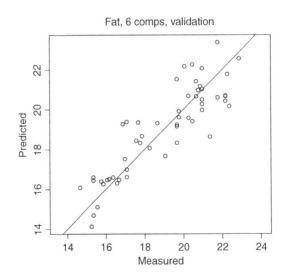

Figure 5.11. *Predicted versus measured plot for soybean data.*

for so many predictors nor is it useful. Instead, their plot versus the frequencies may be useful for interpretive purposes by identifying the frequencies at which the peaks in the plot occur and labeling them as most predictive of the response variable. This plot is shown in Figure 5.12. We see that the peaks in the coefficients plots occur at high frequencies. Prediction of Fat values for new data can be done by using the `predict` function in the usual manner.

Finally, Figure 5.13 gives a plot of the loadings of 88 predictors on LV1 and LV2.

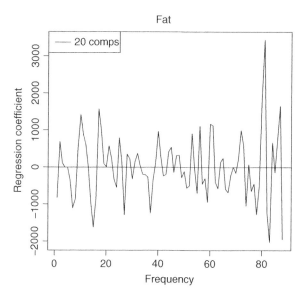

Figure 5.12. *Coefficients plot for soybean data.*

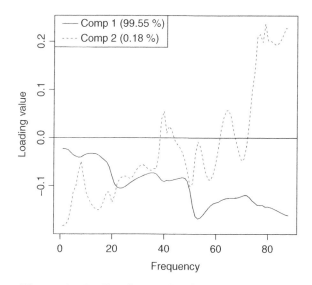

Figure 5.13. *Loadings plot for soybean data.*

Notice that all the loadings on LV1 are negative with the highest negative values at frequencies numbered between 50–60 and 70–88. If we regress Fat on LV1 (i.e., the first column of the scores matrix), we get a positive coefficient. In other words, high values of spectrometric readings at these frequencies are predictive of low values of Fat.   □

5.5   Technical notes*

5.5.1   Properties of ridge estimator

From (5.3) we can write $\widehat{\boldsymbol{\beta}}^{R}(\lambda) = [\boldsymbol{I} + \lambda(\boldsymbol{X}'\boldsymbol{X})^{-1}]^{-1}\widehat{\boldsymbol{\beta}}$ where $\widehat{\boldsymbol{\beta}} = (\boldsymbol{X}'\boldsymbol{X})^{-1}\boldsymbol{X}'\boldsymbol{y}$ is the LS estimator of $\boldsymbol{\beta}$. Hence

$$E[\widehat{\boldsymbol{\beta}}^{R}(\lambda)] = [\boldsymbol{I} + \lambda(\boldsymbol{X}'\boldsymbol{X})^{-1}]^{-1}\boldsymbol{\beta},$$

and so $\widehat{\boldsymbol{\beta}}^{R}(\lambda)$ is biased unless $\lambda = 0$ when the ridge estimator reduces to the LS estimator.

Using the sandwich formula (A.7), the covariance matrix of $\widehat{\boldsymbol{\beta}}^{R}(\lambda)$ for fixed λ is given by

$$\text{Cov}(\widehat{\boldsymbol{\beta}}^{R}(\lambda)) = \sigma^2(\boldsymbol{X}'\boldsymbol{X} + \lambda\boldsymbol{I})^{-1}\boldsymbol{X}'\boldsymbol{X}(\boldsymbol{X}'\boldsymbol{X} + \lambda\boldsymbol{I})^{-1} = \sigma^2\boldsymbol{V} \quad \text{(say).} \quad (5.13)$$

The standard errors of the individual ridge estimates $\widehat{\beta}_{j}^{R}(\lambda)$ are given by $\text{SE}(\widehat{\beta}_{j}^{R}(\lambda)) = s\sqrt{v_{jj}}$, where v_{jj} is the jth diagonal entry of the $\boldsymbol{V}$ matrix defined above and s^2 is an estimate of σ^2 calculated as follows:

The following method of calculating s^2 for ridge regression is given by Cule et al. (2011). First calculate the fitted vector for ridge regression:

$$\widehat{\boldsymbol{y}}^{R}(\lambda) = \boldsymbol{X}\widehat{\boldsymbol{\beta}}^{R}(\lambda) = \boldsymbol{X}(\boldsymbol{X}'\boldsymbol{X} + \lambda\boldsymbol{I})^{-1}\boldsymbol{X}'\boldsymbol{y} = \boldsymbol{H}\boldsymbol{y} \quad \text{(say),}$$

where $\boldsymbol{H} = \boldsymbol{X}(\boldsymbol{X}'\boldsymbol{X} + \lambda\boldsymbol{I})^{-1}\boldsymbol{X}'$ is the hat matrix for ridge regression analogous to the hat matrix for MLR defined in (3.10). Hastie and Tibshirani (1990) defined the **effective error d.f.** as $\nu = n - \text{tr}(2\boldsymbol{H} - \boldsymbol{H}\boldsymbol{H}')$. Note that for MLR $\boldsymbol{H}\boldsymbol{H}' = \boldsymbol{H}$ and hence $\nu = n - \text{tr}(\boldsymbol{H}) = n - p$ if the variables are standardized. For ridge regression, the MSE estimate of σ^2 is given by

$$s^2 = \frac{(\boldsymbol{y} - \widehat{\boldsymbol{y}}^{R}(\lambda))'(\boldsymbol{y} - \widehat{\boldsymbol{y}}^{R}(\lambda))}{n - \text{tr}(2\boldsymbol{H} - \boldsymbol{H}\boldsymbol{H}')}.$$

So the significance of ridge estimates can be tested using $t_j = \widehat{\beta}_{j}^{R}(\lambda)/\text{SE}(\widehat{\beta}_{j}^{R}(\lambda))$ as approximate t-statistics with ν d.f.

The MSE of $\widehat{\boldsymbol{\beta}}^{R}(\lambda)$ is given by

$$\text{MSE}(\widehat{\boldsymbol{\beta}}^{R}(\lambda)) = E[(\widehat{\boldsymbol{\beta}}^{R}(\lambda) - \boldsymbol{\beta})'(\widehat{\boldsymbol{\beta}}^{R}(\lambda) - \boldsymbol{\beta})]$$

$$= \sigma^2 \sum_{j=1}^{p} \frac{\lambda_j}{(\lambda_j + \lambda)^2} + \lambda^2 \boldsymbol{\beta}'(\boldsymbol{X}'\boldsymbol{X} + \lambda\boldsymbol{I})^{-2}\boldsymbol{\beta},$$

where the λ_j are the eigenvalues of $\boldsymbol{X}'\boldsymbol{X}$, and λ is the tuning parameter for ridge regression. The first term represents the sum of the variances of $\widehat{\beta}_j^R(\lambda)$ and the second term represents the sum of the squares of their biases. For the ridge estimator, the variance part is smaller than that of the LS estimator, but the bias part is positive, whereas that for the LS estimator the bias part is 0. One can choose a suitable λ to trade off the bias against the variance and minimize $\text{MSE}(\widehat{\boldsymbol{\beta}}^R(\lambda))$.

5.5.2 Derivation of principal components

Setting the derivative of the objective function in (5.8) w.r.t. $\boldsymbol{u}_1$ equal to $\boldsymbol{0}$, we get

$$2(\boldsymbol{\Sigma} - \lambda_1 \boldsymbol{I})\boldsymbol{u}_1 = \boldsymbol{0}.$$

It follows that $\boldsymbol{u}_1$ is the eigenvector associated with the eigenvalue λ_1. To determine which of the p eigenvalues is λ_1, premultiply the above equation by $\boldsymbol{u}_1'$ resulting in

$$\boldsymbol{u}_1'\boldsymbol{\Sigma}\boldsymbol{u}_1 = \lambda_1 \boldsymbol{u}_1'\boldsymbol{u}_1 = \lambda_1.$$

But $\boldsymbol{u}_1'\boldsymbol{\Sigma}\boldsymbol{u}_1 = \text{Var}(z_1)$, which is being maximized. Therefore, λ_1 is the maximum eigenvalue of $\boldsymbol{\Sigma}$.

The constrained optimization problem to find PC2 is as follows:

$$\max_{\boldsymbol{u}_2} \boldsymbol{u}_2'\boldsymbol{\Sigma}\boldsymbol{u}_2 \quad \text{subject to} \quad \boldsymbol{u}_2'\boldsymbol{u}_2 = 1 \quad \text{and} \quad \boldsymbol{u}_1'\boldsymbol{u}_2 = 0.$$

Using the Lagrangian multiplier λ_2 for the first constraint and μ for the second constraint, the optimization problem becomes

$$\max_{\boldsymbol{u}_2}[\boldsymbol{u}_2'\boldsymbol{\Sigma}\boldsymbol{u}_2 - \lambda_2(\boldsymbol{u}_2'\boldsymbol{u}_2 - 1) - \mu(\boldsymbol{u}_1'\boldsymbol{u}_2)].$$

Setting the derivative of this objective function w.r.t. $\boldsymbol{u}_2$ equal to $\boldsymbol{0}$ (the null vector), we get

$$2(\boldsymbol{\Sigma} - \lambda_2 \boldsymbol{I})\boldsymbol{u}_2 - \mu\boldsymbol{u}_1 = \boldsymbol{0}.$$

As before, premultiplying the above by $\boldsymbol{u}_2'$, we get

$$2\boldsymbol{u}_2'(\boldsymbol{\Sigma} - \lambda_2 \boldsymbol{I})\boldsymbol{u}_2 - \mu\boldsymbol{u}_2'\boldsymbol{u}_1 = 2\boldsymbol{u}_2'\boldsymbol{\Sigma}\boldsymbol{u}_2 - 2\lambda_2 = 0$$

since $\boldsymbol{u}_2'\boldsymbol{u}_1 = 0$ and $\boldsymbol{u}_2'\boldsymbol{u}_2 = 1$. Hence, $\lambda_2 = \boldsymbol{u}_2'\boldsymbol{\Sigma}\boldsymbol{u}_2 = \text{Var}(z_2)$ is the second largest eigenvalue of $\boldsymbol{\Sigma}$. Next, premultiplying the above by $\boldsymbol{u}_1'$ we get

$$2\boldsymbol{u}_1'\boldsymbol{\Sigma}\boldsymbol{u}_2 - \lambda_2\boldsymbol{u}_1'\boldsymbol{u}_2 - \mu\boldsymbol{u}_1'\boldsymbol{u}_1 = 2\boldsymbol{u}_1'\boldsymbol{\Sigma}\boldsymbol{u}_2 - \mu = 0,$$

Hence, $2\boldsymbol{u}'_1\boldsymbol{\Sigma}\boldsymbol{u}_2 = \mu$. However, premultiplying (5.8) by $\boldsymbol{u}'_2$, we get

$$2\boldsymbol{u}'_2\boldsymbol{\Sigma}\boldsymbol{u}_1 - \lambda\boldsymbol{u}'_2\boldsymbol{u}_1 = 2\boldsymbol{u}'_2\boldsymbol{\Sigma}\boldsymbol{u}_1 = 0,$$

Hence, $\mu = 0$ and $\boldsymbol{u}'_2\boldsymbol{\Sigma}\boldsymbol{u}_1 = \mathrm{Cov}(z_1, z_2) = 0$. Thus, PC2 is not only orthogonal to PC1 but is also uncorrelated with it. This process can be continued with higher PCs.

Exercises

Theoretical Exercises

5.1 (Derivation of the ridge estimator) Derive the ridge estimator (5.3) by solving the minimization problem in (5.2).

5.2 (Ridge and lasso regression) This exercise is taken from an example in the book by James et al. (2013, p. 225). Consider a simple set up with $n = p$, no β_0 term and the $\boldsymbol{X}$ matrix is an identity matrix. In that case, the LS estimates are obtained by minimizing $\sum_{i=1}^{n}(y_i - \beta_i)^2$ and so $\widehat{\beta}_i = y_i$ ($i = 1, \ldots, n$).

(a) Show that the ridge estimator is given by

$$\widehat{\beta}_i^R(\lambda) = \frac{\widehat{\beta}_i}{1 + \lambda} \quad (1 \leq i \leq n).$$

(b) Show that the lasso estimator is given by

$$\widehat{\beta}_i^L(\lambda) = \begin{cases} y_i - \lambda/2 & \text{if } y_i > \lambda/2 \\ y_i + \lambda/2 & \text{if } y_i < -\lambda/2 \quad (1 \leq i \leq n) \\ 0 & \text{if } |y_i| < \lambda/2. \end{cases}$$

(c) Discuss the difference between the ridge and lasso estimators in the way they shrink the LS estimators.

5.3 (Ridge estimator of slope for regression through origin) Refer to Exercise 2.1 where the problem of fitting a regression line through the origin was considered.

(a) Show that the ridge estimator $\widehat{\beta}^R$ of the slope parameter β is given by

$$\widehat{\beta}^R = \left(\frac{c}{c + \lambda}\right)\widehat{\beta},$$

where $\widehat{\beta}$ is the LS estimator of β, $\lambda > 0$ is a given ridge parameter and $c = \sum x_i^2$. Hence, compute $E(\widehat{\beta}^R)$. How does the bias of $\widehat{\beta}^R$ change (increase or decrease) as λ increases? In what sense is $\widehat{\beta}^R$ a shrinkage estimator of β?

(b) Give a formula for $\mathrm{Var}(\widehat{\beta}^R)$. How does the variance of $\widehat{\beta}^R$ change (increase or decrease) as λ increases?

(c) Find λ that minimizes $\mathrm{MSE}(\widehat{\beta}^R) = \mathrm{Var}(\widehat{\beta}^R) + \mathrm{Bias}^2(\widehat{\beta}^R)$ and the corresponding minimum $\mathrm{MSE}(\widehat{\beta}^R)$. For what range of values of λ is $\mathrm{MSE}(\widehat{\beta}^R) \le \mathrm{MSE}(\widehat{\beta})$?

(d) Assume $\beta = 2, \sigma^2 = 1$ and $\sum x_i^2 = 10$. Plot $\mathrm{Var}(\widehat{\beta}^R)$, $\mathrm{Bias}(\widehat{\beta}^R)$ and $\mathrm{MSE}(\widehat{\beta}^R)$ for $\lambda = 0(0.1)1.0$. Check that $\mathrm{MSE}(\widehat{\beta}^R)$ is minimized at the optimum λ, and the range of values of λ for which $\mathrm{MSE}(\widehat{\beta}^R) \le \mathrm{MSE}(\widehat{\beta})$.

5.4 **(Principal components of a patterned covariance matrix)** In some applications, the x variables have a so-called spherically symmetric distribution, i.e. they are homoscedastic and equicorrelated, so their covariance matrix is given by

$$\boldsymbol{\Sigma} = \sigma^2 \begin{bmatrix} 1 & \rho & \cdots & \rho \\ \rho & 1 & \cdots & \rho \\ \vdots & \vdots & \ddots & \vdots \\ \rho & \rho & \cdots & 1 \end{bmatrix}.$$

Assume $\sigma^2 = 1$ and $\rho > 0$.

(a) Show that the eigen values of $\boldsymbol{\Sigma}$ are $\lambda_1 = 1 + (p-1)\rho$ and $\lambda_2 = \ldots = \lambda_p = 1 - \rho$.

(b) Show that the eigenvectors are $\boldsymbol{u}_1 = (1/\sqrt{p})(1, 1 \ldots, 1)'$ and $\boldsymbol{u}_2, \ldots, \boldsymbol{u}_p$, which are the loading vectors of the PCs, form an orthonormal system of contrast vectors (i.e. the components of $\boldsymbol{u}_j$ sum to 0, $\|\boldsymbol{u}_j\| = 1$ and $\boldsymbol{u}'_j \boldsymbol{u}_k = 0$ for any $j \ne k$). Show that the following eigenvectors satisfy these conditions:

$$\boldsymbol{u}_j = \left(\frac{1}{\sqrt{(j-1)j}}, \ldots, \frac{1}{\sqrt{(j-1)j}}, \frac{-(j-1)}{\sqrt{(j-1)j}}, 0, \ldots, 0 \right)' \quad (j = 2, \ldots, p).$$

Any orthogonal transformation of these vectors results in another system satisfying the given conditions. The system is not unique because the eigenvalues $\lambda_2, \ldots, \lambda_p$ are equal.

(c) Part (b) shows that the first principal component is proportional to the average of all x's. What is the proportion of variance accounted by it? How does this proportion depend on ρ?

Applied Exercises

5.5 **(Gas mileages of cars: Ridge and lasso regressions)** Perform ridge and lasso regressions on the gas mileage data considered in Exercise 4.11 and compare the results with those of LS regression.

5.6 **(Acetylene data: Ridge and lasso regression)** Perform ridge and lasso regressions on the acetylene data from Exercise 4.12 and compare the results with those of LS regression.

5.7 **(Soybean data: PCR)** Refer to Example 5.8.

   (a) What is the minimum number of PCs needed to explain at least 99.99% of the total variance among the 88 spectroscopic measurements? Note from Example 5.8 that six LVs are needed for this purpose.

   (b) Fit a PCR model with Fat as the response variable using the minimum number of PCs found in Part (a). What proportion of the variance in Fat is explained by this PCR model? Note from Example 5.8 that the PLS model with six LVs explains 81.70% of the variance in Fat. Why is the proportion of the variance explained by the PLS model much higher?

5.8 **(Gasoline data: PLS)** The pls library includes `gasoline` data set, which consists of octane number (octane) as the response variable and near infrared reflectance (NIR) measurements on 60 gasoline samples as the predictor variables. Each NIR spectrum consists of 401 diffuse reflectance measurements (expressed as log(1/Reflectance) taken in the wavelength values in the range 900(2)1700 nm (numbered from 0 to 400). Use data on the first 50 gasoline samples as the training data and the last 10 gasoline samples as the test data.

   (a) How many LVs are needed to achieve most of the reduction in CV and adjCV? Plot CV and adjCV against the number of LVs. Then the number of LVs that achieves most of the reduction in CV and adjCV corresponds to the "elbow" in the plot after which point there is a plateau with the plot becoming essentially flat.

   (b) How many LVs are needed to explain at least 95% of the variance in the NIR measurements? What is the proportion of variance in the octane numbers explained by this number of LVs?

   (c) Make the coefficients plot and the factor loadings plot as in Figures 5.12 and 5.13, respectively. At which wavelengths do the NIR measurements have the highest loadings on LV1 and LV2? At which wavelengths are the NIR measurements most predictive of the octane rating of the gasoline?

Chapter 6

Multiple linear regression: variable selection and model building

Thus far we have assumed that the regression model is fully specified and only its parameters are to be estimated. In practice, often we have a choice of many predictors, and the goal is to select a small subset of them that provides a parsimonious yet well-fitting model with good predictive power. In this chapter, we discuss methods for predictor variable selection for this goal.

We will use two examples to illustrate the methods presented in this chapter. The first example deals with the Hald cement data from Example 4.8. This is a small data set that involves extreme multicollinearity. As we have seen, there are large negative correlations between x_1 and x_3, and x_2 and x_4. However, it is unclear which variable out of each pair should be retained in the model. Ridge regression performed in Example 5.1 is ambivalent on this issue since it shrinks all four regression coefficients. Lasso regression performed in Example 5.2 drops x_3 from the model and keeps the other three variables. We will analyze this data set by applying the methods of variable selection to see which variables should be kept in the model.

The second example deals with the used car prices data from Example 3.12. This is a large data set that involves many predictor variables. Any data set with a large number of variables is likely to have some dependencies among them. Therefore, it is of interest to screen them to find a small subset of predictor variables that gives a good predictive model.

Predictive Analytics: Parametric Models for Regression and Classification Using R,
First Edition. Ajit C. Tamhane.
© 2021 John Wiley & Sons, Inc. Published 2021 by John Wiley & Sons, Inc.
Companion website: www.wiley.com/go/tamhane/predictiveanalytics

6.1   Best subset selection

Denote by m the number of predictor variables, $x_1, \ldots, x_m$, available for selection in the model. Some of them could be functions of the original variables, e.g. x_2 could be x_1^2 or x_3 could be the interaction $x_1 x_2$. Let $p \leq m$ denote the number of predictors in a given model. There are a total of $2^m - 1$ possible models excluding the null model, which has no predictors. (We assume that the intercept is included in every model.) For every $p \leq m$, there are $\binom{m}{p}$ models of size p. The best subset selection method aims at selecting the best model according to a specified criterion by evaluating it for all models. Some commonly used criteria are described below. Efficient algorithms are available to evaluate these criteria without having to actually fit the models. Nevertheless, if m is large, then this method may not be feasible because the total number of possible models can be prohibitively large.

6.1.1   Model selection criteria

A p-variable model consists of a subset of size p from $x_1, \ldots, x_m$. For convenience, we will renumber the p variables in the model as $x_1, \ldots, x_p$, keeping in mind that these are not necessarily the first p variables from $x_1, \ldots, x_m$. To keep the notation simple, we will indicate the dependence of the criterion only on the size p of the model and not on the particular subset of p variables included, e.g. we will denote the R^2 criterion for the model consisting of $x_1, \ldots, x_p$ by R_p^2 rather than by $R^2(x_1, \ldots, x_p)$.

R_p^2 *Criterion*: This criterion compares models based on their R^2 values, which are measures of the goodness of fit. Since R^2 can only increase by adding more variables to the model, it is trivially maximized by including all m variables. Therefore, if this criterion is to be used for model selection, then some condition must be put on it, e.g. choose the smallest model whose R^2 does not increase significantly by adding more variables. The test given in Exercise 6.2 can be used for this purpose. In general, the R_p^2 criterion tends to produce models that are too large. Furthermore, although these models provide good fits, they do not necessarily give good predictions.

We can plot R_p^2 as a function of p as shown in Figure 6.1 for the Hald cement data. We see that $\max R_p^2$ values for each p, where the maximum is taken over all $\binom{m}{p}$ models of size p, increases sharply initially as p increases but soon reaches a plateau with only a marginal increase beyond it. This "elbow" point corresponds to the desired model. The adjusted R^2 criterion, discussed next, can be used to identify it.

Adjusted R_p^2 Criterion: This criterion was introduced in Section 3.2.4 as a modification of R_p^2 in order to incorporate penalty for the number of

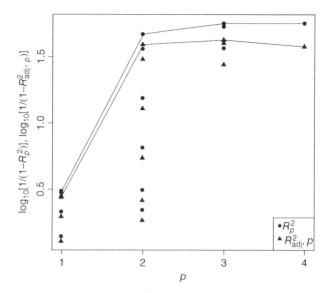

Figure 6.1. $\log_{10}\left(\frac{1}{1-R_p^2}\right)$ and $\log_{10}\left(\frac{1}{1-R_{\mathrm{adj},p}^2}\right)$ as functions of p for the Hald cement data.

variables in the model. It is defined as

$$R_{\mathrm{adj},p}^2 = 1 - \frac{\mathrm{SSE}_p/[n-(p+1)]}{\mathrm{SST}/(n-1)} = 1 - \frac{\mathrm{MSE}_p}{\mathrm{MST}}.$$

Note that MST does not depend on p. As p increases, SSE_p decreases but so do the error degrees of freedom (d.f.) Therefore, their ratio does not necessarily decrease. $R_{\mathrm{adj},p}^2$ reaches a maximum when MSE_p reaches a minimum at some p as shown in Figure 6.1. Thus maximizing $R_{\mathrm{adj},p}^2$ (or equivalently minimizing MSE_p) is a well-defined criterion unlike maximizing R_p^2.

Mallows' C_p Criterion: Mallows' (1973) C_p statistic is an approximately unbiased estimate of the sum of the standardized mean squared error of prediction (MSEP) of all n observations (except for a constant). It is given by

$$C_p = \frac{\mathrm{SSE}_p}{\widehat{\sigma}^2} + 2(p+1) - n,$$

where $\widehat{\sigma}^2$ is an unbiased estimate of σ^2. See Section 6.4.1 for the derivation of this formula. The goal is to minimize C_p.

Usually, the MSE for the full model is used for $\widehat{\sigma}^2$, i.e. $\widehat{\sigma}^2 = \mathrm{MSE}_m = \mathrm{SSE}_m/[n-(m+1)]$. This gives

$$C_m = \frac{\mathrm{SSE}_m}{\mathrm{MSE}_m} + 2(m+1) - n = n - (m+1) + 2(m+1) - n = m + 1.$$

It can be shown that if a p-variable model has zero bias then $E(C_p) \approx p + 1$. Since $C_m = m + 1$, we can say that the full model has zero bias and dropping variables from the full model will increase bias. If we drop variables from a model, p decreases and SSE_p increases but the penalty term $2(p+1)$ decreases. So the minimum C_p is generally attained at some intermediate value of p as shown in Figure 6.2 for the Hald cement data. Note that C_p charges a stiffer penalty for the number of variables than does $R^2_{\mathrm{adj},p}$. Therefore, C_p tends to select a more parsimonious model.

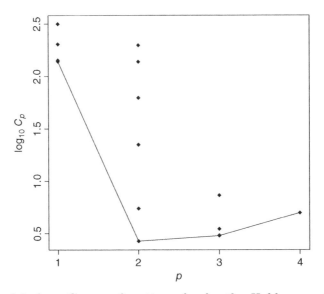

Figure 6.2. $\log_{10} C_p$ *as a function of* p *for the Hald cement data.*

AIC_p and BIC_p criteria: AIC_p stands for **Akaike's information criterion** and is given by

$$\mathrm{AIC}_p = n \ln \mathrm{SSE}_p + 2(p+1) - n \ln n.$$

This is the same formula derived in Equation (9.14) except for the constant term $n \ln 2\pi + n$. AIC_p measures the "information loss" (using the Kullback–Leibler information measure) because of not fitting the true model. Minimizing AIC_p is equivalent to maximizing the expected information in a model subject to a penalty term for the number of variables in the model.

There are many variants of the AIC_p criterion. BIC_p stands for Schwarz's **Bayesian information criterion** and is given by

$$\mathrm{BIC}_p = n \ln \mathrm{SSE}_p + (p+1) \ln n - n \ln n.$$

Note that the multiplying factor $\ln n$ used in the penalty term in the above for the number of variables is greater than 2 used in AIC_p if $n \geq 8$. So the BIC_p criterion charges a stiffer penalty for the number of variables than does AIC_p. Therefore it tends to result in a more parsimonious model.

Example 6.1 (Hald Cement Data: Best Subsets Regression) The values of the above subset selection criteria for the Hald cement data are shown in Table 6.1. For each p, the two best models (in terms of the largest R_p^2) are shown. We see that the $R_{\text{adj},p}^2$, C_p, and the BIC_p criteria choose $\{x_1, x_2\}$ as the best model, whereas the AIC_p criterion chooses $\{x_1, x_2, x_4\}$ as the best model, which was also chosen by lasso regression; see Example 5.2. However, its AIC_p is only marginally smaller than that of the $\{x_1, x_2\}$ model.

Table 6.1. Values of the criteria for two best subsets of each size for the Hald cement data

p	Variables	R_p^2 (%)	$R_{\text{adj},p}^2$ (%)	C_p	AIC_p	BIC_p
1	x_4	67.45	64.50	138.75	58.85	59.98
	x_2	66.62	63.60	142.49	59.18	60.31
2	x_1, x_2	97.87	97.44	2.678	25.42	27.11
	x_1, x_4	97.25	96.69	5.503	28.75	30.44
3	x_1, x_2, x_4	98.23	97.64	3.018	24.97	27.23
	x_1, x_2, x_3	98.23	97.64	3.042	25.01	27.27
4	x_1, x_2, x_3, x_4	98.24	97.36	5.0	26.94	29.76

We illustrate the calculation of the criteria for the $\{x_1, x_2\}$ model. For this model, $\text{SST} = 2715.76, \text{SSE}(x_1, x_2) = 57.90$, and $\text{MSE}(x_1, x_2) = 5.79$. For the full model $\text{SSE}(x_1, x_2, x_3, x_4) = 47.86$, and $\text{MSE}(x_1, x_2, x_3, x_4) = 5.98$. Therefore,

$$R_p^2 = 1 - \frac{57.90}{2715.76} = 97.87\%, \quad R_{\text{adj},p}^2 = 1 - \frac{5.79}{2715.76/12} = 97.44\%,$$

$$C_p = \frac{57.90}{5.98} + 2(2+1) - 13 = 2.678,$$

$$\text{AIC}_p = 13 \ln \left(\frac{57.90}{13} \right) + 2(2+1) = 25.418,$$

and

$$\text{BIC}_p = 13 \ln \left(\frac{57.90}{13} \right) + (2+1) \ln 13 = 27.113.$$

Figure 6.1 shows the plot of $\log_{10}[1/(1 - R_p^2)]$ and $\log_{10}[1/(1 - R_{\mathrm{adj},p}^2)]$ versus p. The log scale is used to clearly bring out the differences between R_p^2 and $R_{\mathrm{adj},p}^2$ near their maximum values for each p.

Figure 6.2 shows the plot of $\log_{10} C_p$ versus p. In this plot, we have chosen the log scale to shrink the wide range of C_p values (from 2.678 to 202.567).

The following R script produces the two best models of each size in terms of the C_p criterion. It uses the library `leaps`.

```
> cement = read.csv("c:/data/cement.csv")
> library(leaps)
> best=leaps(cement[,1:4], cement[,5], method="Cp", nbest=2,
> names=names(cement)[1:4])
> data.frame(size=best$size,Cp=best$Cp,best$which)
```

The output is as follows.

	size	Cp	x1	x2	x3	x4
1	2	138.730833	FALSE	FALSE	FALSE	TRUE
2	2	142.486407	FALSE	TRUE	FALSE	FALSE
3	3	2.678242	TRUE	TRUE	FALSE	FALSE
4	3	5.495851	TRUE	FALSE	FALSE	TRUE
5	4	3.018233	TRUE	TRUE	FALSE	TRUE
6	4	3.041280	TRUE	TRUE	TRUE	FALSE
7	5	5.000000	TRUE	TRUE	TRUE	TRUE

We see that the minimum C_p model is $\{x_1, x_2\}$ with $C_p = 2.678$. $\square$

Example 6.2 *(Used Car Prices: Best Subsets Selection)* We applied the best subsets regression to the used car prices training data set from Example 3.12 with $n = 402$. The model with the smallest $C_p = 13.03$ (this value depends on what is used as the full model and so may vary) includes 12 variables (Mileage, Cylinders, Liter, Cruise, Buick, Cadillac, Chevrolet, SAAB, Convertible, Coupe, Hatchback, Sedan) as shown below.

$$\widehat{\log_{10}(\text{Price})} = 4.082 - 0.0036 \text{ Mileage} - 0.0141 \text{ Cylinder} + 0.1116 \text{ Liter}$$

$$+ 0.0098 \text{ Cruise} + 0.0408 \text{ Buick} + 0.2464 \text{ Cadillac}$$

$$- 0.1292 \text{ Chevrolet} + 0.2794 \text{ SAAB} + 0.0720 \text{ Convertible}$$

$$- 0.0668 \text{ Coupe} - 0.0792 \text{ Hatchback} - 0.0708 \text{ Sedan}.$$

Recall that Mileage is expressed in thousands of miles. The only nonsignificant variable in the above model is Cruise with P-value equal to 0.098. Surprisingly, the Cylinder variable is significant with P-value equal to 0.042 despite its high correlation of 0.958 with the Liter variable. As a comparison, $C_p = 16.05$ for the 10-variable model fitted in Example 3.12 which includes Liter but not the Cylinder variable. We would probably choose that model as it has less number of variables all of which are highly significant ($P < 0.001$), although its C_p is higher. □

6.2 Stepwise regression

As the name suggests, stepwise regression, builds a single model by entering or removing predictors one at a time in a stepwise manner according to a set of rules. Note that stepwise regression does not evaluate all models and does not optimize some well-defined criterion of "bestness" as does the best subset selection method.

There are two basic types of stepwise regression: forward and backward. Forward stepwise regression starts from the null model and enters variables one at a time, while backward stepwise regression starts from the full model and removes variables one at a time. For each type there are two kinds of algorithms. **Forward stepwise algorithm** allows the option of removing the variables entered at previous steps, while the **backward stepwise algorithm** allows the option of entering the variables removed at previous steps. These algorithms can be chosen in the `step` function in R by specifying `direction = "both"`. On the other hand, **forward selection algorithm** proceeds only in one direction with no option of removing the previously entered variables, while **backward elimination algorithm** proceeds also only in one direction with no option of entering the previously removed variables. These algorithms can be chosen in the `step` function in R by specifying `direction = "forward"` or `direction = "backward"`, respectively. In case of multicollinearity, the backward algorithm is recommended. Exercise 6.3 gives a data set which illustrates this point.

Next, we explain the forward and backward stepwise algorithms based on partial F-statistics. Suppose that there are m predictor variables available for selection in the regression model and at a given step of the algorithm there are $p < m$ variables included in the model, labeled as $x_1, \ldots, x_p$. (It is assumed that the intercept term is included in all models.) A critical constant, f_{in}, is specified to decide whether to enter one of the excluded variables, labeled as $x_{p+1}, \ldots, x_m$, into the model while another critical constant, f_{out}, is specified to decide whether to remove one of the included variables, $x_1, \ldots, x_p$, from the model. The forward stepwise algorithm begins with $p = 0$ (i.e. the null model containing only the intercept term) while the backward stepwise algorithm begins with $p = m$ (i.e. the full model). For the "enter" decision, we compute $F_{\text{in}} = \max_{p+1 \leq i \leq m} F_i$ and enter the variable associated with F_{in} if $F_{\text{in}} > f_{\text{in}}$. Here F_i (sometimes referred to as F-**to-enter**) is the extra sum of squares (SS) F-statistic if the ith excluded variable ($i = p + 1, \ldots, m$) is added to a p-variable model. Similarly, for the "remove" decision, we compute $F_{\text{out}} = \min_{1 \le i \le p} F_i$ and remove the variable associated with F_{out} if $F_{\text{out}} < f_{\text{out}}$. Here F_i (sometimes referred to as F-**to-remove**) is the extra SS F-statistic if the ith included variable ($i = 1, \ldots, p$) is removed from a p-variable model. Note that both these F_i-statistics are equal to the squares of the t_i-statistics for that variable when it is included in the regression model along with the other variables.

We require $f_{\text{in}} \ge f_{\text{out}}$, otherwise the algorithm can go into an infinite loop. For example, if $f_{\text{in}} = 3.0$ and $f_{\text{out}} = 4.0$ and suppose $F_i = 3.5$ then x_i will enter the model if $F_{\text{in}} = F_i > 3.0$ but at the next step, x_i will be removed from the model if $F_{\text{out}} = F_i < 4.0$. Thus, the algorithm will keep recycling x_i in and out of the model.

The number of variables selected in the final model depends on the selected values of f_{in} and f_{out}. The smaller values will result in more variables being chosen in the model and vice versa. Finally, it should be noted that the partial F-tests used in these algorithms are not formal F-tests at some pressigned α-level because the tests use not the F_i themselves but their minimum or maximum. Furthermore, they are applied sequentially multiple times with varying error d.f. Hence, f_{in} and f_{out} need not be chosen as critical values of some F-distribution; instead, they are chosen to be some fixed constants. One can vary their values to see how they affect the resulting model.

The forward stepwise algorithm is illustrated in the following example.

Example 6.3 (Hald Cement Data: Forward Stepwise Regression) In this example, we will illustrate the stepwise regression algorithm given above for the Hald cement data from Table 4.3. We will set $f_{\text{in}} = f_{\text{out}} = 4.0$.

Step 1: At Step 1, we decide which variable to enter in the model first. Since there are no variables in the model to begin with, the maximum partial F-statistic corresponds to the maximum absolute bivariate correlation coefficient $|r_{yx_j}|$. The four correlation coefficients are

$$r_{yx_1} = 0.7307, \quad r_{yx_2} = 0.8163, \quad r_{yx_3} = -0.5347, \quad r_{yx_4} = -0.8213,$$

and the corresponding error sum of squares (SSEs) are

$$\text{SSE}(x_1) = 1265.7, \quad \text{SSE}(x_2) = 906.1,$$
$$\text{SSE}(x_3) = 1939.3, \quad \text{SSE}(x_4) = 883.9.$$

Also, SST = $\text{SSE}(\emptyset) = 2715.76$ from the analysis of variance table in Example 4.9, where $\emptyset$ denotes a null model with only the intercept term. The maximum absolute bivariate correlation coefficient is $|r_{yx_4}|$, which corresponds to the minimum SSE (since it reduces the SSE most from the null model) or max F_{in}. Using the extra SS method, we calculate the F_{in} for x_4:

$$F_{\text{in}} = \frac{\text{SSE}(\emptyset) - \text{SSE}(x_4)}{\text{MSE}(x_4)} = \frac{2715.76 - 883.9}{883.9/(13 - 2)} = 22.745 > 4.0.$$

So x_4 enters the model.

Step 2: At Step 2, we choose between x_1, x_2, and x_3 to decide which variable, if any, enters the model next. The variable that gives the smallest SSE will be the candidate. The three SSE values are as follows:

$$\text{SSE}(x_1, x_4) = 74.8, \quad \text{SSE}(x_2, x_4) = 868.9, \quad \text{SSE}(x_3, x_4) = 175.7.$$

Since adding x_1 gives the smallest SSE, it is the candidate for the entry into the model. To test if it meets the entry criterion, we calculate its F_{in} statistic:

$$F_{\text{in}} = \frac{\text{SSE}(x_4) - \text{SSE}(x_1, x_4)}{\text{MSE}(x_1, x_4)} = \frac{883.9 - 74.8}{74.8/(13 - 3)} = 108.17 > 4.0.$$

Therefore x_1 enters the model.

Before going to the next step, we need to check if x_4, which entered at Step 1, should be removed from the model. So we calculate its F_{out} statistic:

$$F_{\text{out}} = \frac{\text{SSE}(x_1) - \text{SSE}(x_1, x_4)}{\text{MSE}(x_1, x_4)} = \frac{1265.7 - 74.8}{74.8/(13 - 3)} = 159.20 > 4.0.$$

Therefore, x_4 cannot be removed from the model.

Step 3: Next, we check whether x_2 or x_3 should be added to the model. Toward this end, we calculate

$$\text{SSE}(x_1, x_2, x_4) = 47.97, \quad \text{SSE}(x_1, x_3, x_4) = 50.84.$$

Since adding x_2 gives the smaller SSE, it is the candidate to enter the model next. To test if it meets the entry criterion, we calculate its F_{in} statistic:

$$F_{\text{in}} = \frac{\text{SSE}(x_1, x_4) - \text{SSE}(x_1, x_2, x_4)}{\text{MSE}(x_1, x_2, x_4)} = \frac{74.8 - 47.97}{47.97/(13 - 4)} = 5.034 > 4.0.$$

Therefore x_2 enters the model.

Before going to the next step, we need to check if either x_1 or x_4 should be removed from the model. So we calculate

$$\text{SSE}(x_1, x_2) = 57.9 \quad \text{and} \quad \text{SSE}(x_2, x_4) = 868.9.$$

Thus, removing x_4 gives a smaller SSE. So we check if it can be removed by calculating its F_{out} statistic:

$$F_{\text{out}} = \frac{\text{SSE}(x_1, x_2) - \text{SSE}(x_1, x_2, x_4)}{\text{MSE}(x_1, x_2, x_4)} = \frac{57.9 - 47.97}{47.97/(13 - 4)} = 1.863 < 4.0.$$

Therefore we remove x_4 from the model. Recall that x_4 was the first variable to enter the model but it is removed at this step.

It is easy to check that neither x_1 nor x_2 can be removed from the model.

Step 4: Since x_4 was just removed from the model, it cannot re-enter (its F_{in} at this step equals its F_{out} at the previous step). Thus, it only remains to check whether x_3 can be entered in the model. We calculate $\text{SSE}(x_1, x_2, x_3) = 48.11$. Hence, its

$$F_{\text{in}} = \frac{\text{SSE}(x_1, x_2) - \text{SSE}(x_1, x_2, x_3)}{\text{MSE}(x_1, x_2, x_3)} = \frac{57.9 - 48.11}{48.11/(13 - 4)} = 1.832 < 4.0.$$

Therefore x_3 cannot enter the model and the algorithm stops with the final model consisting of two variables: x_1 and x_2, which is the same model obtained using the best subsets regression with the C_p criterion.

$\square$

It should be noted that stepwise regression in R uses the AIC criterion and not partial F-tests. Variables are added or removed from the model in a stepwise manner to minimize AIC. The following example illustrates this using the same Hald cement data.

Example 6.4 (Hald Cement Data: Stepwise Regression Using R)
Stepwise regression in R finds the model with the minimum AIC by deleting (in case of backward elimination) or adding (in case of forward selection) one variable at each step. We can run a backward elimination algorithm using the following R script.

```
> cement = read.csv("c:/data/cement.csv")
> fit1 = lm(y ~., cement)
> step(fit1)
```

The resulting output is as follows.

```
Start:   AIC=26.94 y ~ x1 + x2 + x3 + x4

        Df Sum of Sq     RSS     AIC
- x3     1     0.1091 47.973 24.974
- x4     1     0.2470 48.111 25.011
- x2     1     2.9725 50.836 25.728
<none>                  47.864 26.944
- x1     1    25.9509 73.815 30.576

Step:  AIC=24.97
y ~ x1 + x2 + x4

        Df Sum of Sq     RSS     AIC
<none>                  47.97 24.974
- x4     1       9.93  57.90 25.420
- x2     1      26.79  74.76 28.742
- x1     1     820.91 868.88 60.629

Call:
lm(formula = y ~ x1 + x2 + x4, data = cement)

Coefficients:
(Intercept)           x1           x2           x4
    71.6483       1.4519       0.4161      -0.2365
```

We see from the output that the algorithm starts with the full model, deletes one variable at a time and calculates the AIC of the resulting model. The results are arranged in the increasing order of AIC. The full model has AIC = 26.94. The best model of size three is $\{x_1, x_2, x_4\}$ and has AIC = 24.97. From the next step, we see that removing any more variables from this model increases AIC. So the overall best model is $\{x_1, x_2, x_4\}$.

The same result is obtained if we run the forward stepwise algorithm using the following R script:

```
> fit2 = lm(y ~ 1, cement)
> step(fit2, scope=~x1+x2+x3+x4)
```

The resulting output is as follows.

```
Start:  AIC=71.44
y ~ 1

        Df Sum of Sq     RSS     AIC
+ x4     1    1831.90  883.87 58.852
+ x2     1    1809.43  906.34 59.178
```

```
+ x1     1    1450.08 1265.69 63.519
+ x3     1     776.36 1939.40 69.067
<none>                2715.76 71.444
```

```
Step:  AIC=58.85
y ~ x4
```

	Df	Sum of Sq	RSS	AIC
+ x1	1	809.10	74.76	28.742
+ x3	1	708.13	175.74	39.853
<none>			883.87	58.852
+ x2	1	14.99	868.88	60.629
− x4	1	1831.90	2715.76	71.444

```
Step:  AIC=28.74
y ~ x4 + x1
```

	Df	Sum of Sq	RSS	AIC
+ x2	1	26.79	47.97	24.974
+ x3	1	23.93	50.84	25.728
<none>			74.76	28.742
− x1	1	809.10	883.87	58.852
− x4	1	1190.92	1265.69	63.519

```
Step:  AIC=24.97
y ~ x4 + x1 + x2
```

	Df	Sum of Sq	RSS	AIC
<none>			47.97	24.974
− x4	1	9.93	57.90	25.420
+ x3	1	0.11	47.86	26.944
− x2	1	26.79	74.76	28.742
− x1	1	820.91	868.88	60.629

```
Call:
lm(formula = y ~ x4 + x1 + x2, data = cement)
```

```
Coefficients:
(Intercept)           x4            x1            x2
    71.6483       -0.2365        1.4519        0.4161
```

$\square$

Example 6.5 (*Used Car Prices: Stepwise Regression*) We applied stepwise regression with $f_{in} = f_{out} = 4.0$ on the used car prices training data set from Example 3.12 with $n = 402$. The response variable was $\log_{10}(\text{Price})$. The variables are listed in Table 6.2 in the sequence in which they entered the model along with their partial correlation coefficients and the F_{in} statistics. Note that in this example no variables were removed at any step. Thus, the procedure operated as

Table 6.2. Stepwise addition of variables for used car prices regression model

Step	Variable	Partial r	F_{in}	Step	Variable	Partial r	F_{in}
1	Liter	0.590	213.6	6	Wagon	0.390	70.73
2	SAAB	0.785	639.58	7	Chevrolet	-0.244	24.90
3	Cadillac	0.743	491.51	8	Pontiac	-0.155	23.62
4	Mileage	0.457	104.65	9	Saturn	-0.206	17.31
5	Convertible	0.501	132.94	10	Cylinder	-0.142	8.07

if it is a forward selection procedure. Similar to the best model according to the C_p criterion found in Example 6.2, this model also includes the Cylinder variable along with Liter. The fitted equation for this model is

$$\widehat{\log_{10}(\text{Price})} = 4.064 - 0.0035 \text{ Mileage} + 0.1177 \text{ Liter} + 0.2075 \text{ Cadillac}$$

$$-0.0572 \text{ Chevrolet} - 0.0393 \text{ Pontiac} + 0.2406 \text{ SAAB}$$

$$-0.0475 \text{ Saturn} + 0.1423 \text{ Convertible} + 0.0672 \text{ Wagon}$$

$$-0.0183 \text{ Cylinder}.$$

Once again, recall that Mileage is expressed in thousands of miles.

The C_p statistic for this model can be calculated as follows. The full model consists of the following 15 variables: Mileage, Cylinder, Liter, Cruise, Sound, Leather, Buick, Cadillac, Chevrolet, Pontiac, SAAB, Convertible, Coupe, Hatchback, and Sedan. This full model has MSE $= 1.5875 \times 10^{-3}$, and SSE for the above fitted model is 0.6274. Therefore,

$$C_p = \frac{0.6274}{1.5875 \times 10^{-3}} + 2(11) - 402 = 15.2214.$$

□

Example 6.6 *(Used Car Prices: Stepwise Regression Using* R*)* Below we give the R output for stepwise regression for used car prices data. First training and test sets are formed by putting all odd-numbered observations in the former and all even-numbered observations in the latter. Next variable numbers 4 and 5 (Model and Trim) are omitted from the training and test sets to form newtrain and newtest sets. Finally, Make and Type are defined as factors and Price is transformed to log10(Price).

```
> data = read.csv("c:/data/carprices.csv")
> train_indices <- seq(1, nrow(data), by=2)
> train <- data[train_indices,]
```

```
> test <- data[-train_indices,]
> newtrain=train[-c(4:5)]
> newtest=test[-c(4:5)]
> newtrain$Make=factor(newtrain$Make)
> newtrain$Type=factor(newtrain$Type)
> newtrain$Price=log10(newtrain$Price)
> newtest$Make=factor(newtest$Make)
> newtest$Type=factor(newtest$Type)
> newtest$Price=log10(newtest$Price)
> fit0 = lm(Price~.,data = newtrain)
> fit1=step(fit0)
Start:  AIC=-2590.44
Price ~ Mileage + Make + Type + Cylinder + Liter + Doors + Cruise +
    Sound + Leather


Step:  AIC=-2590.44
Price ~ Mileage + Make + Type + Cylinder + Liter + Cruise + Sound +
    Leather

            Df Sum of Sq    RSS     AIC
- Leather    1    0.0003 0.5906 -2592.2
- Cruise     1    0.0018 0.5922 -2591.2
<none>                   0.5903 -2590.4
- Cylinder   1    0.0034 0.5938 -2590.1
- Sound      1    0.0047 0.5950 -2589.3
- Liter      1    0.2903 0.8807 -2431.7
- Mileage    1    0.3244 0.9148 -2416.4
- Type       4    0.4227 1.0130 -2381.4
- Make       5    3.4480 4.0383 -1827.5

Step:  AIC=-2592.25
Price ~ Mileage + Make + Type + Cylinder + Liter + Cruise + Sound

            Df Sum of Sq    RSS     AIC
- Cruise     1    0.0020 0.5926 -2592.9
<none>                   0.5906 -2592.2
- Cylinder   1    0.0033 0.5939 -2592.0
- Sound      1    0.0046 0.5952 -2591.2
- Liter      1    0.2928 0.8835 -2432.4
- Mileage    1    0.3255 0.9161 -2417.8
- Type       4    0.4225 1.0132 -2383.3
- Make       5    3.5181 4.1087 -1822.5

Step:  AIC=-2592.91
Price ~ Mileage + Make + Type + Cylinder + Liter + Sound
```

```
          Df Sum of Sq   RSS      AIC
<none>                  0.5926 -2592.9
- Cylinder  1    0.0032 0.5958 -2592.8
- Sound     1    0.0045 0.5971 -2591.9
- Liter     1    0.2989 0.8915 -2430.7
- Mileage   1    0.3261 0.9187 -2418.7
- Type      4    0.4206 1.0132 -2385.3
- Make      5    4.0403 4.6329 -1776.2
> summary(fit1)

Call:
lm(formula = Price ~ Mileage + Make + Type + Cylinder + Liter +
    Sound, data = newtrain)

Residuals:
     Min       1Q     Median       3Q       Max
-0.140788 -0.024145  0.000808  0.024663  0.116387

Coefficients:
                 Estimate Std. Error t value Pr(>|t|)
(Intercept)     4.184e+00  2.182e-02 191.764  < 2e-16 ***
Mileage        -3.479e-06  2.381e-07 -14.612  < 2e-16 ***
MakeCadillac    1.986e-01  1.011e-02  19.639  < 2e-16 ***
MakeChevrolet  -5.373e-02  7.660e-03  -7.014 1.03e-11 ***
MakePontiac    -3.267e-02  7.945e-03  -4.112 4.79e-05 ***
MakeSAAB        2.462e-01  9.679e-03  25.441  < 2e-16 ***
MakeSaturn     -4.318e-02  1.041e-02  -4.149 4.10e-05 ***
TypeCoupe      -1.373e-01  1.059e-02 -12.962  < 2e-16 ***
TypeHatchback  -1.569e-01  1.241e-02 -12.640  < 2e-16 ***
TypeSedan      -1.412e-01  9.349e-03 -15.101  < 2e-16 ***
TypeWagon      -7.324e-02  1.184e-02  -6.189 1.54e-09 ***
Cylinder       -9.854e-03  6.809e-03  -1.447   0.1487
Liter           1.071e-01  7.652e-03  13.990  < 2e-16 ***
Sound           7.799e-03  4.564e-03   1.709   0.0883 .
---
Signif. codes:  0 '***' 0.001 '**' 0.01 '*' 0.05 '.' 0.1 ' ' 1

Residual standard error: 0.03908 on 388 degrees of freedom
Multiple R-squared:  0.9532,   Adjusted R-squared:  0.9516
F-statistic: 608 on 13 and 388 DF,  p-value: < 2.2e-16

>
> test.pred1 = predict(fit1,newtest)
> SSE1 = sum((newtest$Price-test.pred1)^2)
> SSE1
[1] 0.6418505
```

6.3 Model building

Regression modeling is an iterative process. Below we give suggested steps in this process.

- *Univariate exploration of the data*: This step is generally referred to as **exploratory data analysis** (**EDA**). Examine each variable for outliers, wrong or inconsistent data entries, missing values, etc. If possible, correct outliers and wrong data entries and fill in missing values as appropriate. If outliers and missing values are a small fraction of the total sample then one may discard those observations. It is better to discard a few observations than to discard some variables as they may be important predictors.

  Examine univariate distributions of all variables and in cases of highly skewed distributions make suitable transformations to symmetrize the distributions. Often, the response variable y is skewed with many zero values. In that case, log transformation, $y \to \log(y + 1)$, transforms zeros back to zeros. Transformations may also be made based on subject matter knowledge by consulting subject matter experts. Note that the predictor variables do not need to be normally or even symmetrically distributed (in fact, dummy variables are binary). However, highly skewed predictors are likely to result in influential observations and so it is recommended to symmetrize them.

- *Bivariate exploration of the data*: Make scatter plots and compute correlations between all variables (feasible only when there is a modest number of variables, say, less than 10) to identify any bivariate outliers, influential observations, and highly correlated x's causing multicollinearity. Bivariate scatter plots between y and each x help suggest linearizing transformations for the x's. It is recommended not to transform y at this step unless a common linearizing transformation of y works for all x's or if a subject matter–based transformation exists for y. For example, if one wants to predict the gas mileage (miles per gallon or mpg) of a car from its weight, engine size, etc., then an inverse transformation (gallons per mile or gpm) is preferable since the latter is proportional to the amount of energy spent to move the car one mile which in turn is proportional to its weight, engine size, etc.

- *Feature engineering*: Feature engineering involves creating new predictor variables (features) from the given ones. This includes simple transformations of predictor variables such as log or square-root transformations but mainly includes creating interactions. Generally, only those interactions that are supported by subject matter knowledge should be added. Interactions are often modeled as products but can be ratios.

- *Form training set and test set*: Randomly divide the data into a training set and a test set. The training set should be large enough to build a reliable model and have enough error d.f. If the data set is sufficiently large, use a 50 : 50 split; otherwise, make the training set bigger.

- *Fit several candidate models and compare them*: Use best subsets regression or stepwise regression to fit several models. If there are many variables, say $p \geq 20$, then the best subsets regression approach may not be feasible in which case it may be advisable to first select a subset of variables by using stepwise regression followed by the best subsets regression on that subset. Evaluate the candidate models based on C_p and other criteria such as residual plots and other diagnostics. Choose two or three good candidate models.

- *Selection of the final model*: Choose the final model by comparing the contending models in terms of the total prediction error by applying them to the test set. More generally, the final model selection may be done using m-**fold cross-validation**. In this method, the data set is divided into $m \geq 2$ random subsets. One subset is used as a test set while the remaining $m - 1$ subsets are used as a training set. This process is repeated m times by leaving out one subset as a test set each time and fitting the model on the remaining $m - 1$ subsets. Random splits of the data into m subsets may be generated multiple times. Prediction errors are averaged over all m splits of the data into training and test sets. If m equals the total sample size n then in each iteration of the algorithm, a single observation is set aside as a test set. The model is fitted by deleting one observation at a time and the prediction error for the deleted observation is calculated. This is known as **leave-one-out cross-validation**.

- *Practical checks on the final model*: Check that the model includes the key variables based on subject matter knowledge and their coefficients have the correct signs. The model should pass the standard statistical tests as well as the subject matter criteria.

6.4 Technical notes*

6.4.1 Derivation of the C_p statistic

C_p is an approximately unbiased sample estimate of the **MSEP** except for a constant, using a p-variable model. Let $\widehat{y}_{i,p}$ denote the fitted or predicted value of y_i. The MSEP equals

$$
\begin{aligned}
\text{MSEP}_p &= \sum_{i=1}^{n} E[(\widehat{y}_{i,p} - y_i)^2] \\
&= \sum_{i=1}^{n} E[\{(\widehat{y}_{i,p} - E(y_i)) - (y_i - E(y_i))\}^2] \\
&= \sum_{i=1}^{n} E[(\widehat{y}_{i,p} - E(y_i))^2] + \sum_{i=1}^{n} E[(y_i - E(y_i))^2] \\
&= \sum_{i=1}^{n} E[(\widehat{y}_{i,p} - E(y_i))^2] + n\sigma^2.
\end{aligned}
$$

In the above the cross-product term cancels out since $E[y_i - E(y_i)] = 0$. We can ignore $n\sigma^2$ since it is a constant. Therefore, minimizing MSEP_p is equivalent to minimizing Γ_p, where

$$
\begin{aligned}
\Gamma_p &= \frac{1}{\sigma^2} \sum_{i=1}^{n} E[(\widehat{y}_{i,p} - E(y_i))^2] \\
&= \frac{1}{\sigma^2} \sum_{i=1}^{n} E[\{(\widehat{y}_{i,p} - E(\widehat{y}_{i,p})) + (E(\widehat{y}_{i,p}) - E(y_i))\}^2] \\
&= \frac{1}{\sigma^2} \sum_{i=1}^{n} \{E[(\widehat{y}_{i,p} - E(\widehat{y}_{i,p}))^2] + [(E(\widehat{y}_{i,p}) - E(y_i))^2]\}.
\end{aligned}
$$

In the above, the cross-product term cancels out since $E[(\widehat{y}_{i,p} - E(\widehat{y}_{i,p}))] = 0$. Note that the first term inside the summation is $\text{Var}(\widehat{y}_{i,p})$ and the second term is $[\text{Bias}(\widehat{y}_{i,p})]^2$. In Section 3.8, we have shown that $\text{Cov}(\widehat{\boldsymbol{y}}) = \text{Cov}(\boldsymbol{H}\boldsymbol{y}) = \sigma^2 \boldsymbol{H}$ and $\text{tr}(\boldsymbol{H}) = p + 1$. Hence, $\sum_{i=1}^{n} \text{Var}(\widehat{y}_{i,p}) = \sigma^2 \text{tr}(\boldsymbol{H}) = \sigma^2(p+1)$. Denote the total bias squared term by $B_p = \sum_{i=1}^{n} [(E(\widehat{y}_{i,p}) - E(y_i))^2]$. Then,

$$
\Gamma_p = \frac{1}{\sigma^2} B_p + (p+1).
$$

It can be shown that $E(\text{SSE}_p) = B_p + [n - (p+1)]\sigma^2$. (This equation shows that $\text{MSE}_p = \text{SSE}_p/[n - (p+1)]$ is an unbiased estimator of σ^2 when $B_p = 0$.) Therefore, an approximately unbiased estimator of Γ_p is given by

$$
C_p = \frac{\text{SSE}_p}{\widehat{\sigma}^2} - [n - (p+1)] + (p+1) = \frac{\text{SSE}_p}{\widehat{\sigma}^2} + 2(p+1) - n.
$$

Here, $\hat{\sigma}^2$ is an unbiased estimate of σ^2, which is usually taken to be the MSE of the full model based on the assumption that the full model has zero bias, i.e. $B_m = 0$.

Exercises

Theoretical Exercises

6.1 (Relation between R^2 and R_{adj}^2) Show that

$$1 - R_{\text{adj}}^2 = \left[\frac{n-1}{n - (p+1)} \right] (1 - R^2),$$

and hence $R_{\text{adj}}^2 \leq R^2$.

6.2 (Testing the difference between R_p^2 and R_q^2) Let R_p^2 and R_q^2 denote the R^2's for the full model with p predictors and a partial model with $q < p$ predictors. Show that the extra SS F-statistic equals

$$F = \frac{(R_p^2 - R_q^2)/(p - q)}{(1 - R_p^2)/[n - (p+1)]}.$$

Suppose that $n = 26, q = 3$, and $p = 5$. Further suppose that $R_p^2 = 0.90$ and $R_q^2 = 0.80$. Test whether the increase in R^2 from the partial model to the full model is statistically significant at the 1% level.

Applied Exercises

6.3 (Hamilton data.) Hamilton (1987) gave the data shown in Table 6.3.

(a) Calculate the correlation matrix between y, x_1, and x_2. Notice that the correlations between y and x_1 and y and x_2 are quite small. Figure 6.3 shows three bivariate scatter plots between y, x_1 and x_2. What do you notice? Verify the joint relationship of y with x_1 and x_2 by running a regression.

(b) Why forward stepwise regression would fail but backward stepwise regression would work for this data set?

6.4 (Best subset and stepwise regression) Table 6.4 lists the SSEs obtained by fitting different models involving three predictor variables, x_1, x_2, x_3, based on a total of $n = 20$ observations. (The constant term is included in all models.)

(a) Complete the table by filling in the values of p, error d.f., MSE_p, $R_{\text{adj},p}^2$, C_p, and AIC_p statistics.

Table 6.3. Hamilton data

x_1	x_2	y	x_1	x_2	y
2.23	9.66	12.37	3.04	7.71	12.86
2.57	8.94	12.66	3.26	5.11	10.84
3.87	4.40	12.00	3.39	5.05	11.20
3.10	6.64	11.93	2.35	8.51	11.56
3.39	4.91	11.06	2.76	6.59	10.83
2.83	8.52	13.03	3.90	4.90	12.63
3.02	8.04	13.13	3.16	6.96	12.46
2.14	9.05	11.44			

Source: Hamilton (1987).

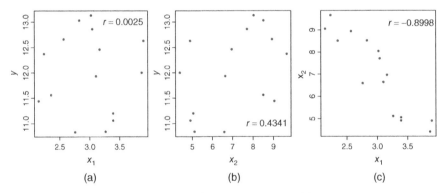

Figure 6.3. *Scatter plots between (a) y versus x_1, (b) y versus x_2, and (c) x_1 versus x_2.*

(b) Which models will be selected as the best using the $R^2_{\text{adj},p}$, C_p, and AIC_p criteria? Which model will you choose and why?

(c) Suppose that stepwise regression is to be carried out with $f_{\text{in}} = f_{\text{out}} = 4.0$. Which variable would be the first to enter the model? What is its F_{in} value?

(d) Which will be the second variable to enter the model? What is its F_{in} value? What is its partial correlation coefficient with respect to y controlling for the first variable that entered the model?

(e) Will the first variable that entered the model be removed upon the entry of the second variable? Check by doing the partial F-test.

Table 6.4. SSEs for all possible models with three predictors

Variables in model	SSE_p	p	Error d.f.	MSE_p	$R^2_{adj,p}$	C_p	AIC_p
None	950						
x_1	720						
x_2	630						
x_3	540						
x_1, x_2	595						
x_1, x_3	425						
x_2, x_3	510						
x_1, x_2, x_3	400						

(f) Will stepwise regression enter the third variable in the model, i.e. will it choose the full model? Check by doing the partial F-test.

6.5 (Partial correlation coefficients) From a data set on 33 adult males on their systolic blood pressure (y) and three predictor variables: age (x_1), smoking history (x_2), and body mass index (x_3), the following correlation matrix was computed:

$$\begin{bmatrix}
y & x_1 & x_2 & x_3 \\
1 & 0.775 & 0.247 & 0.742 \\
0.775 & 1 & -0.140 & 0.803 \\
0.247 & -0.140 & 1 & -0.071 \\
0.742 & 0.803 & -0.071 & 1
\end{bmatrix}.$$

(a) Stepwise regression chooses x_1 at Step 1 because it has the highest correlation with y. Which variable will be chosen next—x_2 or x_3? Determine by calculating $r_{yx_2|x_1}$ and $r_{yx_3|x_1}$.

(b) Explain why $r_{yx_2|x_1} > r_{yx_3|x_1}$ even though $r_{yx_2} < r_{yx_3}$?

(c) Do an F-test on $r_{yx_2|x_1}$ at $\alpha = 0.05$ to check if x_2 should be added to the model that already includes x_1.

6.6 (Backward stepwise regression for the Hald cement data) In Example 6.3, we applied forward stepwise regression to the Hald cement data. In this exercise, apply backward stepwise regression to the same data. You may use the SSE values from the example.

6.7 (Used car prices data: Comparing two models on the test set) In Examples 6.2 and 6.5, we obtained two different models for predicting the price using the training data. Refit these two models (i.e. using the same predictors) on the training data and compute their SSEs for the test data. Which model will you select based on the SSE for the test data and whether all variables included in the model are statistically significant or not at the 5% level.

Chapter 7

Logistic regression and classification

In Chapters 3–6, we studied multiple regression where the response variable was numerical, ideally continuous. In this chapter, we study a regression methodology for categorical responses. The simplest categorical response is **binary** or **dichotomous**, e.g. a treatment outcome is a success or failure; a customer buys or does not buy. More generally, the response may have multiple categories (referred to as **multinomial** or **polytomous response**); furthermore, these categories may be **nominal** or **ordinal**. For example, a customer makes a choice among several nominal alternatives such as make of a car or mode of transportation. On the other hand, some outcomes are intrinsically ordinal such as stock recommendation (buy, hold, or sell) or grade in a course (A, B, C, D, F). Polytomous responses typically have a small, finite range and do not have an underlying interval scale even if integers 1, 2, 3, ... are used to code the ordinal categories. Logistic regression deals with response variables of these types.

When the response variable is categorical, prediction corresponds to **classification** of observations into one of several categorical outcomes. For example, a physician uses a battery of lab tests to determine whether a patient has a certain disease; a fraud detection algorithm classifies an online credit card transaction as fraudulent or legitimate depending on the past transaction history of the customer, origination of transaction etc. Such classifiers are known as **binary classifiers**.

The following two data sets will be analyzed in this chapter.

Example 7.1 (Art Museum Visits: Data) A survey of 2626 Chicago adults was conducted in 1995 by the Kellogg Center for Cultural Marketing

Predictive Analytics: Parametric Models for Regression and Classification Using R.
First Edition. Ajit C. Tamhane.
© 2021 John Wiley & Sons. Inc. Published 2021 by John Wiley & Sons. Inc.
Companion website: www.wiley.com/go/tamhane/predictiveanalytics

at Northwestern University to determine which individual attributes are most predictive of a person visiting an art museum in the Chicago area. Data were collected on the variables listed in Table 7.1 and are stored in file `art.csv`.

Table 7.1. Variables for art museum visits data set

Variable	Description
Visit	At least one visit during the year: 1 (Yes), 0 (No)
Age	1 (18–19 yr), 2 (20–24 yr), ..., 11 (65–69 yr), 12 (70+ yr)
Gender	0 (Male), 1 (Female)
Children	1 (Yes), 0 (No)
Married	1 (Yes), 0 (No)
Education	1 (8th grade or less), 2 (some high school), ..., 8 (post graduate degree)
Income	1 (<$20k), 2($20k–$30k), ... , 9($90k–$100k), 10($100k–$125k), 11($125k–$150k), 12(>$150k)
County of residence	1 (Chicago), 2 (Suburban Cook), 3 (Lake), 4 (DuPage), 5 (Other)

We will use this data set to illustrate both simple logistic regression and multiple logistic regression. In the former case, we will use Education as the only predictor. There are 19 missing observations on Education, so only 2607 complete observations will be used for simple logistic regression. Overall, there are 28 observations with missing data on at least one variable. So for multiple logistic regression, 2598 complete observations will be used. Of these 2598 observations, 1676 are no visits (64.5%) and 922 are visits (35.5%).　　　　　　□

Example 7.2 (MBA Admissions: Data) Johnson and Wichern (2002) have given data on 85 MBA admission decisions (1 = admit, 2 = wait list, 3 = deny admission) by a business school as a function of two predictors: applicant's GPA score and GMAT score. The data are stored in file `MBA.csv`. The labeled scatter plot of the data is shown in Figure 7.1 (with different symbols denoting different admission decisions). It can be seen that the clusters corresponding to the three groups of admission decisions are fairly well-separated. The means of GPA and GMAT of the three groups are calculated as shown below. We notice that the GPA scores are well separated; the mean GMAT score of the "admit" group is much higher than that for the "wait list" and "deny admission" groups, which are nearly equal.

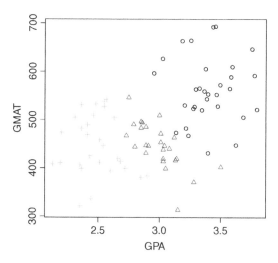

Figure 7.1. *Plot of GMAT versus GPA for MBA admissions data (circle = admit, triangle = wait list, cross = deny admission).*

```
> meanGPA <- sapply(1:3, function(x) mean(MBA$GPA[MBA$admit==x]))
> meanGMAT <- sapply(1:3, function(x) mean(MBA$GMAT[MBA$admit==x]))
> meanGPA
[1] 3.403871 2.992692 2.482500
> meanGMAT
[1] 561.2258 446.2308 447.0714
```

We will use this data set to illustrate multinomial logistic regression, both nominal by disregarding the order among the three decisions and ordinal by taking the order into account. □

7.1 Simple logistic regression

7.1.1 Model

We will begin with the simplest case of a dichotomous response variable y and a single predictor variable x. Following the standard convention, assume that y is coded as 1 or 0 depending on whether the outcome, is a "success" or "failure," respectively. Denote the probability of success by $p(x)$ to indicate its dependence on the predictor x. Just as in multiple regression, we want to model $E(y|x) = P(y = 1|x) = p(x)$ as a function of x. Note that $p(x)$ here is not to be confused with the notation p used to denote the number of predictors.

Why can we not use the simple linear regression model $p(x) = \beta_0 + \beta_1 x$? The main reason is that $p(x)$, being a probability, must lie between 0 and 1 but $\beta_0 + \beta_1 x$ is unconstrained, so predictions made from this model may fall outside the interval $[0, 1]$. Logistic regression addresses this issue by postulating a linear model on the log-transform of the **odds** of success denoted by

$$\psi(x) = \frac{p(x)}{1 - p(x)},$$

which is the ratio of the probability of success to the probability of failure. The log of this ratio is called the **log-odds** or **logit**. The linear model is postulated on this **logistic transform** of $p(x)$:

$$\ln \psi(x) = \ln \left[\frac{p(x)}{1 - p(x)} \right] = \beta_0 + \beta_1 x \qquad (7.1)$$

or equivalently

$$p(x) = \frac{\psi(x)}{1 + \psi(x)} = \frac{\exp(\beta_0 + \beta_1 x)}{1 + \exp(\beta_0 + \beta_1 x)}. \qquad (7.2)$$

The **logistic response function** is an S-shaped curve as shown in Figure 7.2. It has a positive slope if $\beta_1 > 0$ and a negative slope if $\beta_1 < 0$. Just as in linear regression, β_1 has the interpretation of being the change in the log-odds of success for a unit change in x as can be seen from the following:

$$\ln \left[\frac{\psi(x + 1)}{\psi(x)} \right] = \ln \psi(x + 1) - \ln \psi(x) = \beta_0 + \beta_1(x + 1) - (\beta_0 + \beta_1 x) = \beta_1.$$

Thus

$$\frac{\psi(x + 1)}{\psi(x)} = \exp(\beta_1).$$

This is called the **odds ratio**, being the ratio of odds of success for $x + 1$ versus that for x. As an example, suppose that the probability of success for given x is $p(x) = 0.1$ or the odds of success are $\psi(x) = 0.1/0.9 = 0.111$. Furthermore, suppose that $\beta_1 = 0.3$. Then if x is increased by one unit, the odds of success will

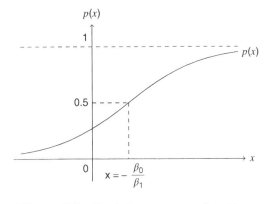

Figure 7.2. *Logistic response function.*

increase by $\exp(0.3) = 1.350$, i.e. the odds will increase to $(1.350)(0.111) = 0.150$. So the probability of success will increase to $0.150/1.150 = 0.130$.

A special case of interest is when the predictor variable is binary, for example, in a clinical trial suppose the control group is coded as $x = 0$ and the treatment group is coded as $x = 1$. Then $\exp(\beta_1)$ gives the odds ratio of success for the treatment compared to the control. If $\beta_1 = 0$ then the odds ratio is 1, which means that the odds of success are the same for the control and the treatment, so the treatment has no effect. On the other hand, if $\beta_1 > 0$ then the treatment is more effective than the control and if $\beta_1 < 0$ then the treatment is less effective than the control.

Logistic transform is just one of many possible functions that can be used to model $p(x)$. Essentially, any cumulative distribution function (c.d.f.) can be used for $p(x)$ since it is a nondecreasing function taking values between 0 and 1 as its argument goes from $-\infty$ to $+\infty$ (logistic function is the c.d.f. of the logistic distribution; see the Technical Notes section). Two examples are as follows:

- *Probit function*: $p(x) = \Phi(\beta_0 + \beta_1 x)$ where $\Phi(x)$ is the standard normal c.d.f. This function is similar in shape to the logistic function shown in Figure 7.2.

- *Complementary log–log function*: $p(x) = 1 - \exp(-\exp(\beta_0 + \beta_1 x))$ or $\ln[-\ln(1 - p(x))] = \beta_0 + \beta_1 x$, which is the c.d.f. of the extreme value distribution. This function is used to model very low probabilities associated with rare events.

These functions are analytically not as tractable as the logistic function and hence are not as commonly used.

7.1.2 Parameter estimation

Suppose that we have n independent observations $y_1, \dots, y_n$ corresponding to the predictor values $x_1, \dots, x_n$, where y_i is the response outcome (0 or 1) for the ith observation. We want to estimate the parameters β_0 and β_1 of the model (7.1) from these data. The least squares (LS) method cannot be used in this case since the observed value of the logistic transform $\ln[y_i/(1 - y_i)]$ is either $-\infty$ if $y_i = 0$ or $+\infty$ if $y_i = 1$ (unless the data are grouped in which case each y_i is a proportion of successes; see Example 7.3). Therefore, we will use the **maximum likelihood estimation** (MLE) method, which is a more general method with certain optimality properties. See Appendix B for a primer on the MLE method.

In the present case, the random data are the y_i, which have Bernoulli distributions with success probabilities

$$p_i = p(x_i) = \frac{\exp(\beta_0 + \beta_1 x_i)}{[1 + \exp(\beta_0 + \beta_1 x_i)]}.$$

This Bernoulli distribution can be written as

$$f(y_i|\beta_0, \beta_1) = (p_i)^{y_i}(1 - p_i)^{1-y_i} = \begin{cases} p_i & \text{if } y_i = 1 \\ 1 - p_i & \text{if } y_i = 0. \end{cases}$$

Hence the likelihood function equals

$$L = L(\beta_0, \beta_1) = \prod_{i=1}^{n}[(p_i)^{y_i}(1 - p_i)^{1-y_i}] = \prod_{i=1}^{n}\left(\frac{p_i}{1 - p_i}\right)^{y_i} \times \prod_{i=1}^{n}(1 - p_i)$$

and the log-likelihood function equals

$$\ln L = \sum_{i=1}^{n} y_i \ln\left(\frac{p_i}{1 - p_i}\right) + \sum_{i=1}^{n} \ln(1 - p_i)$$

$$= \sum_{i=1}^{n} y_i(\beta_0 + \beta_1 x_i) - \sum_{i=1}^{n} \ln[1 + \exp(\beta_0 + \beta_1 x_i)]. \qquad (7.3)$$

The MLEs $\widehat{\beta}_0$ and $\widehat{\beta}_1$ maximize $\ln L$ w.r.t. β_0 and β_1. These maximizing values can be found by setting the partial derivatives of $\ln L$ w.r.t. β_0 and β_1 equal to zero and solving the resulting equations. Now,

$$\frac{\partial \ln L}{\partial \beta_0} = \sum_{i=1}^{n} y_i - \sum_{i=1}^{n} \frac{\exp(\beta_0 + \beta_1 x_i)}{[1 + \exp(\beta_0 + \beta_1 x_i)]}$$

$$= \sum_{i=1}^{n} y_i - \sum_{i=1}^{n} p_i$$

and

$$\frac{\partial \ln L}{\partial \beta_1} = \sum_{i=1}^{n} x_i y_i - \sum_{i=1}^{n} \frac{x_i \exp(\beta_0 + \beta_1 x_i)}{[1 + \exp(\beta_0 + \beta_1 x_i)]}$$

$$= \sum_{i=1}^{n} x_i y_i - \sum_{i=1}^{n} x_i p_i.$$

Since $p_i = E(y_i)$, the equations obtained by setting the above partial derivatives equal to zero can be written as

$$E\sum_{i=1}^{n} y_i = \sum_{i=1}^{n} y_i \quad \text{and} \quad E\sum_{i=1}^{n} x_i y_i = \sum_{i=1}^{n} x_i y_i. \qquad (7.4)$$

Thus, they equate certain expected quantities to the corresponding observed quantities. As we shall see in Chapter 9, this is a common feature of all generalized linear models. We will also see there that an iteratively reweighted least squares algorithm can be used to solve these nonlinear simultaneous equations.

Example 7.3 *(Art Museum Visits: Simple Logistic Regression)* The grouped data for eight increasing levels of Education, coded 1–8, are shown in Table 7.2.

Table 7.2. Art museum visit data

Education	Visit Yes	Visit No	Total
1	7	24	31
2	24	92	116
3	92	408	500
4	53	196	249
5	271	439	710
6	172	277	449
7	107	96	203
8	199	150	349
Total	925	1682	2607

Since the data are grouped, we can calculate the sample fraction f_i of "yes" responses at each Education level $x_i = i$ and the corresponding sample logistic transforms as shown in Table 7.3. Figure 7.3 shows the plot of the logistic transform versus Education level. The plot appears roughly linear with a characteristic "hockey stick" shape. The explanation for this shape is that the Education level must exceed some threshold (e.g. ≥ 3) before its positive effect on the probability of visiting an art museum becomes apparent. The equation of the LS fitted straight line is $\hat{y} = -1.9067 + 0.2587x$, where $\hat{y}$ is the predicted log-odds of a "yes" response for the Education level x.

This preliminary analysis justifies fitting a simple logistic regression model. The following R output shows that the fitted logistic regression model is $\hat{y} = -2.3083 + 0.3281x$, which is qualitatively similar to the LS fitted equation. The odds of visiting an arts museum go up by a factor $\exp(0.3281) = 1.388$ if the Education level goes up one notch.

Table 7.3. Sample proportions and logistic transforms for the art museum visit data

Education	Proportion (f_i)	$\ln[f_i/(1-f_i)]$	$\ln[\hat{p}_i/(1-\hat{p}_i)]$	$\hat{p}_i$
1	0.2258	-1.2321	-1.9802	0.1213
2	0.2069	-1.3437	-1.6521	0.1608
3	0.1840	-1.4895	-1.3240	0.2102
4	0.2129	-1.3078	-0.9959	0.2697
5	0.3817	-0.4824	-0.6678	0.3390
6	0.3831	-0.4765	-0.3397	0.4159
7	0.5271	0.1085	-0.0116	0.4971
8	0.5702	0.2827	0.3165	0.5985

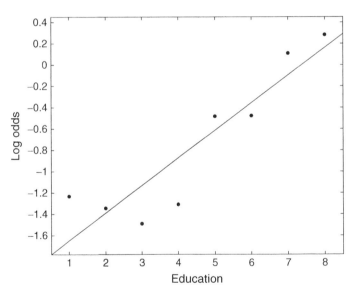

Figure 7.3. *Plot of* $\ln\left(\frac{f_i}{1-f_i}\right)$ *versus Education.*

```
> fit = glm(Visit ~ Education, family=binomial, data = art)

Coefficients:
            Estimate Std. Error z value Pr(>|z|)
(Intercept)  -2.3083     0.1407  -16.41   <2e-16 ***
Education     0.3281     0.0251   13.07   <2e-16 ***
---
```

```
Signif. codes:   0 '***' 0.001 '**' 0.01 '*' 0.05 '.' 0.1 ' ' 1

(Dispersion parameter for binomial family taken to be 1)

    Null deviance: 3391.1  on 2606  degrees of freedom
Residual deviance: 3205.4  on 2605  degrees of freedom
  (19 observations deleted due to missingness)
AIC: 3209.4
```

The deviances in the above output are defined in Section 7.3.1. The estimated probabilities:

$$\widehat{p}_i = \frac{\exp(\widehat{\beta}_0 + \widehat{\beta}_1 x_i)}{1 + \exp(\widehat{\beta}_0 + \widehat{\beta}_1 x_i)} \quad (i = 1 \ldots, n) \tag{7.5}$$

for this model are given in Table 7.3. The following R code and the resulting output show the calculation of these probabilities as well as the classification of observations (using the classification rule: classify the ith response as "Yes" if $\widehat{p}_i > 0.3$, where the cutoff probability 0.3 is chosen arbitrarily for the sake of illustration).

```
> vector1=data.frame(Education=c(1,2,3,4,5,6,7,8))
> probs1=predict(fit1,newdata=vector1,type="response")
> probs1
          1         2         3         4         5         6
0.1213012 0.1608361 0.2101743 0.2697819 0.3390380 0.4159458
          7         8
0.4971766 0.5785568
> pred1=rep("No", 8)
> pred1[probs1>0.3]="Yes"
> pred1
[1] "No"  "No"  "No"  "No"  "Yes" "Yes" "Yes" "Yes"
```

□

7.1.3   Inferences on parameters

To test hypotheses and make confidence intervals on β_0 and β_1, we use the following result: For large n, $(\widehat{\beta}_0, \widehat{\beta}_1)$ is approximately bivariate normal with mean vector (β_0, β_1) and **asymptotic covariance matrix V**, which is the inverse of the **information matrix**:

$$\mathcal{I} = -E \begin{bmatrix} \frac{\partial^2 \ln L}{\partial \beta_0^2} & \frac{\partial^2 \ln L}{\partial \beta_0 \beta_1} \\ \frac{\partial^2 \ln L}{\partial \beta_0 \beta_1} & \frac{\partial^2 \ln L}{\partial \beta_1^2} \end{bmatrix}$$

$$= \begin{bmatrix} \sum_{i=1}^{n} p_i(1 - p_i) & \sum_{i=1}^{n} x_i p_i(1 - p_i) \\ \sum_{i=1}^{n} x_i p_i(1 - p_i) & \sum_{i=1}^{n} x_i^2 p_i(1 - p_i) \end{bmatrix}. \tag{7.6}$$

The derivation of the entries of $\mathcal{I}$ is given in Example B.3. The diagonal entries v_{00} and v_{11} of $\boldsymbol{V} = \mathcal{I}^{-1}$ are the asymptotic variances of $\widehat{\beta}_0$ and $\widehat{\beta}_1$, respectively. In practice, we need to replace the p_i in the information matrix by their sample estimates given by (7.5) and then invert the matrix to obtain the estimated asymptotic covariance matrix $\widehat{\boldsymbol{V}}$ with diagonal entries $\widehat{v}_{00}$ and $\widehat{v}_{11}$. To test $\beta_1 = 0$, we can use $z = \widehat{\beta}_1 / \sqrt{\widehat{v}_{11}}$ as a standard normal test statistic. A large sample $100(1 - \alpha)\%$ CI for β_1 is given by

$$\widehat{\beta}_1 - z_{\alpha/2}\sqrt{\widehat{v}_{11}} \leq \beta_1 \leq \widehat{\beta}_1 + z_{\alpha/2}\sqrt{\widehat{v}_{11}}.$$

An analogous formula is used for the CI on β_0.

Example 7.4 (Art Museum Visits: Simple Logistic Regression Inferences) In this example, we will test a hypothesis and compute a CI on β_1 and on the associated odds ratio. As seen from the R output from Example 7.3, the asymptotic standard error of $\widehat{\beta}_1$ is $\text{SE}(\widehat{\beta}_1) = 0.0251$. We will verify this by hand calculation. Using estimated probabilities $\widehat{p}_i$ given in Table 7.3, we calculate the following entries of the estimated information matrix (7.6):

$$\sum_{i=1}^{8} n_i \widehat{p}_i (1 - \widehat{p}_i) = 555, \sum_{i=1}^{8} n_i x_i \widehat{p}_i (1 - \widehat{p}_i) = 2966, \sum_{i=1}^{8} n_i x_i^2 \widehat{p}_i (1 - \widehat{p}_i) = 17\,435,$$

where $x_i = 1, \ldots, 8$. So the estimated asymptotic covariance matrix equals

$$\begin{bmatrix} 555 & 2966 \\ 2966 & 17435 \end{bmatrix}^{-1} = \begin{bmatrix} 0.01983 & -0.00337 \\ -0.00337 & 0.00063 \end{bmatrix}.$$

Then, $\text{SE}(\widehat{\beta}_1) = \sqrt{0.00063} = 0.0251$. So a large sample 95% CI on β_1 is

$$0.3281 \pm 1.96 \times 0.0251 = [0.2789, 0.3773].$$

Since this interval excludes 0, the null hypothesis $H_0 : \beta_1 = 0$ can be rejected at $\alpha = .05$. The corresponding z-statistic equals $z = 0.3281/0.0251 = 13.07$, which is highly significant. A 95% CI on the odds ratio equals $[\exp(0.2789), \exp(0.3773)] = [1.3217, 1.4583]$, which shows that the odds ratio is significantly greater than 1. $\qquad\square$

7.2　Multiple logistic regression

7.2.1　Model and inference

Extension of simple logistic regression to multiple logistic regression is straightforward. Assume that we have p predictors, $x_1, \ldots, x_p$. Let $\boldsymbol{x} = (1, x_1, \ldots, x_p)'$

denote the predictor vector and let $\boldsymbol{\beta} = (\beta_0, \beta_1, \ldots, \beta_p)'$ denote the regression coefficient vector. Further let $p(\boldsymbol{x})$ denote the probability of success and $\psi(\boldsymbol{x}) = p(\boldsymbol{x})/[1 - p(\boldsymbol{x})]$ denote the odds of success. A straightforward extension of the simple logistic regression model (7.1) gives the multiple logistic regression model:

$$\ln \psi(\boldsymbol{x}) = \ln \left[\frac{p(\boldsymbol{x})}{1 - p(\boldsymbol{x})} \right] = \beta_0 + \beta_1 x_1 + \cdots + \beta_p x_p = \boldsymbol{x}'\boldsymbol{\beta} \qquad (7.7)$$

or equivalently

$$p(\boldsymbol{x}) = \frac{\exp(\boldsymbol{x}'\boldsymbol{\beta})}{1 + \exp(\boldsymbol{x}'\boldsymbol{\beta})}. \qquad (7.8)$$

The regression coefficients β_j have the same interpretation as before, namely $\exp(\beta_j)$ is the odds ratio of success when x_j is increased by one unit, keeping all other x's fixed.

The MLEs of the β_j's can be computed by setting the partial derivatives of the log-likelihood function equal to zero and solving the resulting simultaneous equations. These equations have a relatively simple form, which is a generalization of Equation (7.4). Let $\boldsymbol{X}$ be the $n \times (p + 1)$ model matrix (including the first column of all 1's) as defined in Section 3.1.2 and let $\boldsymbol{p} = (p_1, \ldots, p_n)'$ be the vector of the success probabilities $p_i = E(y_i) = \exp(\boldsymbol{x}_i'\boldsymbol{\beta})/[1 + \exp(\boldsymbol{x}_i'\boldsymbol{\beta})]$. Then the MLE equations are given by

$$\boldsymbol{X}'\boldsymbol{p} = \boldsymbol{X}'\boldsymbol{y} \quad \text{or equivalently} \quad \sum_{i=1}^{n} p_i \boldsymbol{x}_i = \sum_{i=1}^{n} y_i \boldsymbol{x}_i,$$

where $\boldsymbol{x}_i'$ is the ith row of $\boldsymbol{X}$. This is a system of $p + 1$ simultaneous nonlinear equations in $p + 1$ unknowns $\beta_0, \ldots, \beta_p$ (since the p_i are nonlinear functions of $\beta_0, \ldots, \beta_p$).

Similarly, the Hessian matrix of mixed second partial derivatives of the log-likelihood function evaluated at the MLEs of the β_j's gives the observed information matrix which upon inversion yields the estimated asymptotic covariance matrix of the $\widehat{\beta}_j$'s. The following example illustrates these calculations.

Example 7.5 *(Art Museum Visits: Multiple Logistic Regression)*
Before fitting a multiple logistic regression model, it is useful to perform some exploratory analyses to assess the relationships between the proportions of people visiting art museums and the various predictors. Frequency tables for the three binary predictors, Gender, Children, and Married, are shown in Table 7.4. From these tables, we see that females, people with no children and unmarried people are more likely to visit an art museum than their corresponding counterparts.

The frequency table for the County variable is shown in Table 7.5. We see that there is a significant County effect with Chicago Cook county having the largest proportion of art museum visitors compared to collar counties.

Table 7.4. Frequency tables for Gender, Children, and Married variables for art museum visit data

| Predictor | Category | Visit? | | Total | Yes Proportion |
		Yes	No		
Gender	Female	472	749	1221	0.387
	Male	455	938	1393	0.327
Children	No	668	1075	1743	0.383
	Yes	262	621	883	0.297
Married	No	434	617	1051	0.413
	Yes	496	1079	1575	0.315

Table 7.5. Frequency table for County variable for art museum visit data

| County | Visit? | | Total | Yes Proportion |
	Yes	No		
Chicago Cook	106	84	190	0.558
Suburban Cook	434	747	1181	0.368
Lake	121	246	367	0.330
DuPage	90	111	201	0.448
Other	179	508	687	0.261

Finally, Figure 7.4 shows the plots of log-odds of visiting art museums versus Income and Age. (Plot against Education is shown in Figure 7.3.) We see that the relationship with Income is roughly linear but the relationship with Age, if any, is more complex. When included in the model, Age turns out to be a nonsignificant predictor, so we omit it in the following.

The R commands for fitting the multiple logistic regression model are as follows.

```
> fit = glm(Visit ~ Education+Income+Gender+Children+Married,
> factor(County), family=binomial, data = art)
> summary(fit)
```

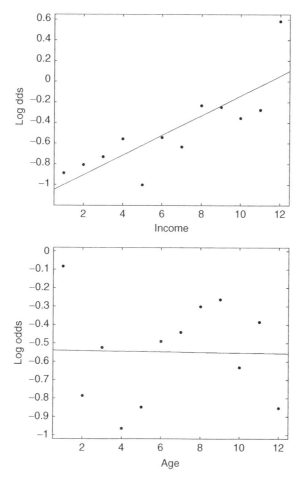

Figure 7.4. *Plots of logistic transforms versus Income and Age.*

The R output is shown below. We see that all the predictors are highly significant. Their regression coefficients represent the change in the log-odds of visiting an art museum if the value of the predictor is increased by one unit for a numerical variable and with respect to the reference category for a categorical variable, keeping all other predictors fixed. The effects of Education and Income are positive, as suggested by the plots, indicating that the odds of visiting an art museum increase as these variables increase in magnitude. The effects of categorical predictors are in agreement with what we saw in the frequency tables given above.

```
Coefficients:
            Estimate Std. Error z value Pr(>|z|)
(Intercept) -1.26621    0.21079  -6.007 1.89e-09 ***
Education    0.30223    0.02768  10.918  < 2e-16 ***
Income       0.07168    0.01627   4.406 1.05e-05 ***
```

```
Gender            -0.42181    0.08958  -4.709 2.49e-06 ***
Children          -0.34137    0.09867  -3.460 0.000541 ***
Married           -0.49891    0.09990  -4.994 5.91e-07 ***
factor(County)2 -0.59726      0.17009  -3.511 0.000446 ***
factor(County)3 -0.74408      0.19613  -3.794 0.000148 ***
factor(County)4 -0.43072      0.21822  -1.974 0.048409 *
factor(County)5 -0.96831      0.18330  -5.283 1.27e-07 ***
---
Signif. codes:  0 '***' 0.001 '**' 0.01 '*' 0.05 '.' 0.1 ' ' 1

(Dispersion parameter for binomial family taken to be 1)

    Null deviance: 3379.6  on 2597  degrees of freedom
Residual deviance: 3068.2  on 2588  degrees of freedom
  (28 observations deleted due to missingness)
AIC: 3088.2
```

To illustrate the use of the coefficients in the above regression output, let us compare the probability of visiting an art museum for a male versus a female, who have all other predictor values the same. Suppose that this probability for a female (denoted by p_f) is 0.30. To find the probability for a male (denoted by p_m), note that the difference in the log-odds for a male versus a female is the coefficient of Gender which is -0.422. Hence

$$\ln\left(\frac{p_m}{1-p_m}\right) - \ln\left(\frac{p_f}{1-p_f}\right) = -0.422,$$

Putting $p_f = 0.30$, we get the log-odds for a female to be $\ln(0.3/0.7) = -0.847$. Hence, the log-odds for a male are $-0.847 - 0.422 = -1.269$. Hence, the odds for a male are $e^{-1.269} = 0.281$ from which we obtain $p_m = 0.219$.   □

7.3 Likelihood ratio (LR) test

The likelihood ratio (LR) test extends the extra sum of squares test discussed in Section 3.3.4 for multiple regression to more general parametric models. Consider a multiple logistic regression model with p predictors, which we refer to as the **full model** (FM), and suppose we want to test if a **partial model** (PM) with a subset of $q < p$ predictors provides an equally good fit. Labeling the q predictors in the partial model as $x_1, \ldots, x_q$, we want to test the null hypothesis $H_0 : \beta_{q+1} = \cdots = \beta_p = 0$, i.e. the extra predictors, $x_{q+1}, \ldots, x_p$, can be dropped from the model without a significant loss in goodness of fit.

Denote the maximums of the likelihood functions for the two models by $L_{\max}(\text{PM})$ and $L_{\max}(\text{FM})$, where the maximums are obtained by substituting the

MLEs of the β_j's under the respective models in the likelihood function. Then, the **LR test statistic** for testing H_0 is given by

$$G^2 = -2\ln\left[\frac{L_{\max}(\text{PM})}{L_{\max}(\text{FM})}\right] = 2[\ln L_{\max}(\text{FM}) - \ln L_{\max}(\text{PM})]. \qquad (7.9)$$

Under H_0, the LR statistic can be shown to be asymptotically (as $n \to \infty$) chi-square with $p - q$ d.f. So we can reject H_0 at level α if $G^2 > \chi^2_{p-q,\alpha}$ and conclude that the full model provides a significantly better fit than the partial model.

Usually, the LR test for comparing PM with FM is conducted by comparing their deviances, defined in the next section.

7.3.1 Deviance

The **deviance** (referred to as **residual deviance** in the R output) of a given model plays the same role in logistic regression (or more generally in generalized linear models; see Chapter 9) as does the SSE in multiple regression. Similar to SSE, the larger the deviance, the poorer the fit of the given model. It is defined as the LR test statistic for comparing the given model (M) with the so-called **saturated model** (SM), which has as many parameters as the number of distinct observations and thus provides an exact fit to the data. In the LR test parlance, the saturated model with n parameters is the full model and the given model with $p + 1 < n$ parameters (p predictor variables) is the partial model. The deviance of the model M equals

$$D^2 = -2[\ln L_{\max}(\text{M}) - \ln L_{\max}(\text{SM})]. \qquad (7.10)$$

We can write

$$\ln L_{\max}(\text{M}) = \sum_{i=1}^{n}[y_i \ln \widehat{p}_i + (1 - y_i)\ln(1 - \widehat{p}_i)],$$

where

$$\widehat{p}_i = \frac{\exp(\boldsymbol{x}_i'\widehat{\boldsymbol{\beta}})}{1 + \exp(\boldsymbol{x}_i'\widehat{\boldsymbol{\beta}})}$$

and $\widehat{\boldsymbol{\beta}}$ is the MLE of $\boldsymbol{\beta}$. Under the saturated model SM, we set $\widehat{p}_i = y_i = 0$ or 1, thus yielding

$$\ln L_{\max}(\text{SM}) = \sum_{i=1}^{n}[y_i \ln y_i + (1 - y_i)\ln(1 - y_i)].$$

Using l'Hospital's rule, it is easy to show that $\ln L_{\max}(\text{SM}) = 0$. This is also explained by the fact that since SM fits the data exactly, $L_{\max}(\text{SM}) = 1$ and hence $\ln L_{\max}(\text{SM}) = 0$. However, this is not true in general as we shall see for some generalized linear models in Chapter 9. Even for binary logistic regression models, if the data are grouped then $L_{\max}(\text{SM}) < 1$ and so $\ln L_{\max}(\text{SM}) < 0$ as Exercise 7.3 asks you to show.

The deviances are mainly useful for carrying out the LR test to compare two models, one (the partial model) being nested under the other model (the full model). Next we consider a test of a given model (the full model) against a null model (the partial model), which has no predictor variables – only the intercept term. The **null deviance** (denoted by D_0^2) is the deviance of the **null model**. Null deviance is analogous to SST in multiple regression. The null model assumes a common success probability for all cases given by

$$p_0 = \frac{\exp(\beta_0)}{1 + \exp(\beta_0)}.$$

It is easy to show that the MLE of p_0 is $\widehat{p}_0 = s/n$, the overall proportion of successes. Therefore,

$$L_{\max}(\text{NM}) = (\widehat{p}_0)^s (1 - \widehat{p}_0)^{n-s} = \left(\frac{s}{n}\right)^s \left(\frac{n-s}{n}\right)^{n-s}.$$

Since $\ln L_{\max}(\text{SM}) = 0$, the **null deviance** equals

$$D_0^2 = -2[s \ln s + (n-s)\ln(n-s) - n \ln n].$$

We can compare the deviance D^2 of the given model with the null deviance D_0^2 to test $H_0 : \beta_1 = \cdots = \beta_p = 0$. This is referred to as the **overall significance test** of the model and is analogous to the ANOVA F-test for multiple regression. From the LR test, it follows that H_0 can be rejected at level α if $D_0^2 - D^2 > \chi_{p,\alpha}^2$. Applying this test to the multiple logistic regression model in Example 7.5, we get $D_0^2 - D^2 = 3379.6 - 3068.2 = 311.4$ on 9 d.f. (corresponding to the nine coefficients under test), which is highly significant. So H_0 can be rejected and we conclude that at least one of the $\beta_j \neq 0$.

Example 7.6 (Art Museum Visits: Comparison of Simple Versus Multiple Logistic Regression Models) Let us compare the simple logistic regression model that uses only Education as the predictor with the multiple logistic regression model that uses six predictors (including Education). A subset of the data consisting of 2598 observations is used to fit the multiple logistic regression model in Example 7.5 as compared to the larger data set consisting of 2607 observations used to fit the simple logistic regression model in Example 7.3 because more observations are missing on additional predictors used in the multiple logistic model. In order for their deviances and d.f. to be comparable, we refitted the simple logistic regression model with Education as the only predictor on the subset data set resulting in the following output.

```
Coefficients:
            Estimate Std. Error z value Pr(>|z|)
(Intercept) -2.30444    0.14088  -16.36   <2e-16 ***
Education    0.32765    0.02515   13.03   <2e-16 ***
---
Signif. codes:  0 '***' 0.001 '**' 0.01 '*' 0.05 '.' 0.1 ' ' 1

(Dispersion parameter for binomial family taken to be 1)

    Null deviance: 3379.6  on 2597  degrees of freedom
Residual deviance: 3195.3  on 2596  degrees of freedom
AIC: 3199.3
```

For this model, $D^2 = 3195.3$ with 2596 d.f., whereas for the multiple logistic regression model, $D^2 = 3068.2$ with 2588 d.f. as seen before. So the test statistic is $3195.3 - 3068.2 = 127.1$ with $2596 - 2588 = 8$ d.f., which is highly significant. So we reject the null hypothesis and conclude that the multiple logistic regression model provides a significantly better fit than the simple logistic regression model.   □

7.3.2 Akaike information criterion (AIC)

The Akaike information criterion (AIC) was introduced in Section 6.1.1 for multiple regression. More generally, the AIC for any model M is given by

$$\text{AIC} = -2 \ln L_{\max}(\text{M}) + 2(p+1), \tag{7.11}$$

where one can regard $2(p+1)$ as penalty for the number of predictors in the model. Since in the case of binary logistic regression, $\ln L_{\max}(\text{SM}) = 0$, we get

$$\text{AIC} = D^2 + 2(p+1).$$

Thus, AIC is the sum of D^2, which is measure of the goodness of fit and $2(p+1)$, which is a measure of the complexity of the model. As a check of this formula, note from the R output in Example 7.5 that $D^2 = 3068.2$ and AIC $= 3068.2 + 2(9+1) = 3088.2$.

7.3.3 Model selection and diagnostics

Best subset and stepwise as well as ridge and lasso regression methods discussed in Chapter 6 can be applied to select the best logistic regression model.

Minimizing AIC is a well-defined model selection criterion. One cannot simply aim at minimizing the deviance as it would lead to the full model just as in multiple regression minimizing SSE (or equivalently maximizing R^2) leads to the full model. Therefore, a predictor variable should be added to a model only if it reduces the deviance more than the increase in the penalty term $2(p+1)$. Furthermore, the decrease in deviance should be statistically significant at a designated level α, i.e. $\Delta D^2 > \chi^2_{1,\alpha}$ (if the predictor variable has 1 d.f.). The following example illustrates the use of `stepAIC` and `bestglm` functions in R for this purpose.

Example 7.7 *(Art Museum Visits: Stepwise Logistic Regression)* In Example 7.5, we fitted a multiple logistic regression model to the art museum visits data. In this example, we apply stepwise logistic regression to the same data. Forward stepwise and backward stepwise both result in the same full model. For the forward stepwise method, the R commands are given below.

```
> fit = glm(Visit ~ 1, binomial, art2)
> stepAIC(fit, scope=~ Income+ Gender + Children + Married
  + Education + County)
```

The R output below gives the best model in terms of the smallest AIC after adding each new variable in a forward stepwise manner. We see that the decrease in deviance at each step exceeds $\chi^2_{1,.05} = 3.843$ (for the addition of the County variable it exceeds $\chi^2_{4,.05} = 9.488$), so the addition of each variable is justified based on the deviance criterion as well.

```
> art1 = read.csv("c:/data/art.csv")
> art1$County = factor(art1$County)
> art2 = art1[complete.cases(art1), -2]
# drop cases with missing values and the Age column
>
> # forward stepwise with stepAIC
> fit = glm(Visit ~ 1, binomial, art2)
> step(fit, scope=~ Income + Gender + Children + Married + Education
> + County,direction = "both")
Start:  AIC=3365.96
Visit ~ 1
```

	Df	Deviance	AIC
+ Education	1	3184.1	3188.1
+ County	4	3291.4	3301.4
+ Income	1	3320.4	3324.4
+ Married	1	3336.2	3340.2
+ Children	1	3342.6	3346.6
+ Gender	1	3354.2	3358.2
<none>		3364.0	3366.0

```
Step:  AIC=3188.09
Visit ~ Education

            Df Deviance    AIC
+ County     4   3132.9 3144.9
+ Married    1   3141.3 3147.3
+ Children   1   3158.7 3164.7
+ Gender     1   3161.3 3167.3
+ Income     1   3181.3 3187.3
<none>           3184.1 3188.1
- Education  1
3364.0 3366.0

Step:  AIC=3144.86
Visit ~ Education + County

            Df Deviance    AIC
+ Married    1   3102.8 3116.8
+ Gender     1   3111.0 3125.0
+ Children   1   3113.2 3127.2
+ Income     1   3130.1 3144.1
<none>           3132.9 3144.9
- County     4   3184.1 3188.1
- Education  1   3291.4 3301.4

Step:  AIC=3116.81
Visit ~ Education + County + Married

            Df Deviance    AIC
+ Gender     1   3086.2 3102.2
+ Income     1   3088.3 3104.3
+ Children   1   3094.0 3110.0
<none>           3102.8 3116.8
- Married    1   3132.9 3144.9
- County     4   3141.3 3147.3
- Education  1   3274.3 3286.3

Step:  AIC=3102.18
Visit ~ Education + County + Married + Gender

            Df Deviance    AIC
+ Income     1   3068.7 3086.7
+ Children   1   3075.8 3093.8
<none>           3086.2 3102.2
- Gender     1   3102.8 3116.8
- Married    1   3111.0 3125.0
- County     4   3125.0 3133.0
- Education  1   3267.3 3281.3
```

```
Step:  AIC=3086.69
Visit ~ Education + County + Married + Gender + Income

            Df Deviance    AIC
+ Children   1   3056.2 3076.2
<none>           3068.7 3086.7
- Income     1   3086.2 3102.2
- Gender     1   3088.3 3104.3
- County     4   3104.1 3114.1
- Married    1   3106.0 3122.0
- Education  1   3193.2 3209.2

Step:  AIC=3076.22
Visit ~ Education + County + Married + Gender + Income + Children

            Df Deviance    AIC
<none>           3056.2 3076.2
- Children   1   3068.7 3086.7
- Income     1   3075.8 3093.8
- Gender     1   3077.9 3095.9
- Married    1   3081.2 3099.2
- County     4   3089.7 3101.7
- Education  1   3179.0 3197.0

Call:  glm(formula = Visit ~ Education + County + Married + Gender +
    Income + Children, family = binomial, data = art2)

Coefficients:
(Intercept)    Education     County2     County3     County4     County5
   -1.22600      0.29913    -0.62642    -0.76893    -0.45761    -0.99383
     Married       Gender      Income    Children
   -0.49895     -0.41604     0.07212    -0.34705

Degrees of Freedom: 2586 Total (i.e. Null);   2577 Residual
Null Deviance:        3364
Residual Deviance: 3056            AIC: 3076
```

□

Next we consider some diagnostics for logistic regression. Two types of residuals are used for this purpose. **Deviance residuals** are defined as

$$d_i = \text{sign}(y_i - \widehat{p}_i)\sqrt{-2[y_i \ln \widehat{p}_i + (1 - y_i) \ln(1 - \widehat{p}_i)]}.$$

Note that the deviance $D^2 = \sum_{i=1}^{n} d_i^2$. These residuals can be used to check for outliers as in the multiple regression case, e.g. if $|d_i| > 3$ then declare the observation to be an outlier.

Influential observations are generally detected by calculating the changes in D^2 by deleting individual observations defined as

$$\Delta D_i^2 = D^2 - D_{(i)}^2, \quad i = 1, 2, \ldots, n,$$

where $D_{(i)}^2$ is the D^2 statistic calculated by deleting the ith observation. Those observations with large values of ΔD_i^2 are identified as influential. There are no formal statistical tests for deciding whether ΔD_i^2 is significantly large. One can make a sequence plot of these changes against the observation index i and identify any spikes in the plot as influential observations. One can use the leverage measures used to detect influential observations and variance inflation factors (VIFs) used to detect multicollinearity in the context of logistic regression as they depend solely on the $\boldsymbol{X}$ matrix.

7.4 Binary classification using logistic regression

7.4.1 Measures of correct classification

Let $\boldsymbol{x} = (1, x_1, \ldots, x_p)'$ denote the vector of predictors for a future observation and let $\widehat{\boldsymbol{\beta}} = (\widehat{\beta}_0, \widehat{\beta}_1, \ldots, \widehat{\beta}_p)'$ denote the estimated parameter vector of the logistic regression model. Then we can calculate the estimated probability of "success" as

$$\widehat{p}(\boldsymbol{x}) = \frac{\exp(\boldsymbol{x}'\widehat{\boldsymbol{\beta}})}{1 + \exp(\boldsymbol{x}'\widehat{\boldsymbol{\beta}})}.$$

Logistic regression is sometimes referred to as a **soft classification** method since it does not classify observations into "success" and "failure" outcomes (which is called **hard classification**) but gives the probabilities of these two outcomes. To make such a hard classification, we need to specify a **cutoff probability** p^*. An observation $\boldsymbol{x}$ is classified as a "success" if $\widehat{p}(\boldsymbol{x}) > p^*$ and as a "failure" if $\widehat{p}(\boldsymbol{x}) \leq p^*$. For example, we may use $p^* = 0.5$ thinking that a better than $50 : 50$ chance is sufficient to predict a "success." However, this may not be the right thing to do for various reasons, one reason being the costs of misclassifying of a true "success" and a true "failure" may be unequal. For instance, the costs of misdiagnosing a patient may be very different depending on the severity of the disease and the cost and side-effects of the treatment. Another reason is that the prevalence of "successes" in the population may be much lower or higher than 50%, and this should be factored into the decision rule. For example, for a rare disease with less than 0.1% prevalence rate, p^* should be correspondingly lower. Therefore, it is worthwhile to study different choices of p^*.

The results of hard classification of n observations can be represented in the form of a 2×2 table, called the **confusion matrix**, shown below: (Here we use the terminology "Negative" and "Positive" instead of "Failure" and "Success".)

Predicted

		Negative	Positive	Row total
	Negative	n_{00}	n_{01}	n_0
Actual				
	Positive	n_{10}	n_{11}	n_1

The proportion of true negatives that are classified as positives, namely n_{01}/n_0, is called the **false positive rate** (FPR) and the proportion of true positives that are classified as negatives, namely n_{10}/n_1, is called the **false negative rate** (FNR). In hypothesis testing terminology, they correspond to the **type I error rate** and the **type II error rate**, respectively. Alternative but equivalent measures can be defined in terms of correct classifications as follows:

$$\text{Specificity} = 1 - \text{FPR} = \frac{n_{00}}{n_0}$$

and

$$\text{Sensitivity} = 1 - \text{FNR} = \frac{n_{11}}{n_1}.$$

These two measures are complementary to each other in that if one increases the other decreases. As p^* is increased, the FPR decreases and hence specificity increases. On the other hand, the FNR increases and hence sensitivity decreases. We can choose p^* to minimize the total number of misclassifications, $n_{01} + n_{10}$. If the costs of misclassifications are unequal, say c_0 is the cost of a false positive and c_1 is the cost of a false negative, then we can minimize the total cost of misclassification, $C = c_0 n_{01} + c_1 n_{10}$. Note that only the cost ratio c_0/c_1 matters, not the actual costs. For example, if the cost of a false positive is twice that of the cost of a false negative then it suffices to minimize $C = 2n_{01} + n_{10}$.

In machine learning and information retrieval, the following variants of the above measures are used:

$$\text{Precision} = P = \frac{n_{11}}{n_{01} + n_{11}}$$

and

$$\text{Recall} = R = \frac{n_{11}}{n_{10} + n_{11}} = \frac{n_{11}}{n_1}.$$

In the context of information retrieval, the true positives and negatives correspond to the relevant and irrelevant items (e.g. documents), respectively. The predicted positives and negatives correspond to the retrieved and non-retrieved items, respectively. Thus Precision is the proportion of the retrieved items that are relevant, while Recall is the proportion of the relevant items that are retrieved. The same kind of complementarity holds between precision and recall as between specificity and sensitivity. Therefore, both cannot be increased simultaneously. A common practice is to maximize the harmonic mean of the two, called the F_1-**score**:

$$F_1 = \left[\frac{1}{2}\left(\frac{1}{P} + \frac{1}{R}\right)\right]^{-1} = \frac{2PR}{P+R}.$$

Example 7.8 *(Art Museum Visits: Determination of Optimum Cutoff Probability)* Suppose that we want to determine the optimum cutoff probability p^* and the associated predictive power of the multiple logistic regression model fitted in Example 7.5. To get unbiased estimates of these quantities, we split the data into a training set and a test set. We did this by taking the training set to be all odd-numbered observations and the test set to be all even-numbered observations. Thus, of the total 2598 complete observations, exactly half, i.e. 1299 observations are in each of the two sets. Of the 1299 observations in the test set, 459 are positives and 840 are negatives.

We used the following R script to form the training and test sets and compute the confusion matrix using a cutoff probability $p^* = 0.5$ for illustration.

```
> art = read.csv("c:/data/art.csv")
> evenrow = seq(2,nrow(art),2)
> oddrow = seq(1,nrow(art),2)
> train = art[oddrow,]
> test = art[evenrow,]
> #Fit a multiple logistic regression model
> fit = glm(Visit ~ Education+Income+Gender+Children+Married,
> factor(County),family=binomial, data = train)
> summary(fit)
> testpredict=predict(fit,newdata=test)
> tab=table(test$Visit, testpredict>.50)
> tab

    FALSE TRUE
  0   802   38
  1   391   68
```

From the confusion matrix, we can calculate the **correct classification rate** (CCR) by summing the diagonal entries, n_{00} and n_{11};, and dividing by the total sample size. From the following, we see that CCR $= 67\%$.

```
> CCR=sum(diag(tab))/sum(tab)
> CCR
[1] 0.669746
```

The CCRs for $p^* = 0$ and $p^* = 1$ serve as useful limiting values. If $p^* = 0$, then all true positives are correctly classified but all true negatives are misclassified, so CCR $= 459/1299 = 0.3534$. Similarly, if $p^* = 1$, then all true negatives are correctly classified but all true positives are misclassified, so CCR $= 840/1299 = 0.6467$.

By a numerical search, we find the optimum $p^* = 0.44$ corresponding to the confusion matrix shown below which gives maximum CCR $= (797 + 75)/1299 = 0.6713$, which is only slightly greater than 0.6697 obtained using $p^* = 0.5$.

```
    FALSE TRUE
 0   797   43
 1   384   75
```

```
[1] 0.6712856
```

In this example, CCR is a fairly flat function of p^* in the search interval $[0.40, 0.60]$, so almost any value in this interval is equally good.

The values of the other measures defined above for $p^* = 0.44$ are as follows:

$$\text{Sensitivity} = \frac{75}{384 + 75} = 0.1634, \quad \text{Specificity} = \frac{797}{797 + 43} = 0.9488,$$

$$\text{Precision} = \frac{75}{75 + 43} = 0.6356, \quad \text{Recall} = \frac{75}{384 + 75} = 0.1634.$$

The F_1-score equals

$$F_1 = \frac{2(0.6356)(0.1634)}{0.6356 + 0.1634} = 0.2600.$$

We see that this model is good for predicting non-visitors (its specificity is high) but not so good for predicting visitors (its sensitivity is low). Recalling that in the entire sample, there were 35.5% visitors (see Example 7.1), sensitivity is less than half the true proportion of visitors. If we think of visitors as relevant items and non-visitors as irrelevant items, then 63.56% of the retrieved items are relevant but only 16.34% of the relevant items are retrieved. The F_1-score is correspondingly low. $\square$

7.4.2   Receiver operating characteristic (ROC) curve

The receiver operating characteristic (ROC) curve is used to assess the performance of a binary classifier for the entire range $[0, 1]$ of the cutoff probabilities p^*. It is obtained by plotting Sensitivity versus $1-$ Specificity by varying p^*.

Denote by $\widehat{p}_i = \widehat{p}(\boldsymbol{x}_i)$ the estimated probability of a "positive" outcome (or "success") for the ith observation with the predictor vector $\boldsymbol{x}_i = (1, x_{i1}, \ldots, x_{ip})'$, $i = 1, \ldots, n$. For simplicity, assume that all $\widehat{p}_i$ are distinct. Suppose that we start with $p^* = 1$. Then all $\widehat{p}_i$'s are $< p^*$, so all observations are classified as "negatives" (or "failures"). Therefore Sensitivity $= 0$ and Specificity $= 1$ or $1-$ Specificity $= 0$. This results in the lower left hand corner point or the origin in

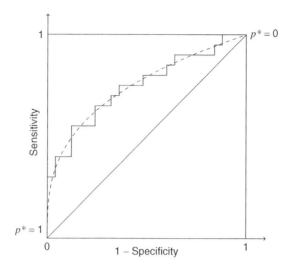

Figure 7.5. *Typical ROC curve.*

Figure 7.5. The reverse happens when $p^* = 0$ and we get the upper right hand corner point. As p^* is decreased from 1 to 0, every time it crosses one of the calculated $\widehat{p}_i$ values we get one more observation classified as a "positive." If it is a true "positive," then Sensitivity goes up by $1/n_1$; if it is a true "negative," then Sensitivity remains unchanged but Specificity goes down by $1/n_0$ or $1-$ Specificity goes up by $1/n_0$. Thus the ROC curve is a nondecreasing step function as shown in Figure 7.5 along with a smooth approximation to it.

The ideal binary classifier perfectly discriminates between the true "positives" and the true "negatives." This will happen when the estimated probabilities $\widehat{p}_i$ for all true negatives are strictly less than those for all true positives, so that there is a perfect separation. In that case it is easy to check that the ROC curve traverses the path along the left vertical edge and then along the top horizontal edge of the square in Figure 7.5 as p^* decreases from 1 to 0. On the other hand, if the $\widehat{p}_i$'s for the true negatives and the true positives are uniformly mixed, then the ROC curve traverses the path along the 45° line. A typical ROC curve falls between these two extremes.

The discriminating power of a binary classification model can be quantified by the **area under curve** (AUC) of the ROC. Note that for the perfectly separated data, AUC $= 1$ and for perfectly mixed data, the ROC curve is the 45° line with AUC $= 1/2$.

The AUC can be shown to be equal to the so-called **concordance index** γ, which is defined as follows. Let n_0 be the number of observations with $y_i = 0$ and n_1 be the number of observations with $y_i = 1$. The number of pairs of observations (i, j) with $y_i = 0, y_j = 1$ is $N = n_0 n_1$. Let $(\widehat{p}_i, \widehat{p}_j)$ denote the estimated success probabilities using a given model for the pair (i, j). If $\widehat{p}_i < \widehat{p}_j$ then the pair is said to be concordant, if $\widehat{p}_i > \widehat{p}_j$ then the pair is said to be discordant, and if $\widehat{p}_i = \widehat{p}_j$

then the pair is said to be tied. Let N_c, N_d, N_t be the numbers of concordant, discordant, and tied pairs with $N_c + N_d + N_t = N$. Then, the concordance index is defined as $\gamma = (N_c + 0.5N_t)/N$.

Example 7.9 (Art Museum Visits: AUC Calculation for Simple Logistic Regression) In general, the AUC calculation is difficult and requires a computer program. However, it is relatively straightforward when there is a single ordered categorical predictor, as in the case of Education in Example 7.3 so that the $\hat{p}_i$ are naturally ordered as seen in Table 7.3. Let n_{i0} and n_{i1} denote the number of "no" and "yes" responses for the ith Education level $(i = 1, \ldots, 8)$. Here $\sum_{i=1}^8 n_{i0} = n_0 = 1682$ and $\sum_{i=1}^8 n_{i1} = n_1 = 925$. The n_{i0} and n_{i1} for each i are given in Table 7.2. Then since $\hat{p}_i < \hat{p}_j$ for $i < j$, the sum of the products $n_{i0}n_{j1}$ gives the total number of concordances and the sum of the products $n_{i1}n_{j0}$ gives the total number of discordances for $i < j$. Thus,

$$N_c = \sum_{i=1}^7 \sum_{j=i+1}^8 n_{i0}n_{j1} = \sum_{i=1}^7 n_{i0}\left(\sum_{j=i+1}^8 n_{j1}\right)$$

$$= 24(24 + \cdots + 199) + 92(92 + \cdots + 199) + \cdots + 96(199) = 892{,}008,$$

$$N_d = \sum_{i=1}^7 \sum_{j=i+1}^8 n_{i1}n_{j0} = \sum_{i=1}^7 n_{i1}\left(\sum_{j=i+1}^8 n_{j0}\right)$$

$$= 7(92 + \cdots + 150) + 24(408 + \cdots + 150) + \cdots + 107(150) = 406{,}807$$

and

$$N_t = N - N_c - N_d = 1682 \times 925 - 892{,}008 - 406{,}807 = 257{,}035.$$

So AUC equals

$$\text{AUC} = \gamma = \frac{N_c + 0.5N_t}{N} = \frac{892{,}008 + 0.5 \times 257{,}035}{1682 \times 925} = 0.6559.$$

$\square$

Example 7.10 (Art Museum Visits: ROC Curves for Simple and Multiple Logistic Regression Models) The ROC curves can be plotted and their AUCs can be computed using the R function `plot.roc`, which requires the library pROC. Here is the R code:

```
> library(pROC)
> art = read.csv("c:/data/art.csv")
> summary(art)
> artnomiss1=na.omit(art[,c(1,6)])
> artnomiss2=na.omit(art[,-2])
```

```
> fit1 = glm(Visit ~ Education, family=binomial, data = artnomiss1)
> plot.roc(artnomiss1$Visit, fit1$fitted.values,print.auc=T,xlab="Specificity",
> ylab="Sensitivity")
> fit2 = glm(Visit ~ Education+Income+Gender+Children+Married,
>   factor(County)+ family=binomial, data = artnomiss2)
> plot.roc(artnomiss2$Visit, fit2$fitted.values,print.auc=T,xlab="Specificity",
> ylab="Sensitivity")
```

The ROC curves produced by this code are shown in Figure 7.6. Note that the AUC for the simple logistic regression model is 0.656 as calculated in Example 7.9 and that for the multiple logistic regression model is 0.700. Thus, there is an increase in AUC due to inclusion of additional predictors, which helps to better discriminate between visitors and non-visitors. Also note that the horizontal axis in these plots is marked "Specificity" instead of "$1-$ Specificity" but the labels on the axis are reversed and go from 1 to 0 instead of 0 to 1. Thus what is plotted is actually $1-$ Specificity.

The ROC curve for the simple logistic regression model does not appear to be a step function. This is due to the fact that the data are grouped at each of the eight levels of Education, so both sensitivity and specificity change at each level of Education, Exercise 7.11 asks you to verify the plot of this ROC curve.

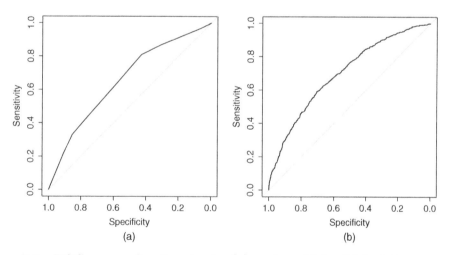

Figure 7.6. *ROC curves for the simple (a) and multiple (b) logistic regression models from Examples 7.3 and 7.5.* □

7.5 Polytomous logistic regression

As mentioned in the introduction of this chapter, polytomous responses may be nominal or ordinal. Different models are needed for these two types of responses. When the response is binary, it makes no difference whether it is nominal or ordinal, so we use the regular binary logistic regression. From here on, assume that the number of possible categorical responses is $m > 2$, and the responses are coded numerically as $y = 1, \dots, m$. Let $\boldsymbol{x} = (1, x_1, \dots, x_p)'$ denote the vector of predictor variables. The probability of response $y = k$ is denoted by $p_k = P(y = k|\boldsymbol{x}) = p_k(\boldsymbol{x})$.

7.5.1   Nominal logistic regression

In binary logistic regression, we use a single logit, which compares the probability of success with the probability of failure. For general m, we use $m - 1$ logits, which use any one response as the reference and compare the probabilities of the remaining $m - 1$ responses to it. Suppose we choose the last response as reference and consider the logits $\ln(p_k/p_m)$, $k = 1, \dots, m - 1$. This results in $m - 1$ logistic models:

$$\ln\left(\frac{p_k}{p_m}\right) = \beta_{0k} + \beta_{1k}x_1 + \cdots + \beta_{pk}x_p = \boldsymbol{x}'\boldsymbol{\beta}_k, \quad (k = 1, \dots, m - 1), \quad (7.12)$$

where $\boldsymbol{\beta}_k = (\beta_{0k}, \beta_{1k}, \dots, \beta_{pk})'$ is an unknown parameter vector to be estimated for each response $k = 1, \dots, m - 1$ and $\boldsymbol{x} = (1, x_1, \dots, x_p)'$ is the vector of predictor variables.

The interpretation of the coefficient β_{jk} is similar to that of the β_j coefficient for binary logistic regression. It is the change in the log-odds of response k relative to that of the reference response m when the predictor variable x_j is increased by one unit keeping all other variables fixed. As an example, suppose $\beta_{jk} = 0.5$ then $\exp(0.5) = 1.649$. Hence the odds of outcome k versus outcome m increase by a factor of 1.649 if x_j is increased by one unit keeping all other variables fixed.

From the logistic model (7.12), it follows that

$$p_k(\boldsymbol{x}) = p_m(\boldsymbol{x}) \exp(\boldsymbol{x}'\boldsymbol{\beta}_k), \quad k = 1, \dots, m - 1.$$

Hence,

$$\sum_{k=1}^{m} p_k(\boldsymbol{x}) = p_m(\boldsymbol{x}) \sum_{k=1}^{m-1} \exp(\boldsymbol{x}'\boldsymbol{\beta}_k) + p_m(\boldsymbol{x}) = 1.$$

Solving for $p_m(\boldsymbol{x})$, we get

$$p_m(\boldsymbol{x}) = \frac{1}{1 + \sum_{j=1}^{m-1} \exp(\boldsymbol{x}'\boldsymbol{\beta}_j)}$$

and substituting back in the formula for $p_k(\boldsymbol{x})$, we get

$$p_k(\boldsymbol{x}) = \frac{\exp(\boldsymbol{x}'\boldsymbol{\beta}_k)}{1 + \sum_{j=1}^{m-1} \exp(\boldsymbol{x}'\boldsymbol{\beta}_j)}.$$

Estimation

The maximum likelihood method is used to estimate the parameters of the model (7.12). Consider the data $(\boldsymbol{x}_i, y_i)$, $i = 1, \ldots, n$, where $\boldsymbol{x}_i = (1, x_{i1}, \ldots, x_{ip})'$ is the vector of x variables for the ith observation and $y_i = k$ is the corresponding response for $k = 1, \ldots, m$. Let z_{ik} be the indicator variable of the outcome y_i, i.e. $z_{ik} = 1$ if $y_i = k$ and $z_{ij} = 0$ for $j \neq k$. Let $p_{ik} = p_k(\boldsymbol{x}_i) = P(z_{ik} = 1 | \boldsymbol{x}_i) = P(y_i = k | \boldsymbol{x}_i)$. Note that $\sum_{k=1}^{m} p_{ik} = 1$ for all observations $i = 1, \ldots, n$.

The likelihood function is given by

$$L = \prod_{i=1}^{n} \prod_{k=1}^{m} (p_{ik})^{z_{ik}},$$

where the inside product includes only one term, p_{ik} for which $z_{ik} = 1$. Hence, the log-likelihood function equals

$$\ln L = \sum_{i=1}^{n} \sum_{k=1:z_{ik}=1}^{m} z_{ik} \ln p_{ik}.$$

The MLE $\widehat{\boldsymbol{\beta}}_k = (\widehat{\beta}_{0k}, \widehat{\beta}_{1k}, \ldots, \widehat{\beta}_{pk})'$ maximizes $\ln L$. The estimated asymptotic variance–covariance matrix of $\widehat{\boldsymbol{\beta}}$ is given by the inverse of the observed information matrix, which is the negative of the matrix of second partial derivatives of $\ln L$ with respect to the β's with the β's replaced by their MLEs. The elements of this matrix can be used to make inferences on the parameters of the model.

Classification

To classify the outcome for any observation with given $\boldsymbol{x} = (1, x_1, \ldots, x_p)'$ vector, we can calculate the estimated probabilities of all m outcomes:

$$\widehat{p}_k(\boldsymbol{x}) = \exp(\boldsymbol{x}'\widehat{\boldsymbol{\beta}}_k)\widehat{p}_m(\boldsymbol{x}) \quad (k = 1, \ldots, m - 1),$$

where

$$\widehat{p}_m(\boldsymbol{x}) = \frac{1}{1 + \sum_{j=1}^{m-1} \exp(\boldsymbol{x}'\widehat{\boldsymbol{\beta}}_j)}.$$

Then we classify the outcome to be that k for which $\widehat{p}_k(\boldsymbol{x})$ is maximum. We refer to this as the **maximum probability classifier**.

Example 7.11 (*MBA Admissions: Nominal Logistic Regression Model*) We use the library `nnet` to perform nominal logistic regression, which is more convenient to use than the library `mlogit`. The R script is shown below. Note that we specified the maximum number of iterations to be 1000 in the `multinom` function; the default is 100.

```
> library(nnet)
> MBA=read.csv("c:/data/MBA.csv")
> fit1=multinom(admit~GPA + GMAT, data = MBA,maxit=1000)
# weights:  12 (6 variable)

> summary(fit1)
```

The output is as follows.

```
Call:
multinom(formula = admit ~ GPA + GMAT, data = MBA, maxit = 1000)

Coefficients:
  (Intercept)       GPA        GMAT
2    155.2575  -28.86054  -0.1350340
3    424.2290 -101.90704  -0.2830152

Std. Errors:
  (Intercept)      GPA        GMAT
2   0.2360282 1.706583 0.01210057
3   0.6922946 2.588929 0.01785585

Residual Deviance: 11.16483
AIC: 23.16483
```

The `multinom` function took 390 iterations to converge because the MBA admissions data form nearly separated clusters as seen in Figure 7.1; as a result, fitting the nominal logistic regression model is more difficult. This may seem counterintuitive, but it can be understood from the following example. Consider fitting a simple logistic regression model to the data in which all x-values corresponding to $y = 0$ are less than those corresponding to $y = 1$, so the two clusters are completely separated. Then the MLE of the slope β_1 of the logistic response curve must approach ∞ and so its MLE does not converge. Exercise 7.2 gives a small data set to illustrate this phenomenon.

Note that `admit` = 1 (admit) is used as the reference level by default which explains the negative coefficients on GPA and GMAT. Denoting the predicted probabilities of the three admission decisions by $\widehat{p}_1, \widehat{p}_2$, and $\widehat{p}_3$, the models are

$$\ln\left(\frac{\widehat{p}_2}{\widehat{p}_1}\right) = 155.2507 - 28.8605 \times \text{GPA} - 0.1350 \times \text{GMAT}$$

and

$$\ln\left(\frac{\widehat{p}_3}{\widehat{p}_1}\right) = 424.2290 - 101.9070 \times \text{GPA} - 0.2830 \times \text{GMAT}.$$

By computing the ratios of the estimated regression coefficients to their standard errors, all the coefficients can be seen to be highly significant.

From the coefficient of -0.135 of GMAT in the model for $\ln(\widehat{p}_2/\widehat{p}_1)$ we see that if an applicant's GMAT score is 10 points higher then $\widehat{p}_1/\widehat{p}_2$, i.e. the odds of the applicant getting admitted versus wait-listed increase by a factor of $\exp(1.350) = 3.86$ If the original odds were 50:50 (i.e, the ratio of their probabilities was 1), then the odds would increase to 3.86 to 1.

Next let us consider predictions using this model. First, we compute predicted probabilities for an applicant with GPA $= 3.2$ and GMAT $= 450$.

```
> predicted=predict(fit1,type='probs',newdata=data.frame(GPA=3.20,GMAT=450))
> predicted
            1            2            3
1.054139e-01 8.945861e-01 2.129714e-14
```

Thus, we get $\widehat{p}_1 \approx 0.1054, \widehat{p}_2 \approx 0.8946$, and $\widehat{p}_3 \approx 0$. Since $\widehat{p}_2$ is the largest, this applicant will most likely be wait-listed (`admit = 2`) with almost 90% chance.

We can check the values $\widehat{p}_1, \widehat{p}_2$, and $\widehat{p}_3$ by hand calculation as follows. We have

$$\ln\left(\frac{\widehat{p}_2}{\widehat{p}_1}\right) = 155.2507 - 28.8605 \times 3.20 - 0.1350 \times 450 = 2.1539$$

and

$$\ln\left(\frac{\widehat{p}_3}{\widehat{p}_1}\right) = 424.2290 - 101.9070 \times 3.20 - 0.2830 \times 450 = -29.223.$$

Since $e^{-29.223} \approx 0$, we ignore $\widehat{p}_3$ in the following calculation. Then

$$\widehat{p}_1 = \frac{1}{1 + e^{2.1539}} = 0.1040, \quad \widehat{p}_2 = \frac{e^{2.1539}}{1 + e^{2.1539}} = 0.8960.$$

These hand-calculated probabilities match closely those calculated in the R output above.

Next we calculate the confusion matrix by applying the maximum probability rule for classification.

```
> Y.prob.1 = fitted(fit1, outcome= FALSE);
> Y.hat.1 = rep(0,n);
> for(i in 1:n){if(max(Y.prob.1[i,]) == Y.prob.1[i,1]){Y.hat.1[i]=1;}
> else if(max(Y.prob.1[i,]) == Y.prob.1[i,2]){Y.hat.1[i]=2;}
> else if(max(Y.prob.1[i,]) == Y.prob.1[i,3]){Y.hat.1[i]=3;}}
>
> ctable1 = table(MBA$admit, Y.hat.1);
> ctable1;
    Y.hat.1
```

```
       1  2  3
  1 30  1  0
  2  2 23  1
  3  0  0 28
> correct.rate1 = sum(diag(ctable1)[1:3])/n;
> correct.rate1
[1] 0.9529412
```

From the confusion matrix we see that the CCR is $(30 + 23 + 28)/85 = 95.29\%$. $\qquad\qquad\qquad\qquad\qquad\qquad\qquad\qquad\qquad\qquad\qquad\qquad\qquad\qquad$ $\square$

7.5.2 Ordinal logistic regression

Now suppose the responses are ordered: $1 < 2 < \cdots < m$. Then it makes sense to define cumulative probabilities $P(y \le k)$, $k = 1, \ldots, m$. Note that $P(y \le m) = 1$. Next define **cumulative logits**:

$$\ln \left[\frac{P(y \le k)}{P(y > k)} \right], \quad k = 1, \ldots, m - 1.$$

A linear model is postulated on these cumulative logits. The part of the model that depends on the predictor variables will be assumed to be common to all cumulative logits. Thus, let $\boldsymbol{\beta} = (\beta_1, \ldots, \beta_p)'$ denote a common parameter vector and $\boldsymbol{x} = (x_1, \ldots, x_p)'$ denote the predictor variable vector. The final model is given by

$$\ln \left[\frac{P(y \le k)}{P(y > k)} \right] = \beta_{0k} + \boldsymbol{x}'\boldsymbol{\beta}, \quad k = 1, \ldots, m - 1, \qquad (7.13)$$

where $\beta_{01} < \beta_{02} < \cdots < \beta_{0,m-1}$. Note that the intercept term β_{0k} is different for each $k = 1, \ldots, m - 1$. This model is equivalent to

$$P(y \le k) = \frac{\exp(\beta_{0k} + \boldsymbol{x}'\boldsymbol{\beta})}{1 + \exp(\beta_{0k} + \boldsymbol{x}'\boldsymbol{\beta})}, \quad k = 1, \ldots, m - 1. \qquad (7.14)$$

The β_{0k} terms are constrained to be nondecreasing to ensure that the cumulative logits (and hence the cumulative probabilities) are nondecreasing.

Note from (7.13) that the difference in log-odds of two individuals i and j with covariate vectors $\boldsymbol{x}_i$ and $\boldsymbol{x}_j$ for any category k is

$$\ln \left[\frac{P(y \le k | \boldsymbol{x}_i)}{P(y > k | \boldsymbol{x}_i)} \right] - \ln \left[\frac{P(y \le k | \boldsymbol{x}_j)}{P(y > k | \boldsymbol{x}_j)} \right] = (\boldsymbol{x}_i - \boldsymbol{x}_j)'\boldsymbol{\beta},$$

which is independent of $k = 1, \ldots, m - 1$. Because of this property, this model is referred to as the **proportional odds model**.

The derivation of model (7.14) is given in Section 7.7. The derivation assumes that there is a latent (unobservable) continuous response z. Also, there are m intervals defined by unknown cutpoints $\beta_{01} < \cdots < \beta_{0,m-1}$ such that if z falls in the kth interval, $\beta_{0,k-1} \le z < \beta_{0k}$, then we observe the outcome $y = k$ as shown in Figure 7.14. The latent response z is assumed to have a logistic distribution with with mean $\mu = \boldsymbol{x}'\boldsymbol{\beta} = \beta_1 x_1 + \cdots + \beta_p x_p$.

Instead of the logistic distribution if we assume that $z \sim N(\mu = \boldsymbol{x}'\boldsymbol{\beta}, 1)$ then

$$P(y \leq k) = P(z \leq \beta_{0k}) = \Phi(\beta_{0k} - \boldsymbol{x}'\boldsymbol{\beta})$$

or equivalently $\Phi^{-1}[P(y \leq k)] = \beta_{0k} - \boldsymbol{x}'\boldsymbol{\beta}$, where $\Phi(\cdot)$ is the standard normal c.d.f. and $\Phi^{-1}(\cdot)$ is its inverse. This is called the **probit model**. However, this model is not analytically tractable since $\Phi^{-1}(\cdot)$ cannot be expressed in a closed algebraic form. On the other hand, the standard logistic c.d.f. $F(x) = \exp(x)$ $/[1 + \exp(x)] = p(x)$ (say) and its inverse $F^{-1}(p(x)) = \ln[p(x)/(1 - p(x))]$ have simple closed algebraic forms. As shown in Figure 7.13, the standard normal and standard logistic distributions are quite close. Now put $p(x) = P(y \leq k)$ and $x = \beta_{0k} - \boldsymbol{x}'\boldsymbol{\beta}$ in these expressions, which yields the model (7.13), except for the "–" sign on $\boldsymbol{x}'\boldsymbol{\beta}$ instead of the "+" sign. This sign is a matter of convention. R uses the $-$ sign on the $\boldsymbol{x}'\boldsymbol{\beta}$ term.

Estimation

Once again, we use the maximum likelihood method to estimate the parameters of the model (7.13). Assume that the data are in the same format as for nominal logistic regression. We have

$$
\begin{aligned}
p_{ik} &= P(y_i = k) \\
&= P(y_i \leq k) - P(y_i \leq k - 1) \\
&= \frac{\exp(\beta_{0k} + \boldsymbol{x}_i'\boldsymbol{\beta})}{1 + \exp(\beta_{0k} + \boldsymbol{x}_i'\boldsymbol{\beta})} - \frac{\exp(\beta_{0,k-1} + \boldsymbol{x}_i'\boldsymbol{\beta})}{1 + \exp(\beta_{0,k-1} + \boldsymbol{x}_i'\boldsymbol{\beta})}.
\end{aligned}
$$

The likelihood function is given by

$$L = \prod_{i=1}^{n} \prod_{k=1}^{m} (p_{ik})^{z_{ik}},$$

where the inside product includes only one term, p_{ik} for which $z_{ik} = 1$. The log-likelihood function equals

$$\ln L = \sum_{i=1}^{n} \sum_{k=1:z_{ik}=1}^{m} z_{ik} \ln p_{ik}.$$

The MLEs $\widehat{\beta}_{01}, \ldots, \widehat{\beta}_{0,m-1}$ and $\widehat{\boldsymbol{\beta}} = (\widehat{\beta}_1, \ldots, \widehat{\beta}_p)'$ maximize $\ln L$. The estimated asymptotic variance–covariance matrix of $\widehat{\boldsymbol{\beta}}$ is given by the inverse of the observed information matrix, which is the negative of the matrix of second partial derivatives of $\ln L$ with respect to the β's with the β's replaced by their MLEs. The elements of this matrix can be used to make inferences on the parameters of the model.

Classification

To classify the outcome for any observation with given $\boldsymbol{x} = (x_1, \ldots, x_p)'$ vector, we can calculate the estimated probabilities of all m outcomes:

$$\widehat{p}_k(\boldsymbol{x}) = \frac{\exp(\widehat{\beta}_{0k} + \boldsymbol{x}'\widehat{\boldsymbol{\beta}})}{1 + \exp(\widehat{\beta}_{0k} + \boldsymbol{x}'\widehat{\boldsymbol{\beta}})} - \frac{\exp(\widehat{\beta}_{0,k-1} + \boldsymbol{x}'\widehat{\boldsymbol{\beta}})}{1 + \exp(\widehat{\beta}_{0,k-1} + \boldsymbol{x}'\widehat{\boldsymbol{\beta}})} \quad (k = 1 \ldots, m-1)$$

and

$$\widehat{p}_m(\boldsymbol{x}) = 1 - \frac{\exp(\widehat{\beta}_{0,m-1} + \boldsymbol{x}'\widehat{\boldsymbol{\beta}})}{1 + \exp(\widehat{\beta}_{0,m-1} + \boldsymbol{x}'\widehat{\boldsymbol{\beta}})}.$$

Then, we classify the outcome to be that k for which $\widehat{p}_k(\boldsymbol{x})$ is maximum.

Example 7.12 *(MBA Admissions: Ordinal Logistic Regression Model)* We will use the MBA admissions data from Example 7.12, but we will take the order among the three groups into account to fit an ordinal logistic regression model using the following R script.

```
> library(ordinal)
> MBA$admit.ordered= as.ordered(MBA$admit)
> fit2=clm(admit.ordered~GPA+GMAT,data=MBA)
> summary(fit2)
```

The output is shown below.

```
Coefficients:
      Estimate Std. Error z value Pr(>|z|)
GPA   -26.20487    7.46633  -3.510 0.000449 ***
GMAT   -0.05928    0.01932  -3.069 0.002150 **
---
Signif. codes:  0 '***' 0.001 '**' 0.01 '*' 0.05 '.' 0.1 ' ' 1

Threshold coefficients:
    Estimate Std. Error z value
1|2  -111.24      32.10  -3.465
2|3   -99.61      28.81  -3.458
```

The negative coefficients of GPA and GMAT seem counterintuitive since increasing either one or both should increase the probability of an applicant to be admitted or wait-listed, i.e. increase $P(y \leq 1)$ and $P(y \leq 2)$. The explanation is that R fits the model

$$\ln\left[\frac{P(y \leq k)}{P(y > k)}\right] = \beta_{0k} - \boldsymbol{x}'\boldsymbol{\beta}, \quad k = 1, \ldots, m-1,$$

so the signs of the regression coefficients are reversed. Thus the actual fitted models are as follows:

$$\ln\left[\frac{P(\text{admit})}{P(\text{wait-list or deny admission})}\right] = -111.24 + 26.205 \times \text{GPA} + 0.059 \times \text{GMAT}$$

and

$$\ln\left[\frac{P(\text{admit or wait-list})}{P(\text{deny admission})}\right] = -99.61 + 26.205 \times \text{GPA} + 0.059 \times \text{GMAT}.$$

Suppose we want to predict the admission decision for an applicant whose GPA $= 3.20$ and GMAT $= 450$. Then using the function

```
> predict(fit2,newdata=data.frame(GPA=3.20,GMAT=450))
```

we get the predicted probabilities of the three outcomes as

```
          1          2            3
1 0.3307963 0.6691856 1.810068e-05
```

Thus, there is a 33% chance that this student will be admitted and 67% chance that the student will be wait-listed; there is very little chance that the student will be denied admission.

For illustration purposes, we check the above probabilities by hand calculation. Toward this end, we first calculate two scores for this student:

$$X_1 = -111.24 + 26.205 \times 3.20 + 0.059 \times 450 = -0.699$$

and

$$X_2 = -99.61 + 26.205 \times 3.20 + 0.059 \times 450 = 10.931.$$

Then

$$P(y \leq 1) = \frac{\exp(-0.699)}{1 + \exp(-0.699)} = 0.331 \quad \text{and} \quad P(y \leq 2) = \frac{\exp(10.931)}{1 + \exp(10.931)} \approx 1.$$

Thus

$$\widehat{p}_1 = 0.331, \widehat{p}_2 \approx 1 - 0.331 = 0.669 \quad \text{and} \quad \widehat{p}_3 \approx 0.$$

$\square$

7.6 Modern extensions[*]

In this section, we introduce two nonparametric classifiers: **classification trees** and **support vector machines**. Both are hard classifiers.

7.6.1 Classification trees

Classification trees are similar to regression trees discussed in Section 3.7.1 except that the outcome is now one of $m \geq 2$ classes. They are constructed in the same way in a top-down manner by recursive binary splitting on one variable at a time resulting in partitioning of the feature space $\mathcal{X}$ into $N \geq 2$ disjoint regions (leaf nodes), $R_1, \ldots, R_N$. The predicted outcome for the jth region is the class that has the highest fraction of observations $f_j^* = \max_{1 \leq k \leq m} f_{jk}$ in that region where f_{jk} is the fraction of observations belonging to class k in region R_j. Since the outcome is categorical, we cannot use the SSE as the criterion to minimize. The `tree` function in R minimizes the deviance at each new split which is equivalent to minimizing the **entropy** at each new leaf node defined as

$$E_j = -2 \sum_{k=1}^{m} f_{jk} \ln f_{jk}.$$

The entropy is a measure of diversity and if the node is "pure," i.e. if all observations in the node belong to one class, in other words, if only one $f_{jk} = 1$ and all other $f_{jk} = 0$, then E_j can be shown to be equal to 0. Thus the goal is to make the leaf nodes as "pure" as possible. Obviously, this can be done by growing the tree to its fullest extent so that each node consists of a single observation. But such a tree is worthless for prediction purposes since it just replicates the training data, i.e. it represents a saturated model. Therefore, minimization of the entropy (or any other loss function) must be constrained subject to a penalty on the complexity of the tree. Another measure having similar properties as the entropy is the **Gini index** defined as

$$G_j = \sum_{k=1}^{m} f_{jk}(1 - f_{jk}).$$

It is immediately seen that $G_j = 0$ for a "pure" node.

Example 7.13 *(GPA Data: Classification Tree)* To illustrate a classification tree, we use the GPA data used in many examples in Chapter 3. GPA is a continuous variable, but we created three outcome classes: $(1,2] = \{1 < \text{GPA} \leq 2\}, (2,3] = \{2 < \text{GPA} \leq 3\}$, and $(3,4] = \{3 < \text{GPA} \leq 4\}$. The classification tree is shown in Figure 7.7. Notice the similarity of this tree with the regression tree in Figure 3.4. Both use similar cutpoints, but because of the coarseness of the classes, the classification tree has only four leaf nodes while the regression tree has six. The partition plot for the classification tree is shown in Figure 7.8.

The confusion matrix using this decision tree is shown below.

```
                      predicted
                  (1,2] (2,3] (3,4]
          (1,2]     5     1     0
Actual    (2,3]     3    15     0
          (3,4]     0     1    15
```

We see that 35 out of 40 observations are correctly classified, so the CCR is $35/40 = 87.5\%$. $\square$

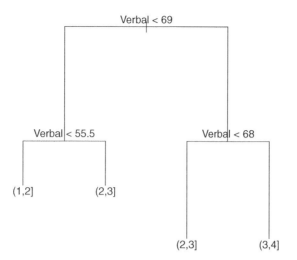

Figure 7.7. *Classification tree for predicting three GPA classes from Verbal and Math scores (left branch at each node corresponds to "yes" and the right branch corresponds to "no").*

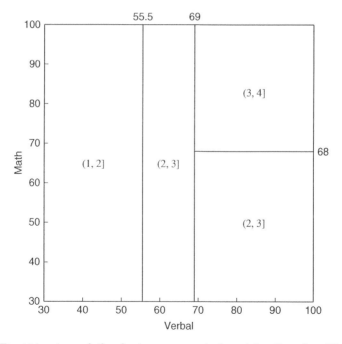

Figure 7.8. *Partitioning of the feature space induced by the classification tree in Figure 7.7.*

7.6.2　Support vector machines

The idea of a **support vector machine** (SVM) is derived from separating hyperplanes. Suppose we have observations $\boldsymbol{x}_i$ belonging to two classes that are completely separated in the feature space. Then it is clear that one can pass an infinite number of hyperplanes that will separate the two clusters of data. A hyperplane is defined by a linear equation: $f(\boldsymbol{x}) = \beta_0 + \boldsymbol{x}'\boldsymbol{\beta} = \beta_0 + \beta_1 x_1 + \cdots + \beta_p x_p$ such that if the observation $\boldsymbol{x}_i$ falls on one side of the hyperplane (e.g. $f(\boldsymbol{x}_i) > 0$) then it belongs to one class and if it falls on the other side of the hyperplane (e.g. $f(\boldsymbol{x}_i) < 0$) then it belongs to the other class. Logistic regression fails in this case because of nonconvergence of the MLEs of the model parameters as discussed in Example 7.11. On the other hand, a separating hyperplane does perfect classification.

How can we choose an optimal separating hyperplane from the infinitely many available when data clusters are fully separated? Let d_i be the shortest distance from an observation $\boldsymbol{x}_i$ to a given hyperplane, which is the length of the projection of $\boldsymbol{x}_i$ on the hyperplane. Then the optimal hyperplane is defined to be the one that maximizes $\min d_i$ where the minimum is taken over all observations and the maximum is taken over all separating hyperplanes. This optimal hyperplane is called the **maximal margin hyperplane**. The width of the margin associated with this hyperplane is $M = \max \min d_i$. This is the widest margin such that no data points fall within it. By choosing the separating hyperplane with the widest margin for the training data, we hope to achieve good classifications for the test data as well. This hyperplane along with its margin is shown in Figure 7.9. Its calculation involves solving a quadratic programming problem.

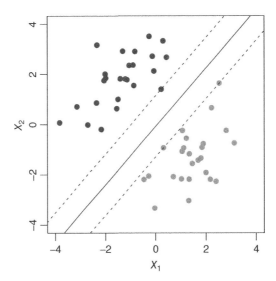

Figure 7.9. *Maximal margin SVM classifier for two fully separated clusters with three support vectors.*

Notice from Figure 7.9 that there are three points that fall on the margin boundaries and hence are at a minimum distance from the separating hyperplane (which is a straight line in the case of two predictors). Thus the maximal margin separating hyperplane for this data set depends only on these three points in the sense that even if other points are perturbed, as long as they do not cross the margin, the maximal margin hyperplane does not change. The data vectors corresponding to these three points are called **support vectors,** and the set of support vectors is denoted by $\mathcal{S}$. In general, it can be shown that there are at least two support vectors.

Thus far we have considered perfectly separated data clusters. Real data almost always have overlapped clusters, so some misclassifications are inevitable. Of course, one would like to have as few misclassifications as possible, which requires making the margin as narrow as possible. However, this leads to overfitting which minimizes misclassifications for the training data but inflates misclassifications for the test data. Another way to state this is that narrow margins have a low bias but high variance and wide margins have a high bias but low variance. To balance this bias-variance trade-off, an upper bound is specified on the number of misclassifications and the maximal margin separating hyperplane is determined subject to this constraint. The higher this upper bound, the wider is the margin and more the number of support vectors and vice versa. Finding this margin and the associated separating hyperplane involves solving another quadratic programming problem. The upper bound on the number of misclassifications is a tuning parameter, which is determined by cross-validation. The number of support vectors for the resulting separating hyperplane is the number of data points that fall within the margins. An example with 13 support vectors is shown in Figure 7.10. Note that classifications are still done depending on which side of the hyperplane the observation falls.

While a separating hyperplane is a linear function of the predictors, in general, SVM uses a class of nonlinear functions, which result in nonlinear decision boundaries. To see how separating hyperplanes are generalized in this manner, we note that the solution for linear decision boundaries has the form

$$\widehat{f}(\boldsymbol{x}) = \widehat{\beta}_0 + \boldsymbol{x}'\widehat{\boldsymbol{\beta}} = \widehat{\beta}_0 + \sum_{\boldsymbol{x}_i \in \mathcal{S}} \widehat{\alpha}_i \boldsymbol{x}' \boldsymbol{x}_i = \widehat{\beta}_0 + \sum_{\boldsymbol{x}_i \in \mathcal{S}} \widehat{\alpha}_i \langle \boldsymbol{x}, \boldsymbol{x}_i \rangle,$$

where the $\widehat{\alpha}_i$ are the estimated regression coefficients in the transformed feature space consisting of the **inner products** $\langle \boldsymbol{x}, \boldsymbol{x}_i \rangle = \boldsymbol{x}' \boldsymbol{x}_i$ (which are the usual dot products in this special case). More generally, SVM fits the model

$$\widehat{f}(\boldsymbol{x}) = \widehat{\beta}_0 + \sum_{\boldsymbol{x}_i \in \mathcal{S}} \widehat{\alpha}_i K(\boldsymbol{x}, \boldsymbol{x}_i),$$

where $K(\cdot, \cdot)$ is called a **kernel function**. These are functions of the inner products. One of the simplest nonlinear kernel functions is the polynomial kernel

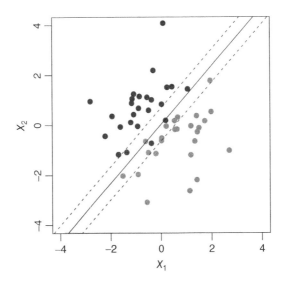

Figure 7.10. *SVM classifier for two overlapping clusters with 13 support vectors.*

given by

$$K(\boldsymbol{x}, \boldsymbol{y}) = (1+ < \boldsymbol{x}, \boldsymbol{y} >)^d,$$

where $d > 1$ is the degree of the polynomial. A popular kernel function is the radial kernel given by $K(\boldsymbol{x}, \boldsymbol{y}) = \exp(-\delta||\boldsymbol{x} - \boldsymbol{y}||)$ for some $\delta > 0$.

Example 7.14 *(GPA Data: Support Vector Classification)* The linear boundaries for three GPA classes, $(1, 2]$, $(2, 3]$, and $(3, 4]$ (the corresponding data points are colored black, dark gray, and light gray, respectively), are shown in Figure 7.11.

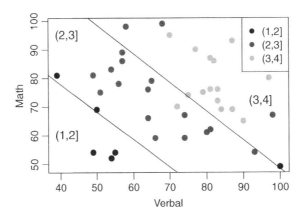

Figure 7.11. *Linear decision boundaries for classifying GPA data into three classes.*

The equations for the linear boundaries are as follows:

$$L_1 = -17.8872 + 0.1481\text{Verbal} + 0.1538\text{Math}$$

and

$$L_2 = -110.876 + 0.7491\text{Verbal} + 0.7448\text{Math}.$$

Note that the boundaries are nearly parallel. If $L_1 < 0$ for a given observation then that observation is classified to the GPA class (1,2]. If $L_1 > 0$ and $L_2 < 0$ then that observation is classified to the GPA class (2,3]. Finally, if $L_2 > 0$ then the observation is classified to the GPA class (3,4]. The confusion matrix for this classification rule is

```
                     predicted
                 (1,2]  (2,3]  (3,4]
         (1,2]     3      2      1
Actual   (2,3]     1     15      3
         (3,4]     0      1     15
```

Seven observations are misclassified resulting in CCR $= 33/40 = 82.5\%$.

The most extreme misclassification occurs for the observation (Verbal=100, Math=49) with GPA=1.54, which is classified to the GPA class (3,4], so we may want to check it. Substituting these Verbal and Math values in the equations for L_1 and L_2, we get $L_1 = 4.459 > 0$ and $L_2 = 0.529 > 0$. Thus, the classification rule is correctly applied but the observation is misclassified.

Next we give an illustration of SVM with nonlinear decision boundaries. Figure 7.12a shows the decision boundaries if the cost of a misclassification is set at $C = 0.5$, while Figure 7.12b shows the decision boundaries if $C = 2.0$. Figure 7.12a has two decision boundaries with four misclassifications (three black dots are misclassified to the light gray dots class and one dark gray dot is misclassified to the light gray dots class), while Figure 7.12b has three decision boundaries with one misclassification (a dark gray dot is misclassified to the light gray dots class). The latter overfits the data with one boundary being essentially an island to correctly classify a single extreme observation (Verbal=100, Math=49).

How these nonlinear decision boundaries are calculated is beyond the scope of this short review of SVM. We may, however, note that for $C = 2.0$, the decision boundary between the black dots class and the dark gray dots class has 18 support vectors (i.e. the decision boundary is a function of 18 data vectors out of a total of 40), while the decision boundary between the dark gray dots class and the light gray dots class has 20 support vectors. The same numbers for the decision boundary for $C = 0.5$ are 18 and 28, respectively. □

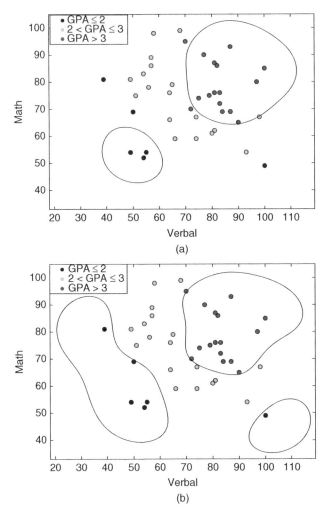

Figure 7.12. *Nonlinear decision boundaries for classifying GPA data into three classes ((a) $C = 0.5$ and (b) $C = 2.0$).*

7.7   Technical notes*

To derive the ordinal logistic regression model, we first introduce the **logistic distribution**. If x has a logistic distribution with mean (location parameter) μ and scale parameter τ (the standard deviation of x equals $\sigma = \pi\tau/\sqrt{3}$), then the p.d.f. and the c.d.f. of the standardized variable $z = (x - \mu)/\tau$ are given by

$$f(z) = \frac{\exp(z)}{[1 + \exp(z)]^2} \quad \text{and} \quad F(z) = \frac{\exp(z)}{[1 + \exp(z)]},$$

respectively. In Figure 7.13 we have plotted the $N(0, 1)$ p.d.f. (shown by a dashed curve) and the logistic p.d.f. (shown by a solid curve) also with $\mu = 0$ and $\sigma = 1$. As can be seen, the two p.d.f. curves match quite closely except in the center.

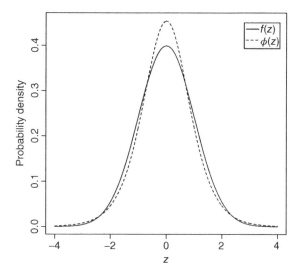

Figure 7.13. *Plots of logistic density $f(z)$ (solid curve) and the standard normal density $\phi(z)$ (dashed curve) both with $\mu = 0$ and $\sigma = 1$.*

We can think of the observable ordinal outcome, $y = 1, 2, \ldots, m$, as resulting from an underlying unobservable continuous **latent variable** z as follows. Suppose that the real axis is divided into m intervals by unknown cutpoints $-\infty < \beta_{01} < \beta_{02} < \cdots < \beta_{0,m-1} < \infty$ such that if z falls in the kth interval, $[\beta_{0,k-1}, \beta_{0k})$, then we observe the discrete outcome $y = k$, as shown in Figure 7.14.

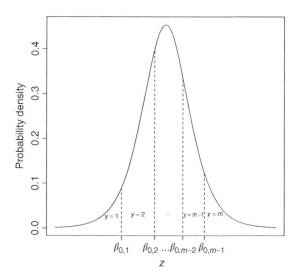

Figure 7.14. *Latent variable distribution with cutpoints resulting in ordinal outcomes.*

Next we assume that z has a logistic distribution with parameters μ and $\tau = 1$ and μ depends on the predictor variable vector $\boldsymbol{x} = (x_1, \ldots, x_p)'$ through a linear predictor, $\mu = \beta_1 x_1 + \cdots + \beta_p x_p = \boldsymbol{x}' \boldsymbol{\beta}$, where $\boldsymbol{\beta} = (\beta_1, \ldots, \beta_p)'$ is an unknown parameter vector. Then we have

$$P(y \leq k) = P(z \leq \beta_{0k}) = \frac{\exp(\beta_{0k} - \mu)}{1 + \exp(\beta_{0k} - \mu)} = \frac{\exp(\beta_{0k} - \boldsymbol{x}' \boldsymbol{\beta})}{1 + \exp(\beta_{0k} - \boldsymbol{x}' \boldsymbol{\beta})},$$

and hence
$$\ln \left[\frac{P(y \leq k)}{P(y > k)} \right] = \beta_{0k} - \boldsymbol{x}' \boldsymbol{\beta}, \quad k = 1, \ldots, m - 1.$$

This is the same model as given in (7.14) except for a change of sign of the $\boldsymbol{x}' \boldsymbol{\beta}$ term. This reduces to the simple logistic regression model (7.1) if $m = 2$ in which case there is only one cutpoint $\beta_{01} = \beta_0$.

Exercises

Theoretical Exercises

7.1 (2×2 contingency table) Consider a clinical trial to compare a new treatment (coded as $x = 1$) with a control (coded as $x = 0$) for some disease. Denote their success probabilities by p_1 and p_0. Suppose that there are n_0 patients in the control group of whom s_0 are successes and n_1 patients in the treatment group of whom s_1 are successes. The MLEs of p_0 and p_1 can be shown to be the sample proportions of successes $\widehat{p}_0 = s_0/n_0$ and $\widehat{p}_1 = s_1/n_1$.

(a) Show that

$$\widehat{\beta}_1 = \ln \widehat{\psi} = \ln[\{\widehat{p}_1/(1 - \widehat{p}_1)\}/\{\widehat{p}_0/(1 - \widehat{p}_0)\}],$$

where $\widehat{\beta}_1$ is the MLE of β_1 in the logistic response model and $\ln \widehat{\psi}$ is the sample log-odds ratio.

(b) Use the information matrix (7.6) to derive the formula:

$$\widehat{\mathrm{Var}}(\ln \widehat{\psi}) \approx \frac{1}{n_0 \widehat{p}_0 (1 - \widehat{p}_0)} + \frac{1}{n_1 \widehat{p}_1 (1 - \widehat{p}_1)}.$$

7.2 (Nonconvergence of MLEs in logistic regression) This exercise is based on Allison (2008). Consider completely separated data in the following table.

x	-5	-4	-3	-2	-1	$+1$	$+2$	$+3$	$+4$	$+5$
y	0	0	0	0	0	1	1	1	1	1

Because these data are symmetric, it can be shown that β_0 in the simple logistic regression model can be taken to be zero. So the likelihood function can be treated as a function only of the slope parameter β_1.

(a) Write the log-likelihood function and plot it versus β_1 for these data and check that it approaches the maximum value of 0 (i.e. the likelihood function approaches the maximum value of 1) as $\beta_1 \to \infty$. So the MLE of β_1 does not exist and the algorithm to find it does not converge.

(b) Next consider quasi-separated data obtained by adding two observations $(x, y) = (0, 0)$ and $(x, y) = (0, 1)$ to the above data set and repeat the exercise. Check that the log-likelihood function approaches a number less than 0 as $\beta_1 \to \infty$. So again the MLE of β_1 does not exist and the algorithm to find it does not converge.

7.3 (Deviance for grouped data) Derive the following formula for the deviance of logistic regression model for grouped data with $g \geq 2$ groups:

$$D^2 = -2 \sum_{i=1}^{g} [\{s_i \ln \widehat{p}_i + (n_i - s_i) \ln(1 - \widehat{p}_i)\} - \{s_i \ln f_i + (n_i - s_i) \ln(1 - f_i)\}]$$

$$= -2 \sum_{i=1}^{g} \left[s_i \ln \left(\frac{\widehat{p}_i}{f_i} \right) + (n_i - s_i) \ln \left(\frac{1 - \widehat{p}_i}{1 - f_i} \right) \right].$$

How many d.f. does D^2 have? (*Hint*: The MLEs of the success probabilities p_i for grouped data under the saturated model are the sample proportions of successes $f_i = s_i/n_i$.)

7.4 (Derivation of the logistic regression model assuming the covariate x is Poisson) The simple logistic regression model can be derived as a conditional probability model, $P(y = 1|x)$, assuming that the covariate x has a Poisson distribution. As an example, consider automated screening of emails for phishing attempts. Suppose there is a set of key words that are identified with phishing attempts and we want to use the count x of these key words as a covariate. Assume that the distributions of x conditioned on y ($y = 0$ for a non-phishing email, $y = 1$ for a phishing email) are

$$f_0(x) = f(x|y = 0) = \frac{e^{-\mu_0} \mu_0^x}{x!} \quad \text{and} \quad f_1(x) = f(x|y = 1) = \frac{e^{-\mu_1} \mu_1^x}{x!}$$

for $x = 0, 1, \ldots$.

Also assume prior probabilities $\pi_0 = 1 - \pi$ and $\pi_1 = \pi$ for non-phishing and phishing emails, respectively. Apply the Bayes formula (see Section 8.4.3) to show that the posterior probability of $y = 1$ conditioned on x is given by

$$P(y = 1|x) = \frac{1}{1 + \exp\left\{-\left[(\mu_0 - \mu_1)) + \ln\left(\frac{\pi}{1-\pi}\right) + x\ln\left(\frac{\mu_1}{\mu_0}\right)\right]\right\}}$$

$$= \frac{1}{1 + \exp\{-(\beta_0 + \beta_1 x)\}},$$

which is the logistic regression model. Here

$$\beta_0 = (\mu_0 - \mu_1) + \ln\left(\frac{\pi}{1 - \pi}\right) \quad \text{and} \quad \beta_1 = \ln\left(\frac{\mu_1}{\mu_0}\right),$$

Under what condition on μ_1 and μ_0 is $P(y = 1|x)$ an increasing function of x?

7.5 (Derivation of the logistic regression model assuming the covariate x is exponential) The diagnoses of some diseases are predicated on high or low values of the count of some blood chemical. For example, high values of the white blood cell (WBC) count is indicative of leukemia. Suppose we assume that the WBC count is exponentially distributed and derive a model for the conditional probability of leukemia given the WBC count (x). Assume that the distributions of x conditioned on y ($y = 0$ for a non-leukemia patient, $y = 1$ for a leukemia patient) are

$$f_0(x) = f(x|y = 0) = \lambda_0 e^{-\lambda_0 x} \quad \text{and} \quad f_1(x) = f(x|y = 1) = \lambda_1 e^{-\lambda_1 x}$$

for $x \geq 0$.

The means of x for the two groups of patients are $\mu_0 = 1/\lambda_0$ and $\mu_1 = 1/\lambda_1$. Assume prior probabilities $\pi_0 = 1 - \pi$ and $\pi_1 = \pi$ for non-leukemia and leukemia patients, respectively. Apply the Bayes formula (see Section 8.4.3) to show that the posterior probability of $y = 1$ conditioned on x is given by

$$P(y = 1|x) = \frac{1}{1 + \exp\left\{-\left[\ln\left(\frac{\lambda_1}{\lambda_0}\right) + \ln\left(\frac{\pi}{1-\pi}\right) + (\lambda_0 - \lambda_1)x\right]\right\}}$$

$$= \frac{1}{1 + \exp\{-(\beta_0 + \beta_1 x)\}},$$

which is the logistic regression model. Here

$$\beta_0 = \ln\left(\frac{\lambda_1}{\lambda_0}\right) + \ln\left(\frac{\pi}{1 - \pi}\right) \quad \text{and} \quad \beta_1 = \lambda_0 - \lambda_1,$$

Under what condition on λ_1 and λ_0 is $P(y = 1|x)$ an increasing function of x?

Applied Exercises

7.6 (2×2 contingency table) In a clinical trial to compare prednisone therapy (active control) with prednisone + VCR therapy (treatment) for leukemia, the following data were obtained.

Outcome

Therapy	Success	Failure	Row total
Prednisone	14	7	21
Prednisone + VCR	38	4	42

Source: Tamhane and Dunlop (2000, Example 9.14).

(a) Calculate the sample log-odds ratio and show that it is significantly different from zero. Use the formula for $\widehat{\text{Var}}(\ln \widehat{\psi})$ from Exercise 7.1 to perform the test.

(b) Do a large sample z-test of $H_0 : p_0 = p_1$ using the test statistic

$$z = \frac{\widehat{p}_1 - \widehat{p}_0}{\sqrt{\widehat{p}(1 - \widehat{p})[1/n_0 + 1/n_1]}},$$

where $\widehat{p}_0$ and $\widehat{p}_1$ are the sample proportions of successes for the Prednisone (Control) and Prednisone + VCR (Treatment) groups, respectively, and $\widehat{p} = (n_0\widehat{p}_0 + n_1\widehat{p}_1)/(n_0 + n_1)$ is the pooled sample proportion of successes. Compare the result with that of the test on the odds ratio from Part (a).

7.7 (Simpson's Paradox) Data are available in the data file `UCBAdmissions.csv` (derived from the data file `UCBAdmission` in the R library) on 4526 applicants (2691 men and 1835 women) who applied for admission to six departments in a university. Table 7.6 shows the data in a summary form. The admission rate for men was 44.5% (1198/2691) and that for women was 30.4% (557/1835). This naturally raised the question of sex discrimination. Although the overall admission rate was 14.1% lower for women than that for men, the admission rate for women was actually higher than that for men in 4 out of 6 departments, as can be seen from Table 7.6. This is called Simpson's paradox.

(a) Explain why Simpson's paradox occurs for these data.

(b) Calculate the sample log odds ratio for Men versus Women given their admission rates at the bottom of Table 7.6. Fit a logistic regression model to the data using only Gender (1 = Men, 0 = Women) as the predictor. Check that this sample log odds ratio is equal to the estimated β coefficient for Gender.

Table 7.6. Admissions data for men and women

Department	Men			Women			Total		
	# Apply	# Admit	Admit (%)	# Apply	# Admit	Admit (%)	# Apply	# Admit	Admit (%)
A	825	512	62.1	108	89	82.4	993	620	64.4
B	560	353	63.0	25	17	68.0	585	577	63.2
C	325	120	36.9	593	202	34.1	918	322	35.1
D	417	138	33.1	375	131	34.9	792	269	34.0
E	191	53	27.7	393	94	23.9	584	147	25.2
F	373	22	5.9	341	24	7.0	714	46	6.4
Total	2691	1198	44.5	1835	557	30.4	4526	1755	38.8

(c) Next fit a second logistic regression model by including the Department as an additional predictor. Explain why the Gender coefficient changes sign from positive to negative, and how this illustrates Simpson's paradox.

7.8 (Radiation therapy) Twenty-four cancer patients were treated with radiation therapy for different number of days (x) and the presence $(y = 0)$ or absence $(y = 1)$ of tumor was observed.

Days (x)	Response (y)	Days (x)	Response (y)
21	1	51	1
24	1	55	1
25	1	25	0
26	1	29	0
28	1	43	0
31	1	44	0
33	1	46	0
34	1	46	0
35	1	51	0
37	1	55	0
43	1	56	0
49	1	58	0

Source: Tanner (1996, p. 28).

(a) Fit a binary logistic regression model to the data.

(b) Calculate a 95% confidence interval for the odds of absence of tumor versus presence of tumor if the number of days of therapy is increased by five days.

(c) Calculate the estimated success probabilities $\widehat{p}_i$ for the 24 patients in the sample. Find the optimum cutoff probability p^* that maximizes the correct classification rate (CCR). Calculate sensitivity, specificity, and the F_1-score for this p^*.

7.9 (Odds ratios for coronary disease data) Logistic regression analysis was done on data from a random sample of patients in a hospital about half of whom had coronary disease. The following coefficients were estimated along with their standard errors for three predictors: Age ($\widehat{\beta} = 0.0906$, SE $=$ 0.0184), Cholesterol ($\widehat{\beta} = 0.0755$, SE $= 0.0136$), and Sex: Female $= 0$, Male $= 1$ ($\widehat{\beta} = 0.035$, SE $= 0.0148$).

(a) What is the odds ratio of coronary disease for males versus females. Calculate a 95% confidence interval for it.

(b) If the odds of coronary disease for a female with Age $= 50$ and Cholesterol $= 180$ are 1 in 10, what are the odds for a male with Age $= 60$ and Cholesterol $= 200$? What is the corresponding probability of coronary disease for that male?

7.10 (Measures of accuracy of classification) Consider the following two confusion matrices arising in a document retrieval system where a binary classifier is used to classify 10,000 documents as relevant or irrelevant.

Classified

		Irrelevant	Relevant	Row total
Actual	Irrelevant	8100	900	9000
	Relevant	100	900	1000
Column	Total	8200	1800	10,000

Classified

		Irrelevant	Relevant	Row total
Actual	Irrelevant	8910	990	9900
	Relevant	10	90	100
Column	Total	8920	1080	10,000

Calculate CCR, Sensitivity, Specificity, Precision, Recall, and F_1-score for the two matrices. What do you conclude?

7.11 (ROC curve for simple logistic regression model for the art museum visits data)

(a) Make a table of Sensitivity versus $1-$ Specificity using the art museum visit data in Table 7.2.

(b) Plot Sensitivity versus $1-$ Specificity to obtain the ROC curve and check that it matches with the one in Figure 7.6a.

7.12 (Pregnancy duration) Kutner et al. (2005) gave data on 102 women whose pregnancy durations were classified as $1 =$ Preterm (less than 36 weeks), $2 =$ Intermediate term (36–37 weeks), and $3 =$ Full term (more than 37 weeks). The predictor variables are Nutrition (higher scores mean better nutrition), Alcohol use ($1 = $ yes, $0 = $ no), Smoking history ($1 = $ yes, $0 = $ no), and Age ($1 = $ less than 20 years, $2 = $ 21–30 years, $3 = $ greater than 30 years). Treat Age as a categorical variable and use Age $= 2$ as the reference category since mothers in this age group are known to have the lowest risk of preterm delivery. Divide the data by putting all odd-numbered observations into the training set and all even-numbered observations into the test set.

a) Fit a nominal logistic regression model to the training set and make predictions for the test set using the maximum probability rule. What is the correct classification rate and how does it break down among the three categories?

b) Repeat the above exercise by fitting an ordinal logistic regression model. Do you get better predictions?

7.13 **(Mammography testing history)** Hosmer and Lemeshow (1989) gave data on 412 women who were asked about their mammography testing history with possible responses (y): Never $= 0$ (234 responses), Within the past year $= 1$ (104 responses), and More than one year ago $= 2$ (74 responses). There are two predictors: family history (HIST) of breast cancer (mother or sister) with values No $= 0$ and Yes $= 1$ and perception of benefit (PB) of mammography on a scale of 5–20 with low values representing high perception of benefit. Divide the data by putting all odd-numbered observations into the training set and all even-numbered observations into the test set.

(a) Fit a nominal logistic regression model to the training set and make predictions for the test set using the maximum probability rule. What is the correct classification rate and how does it break down among the three categories?

(b) Repeat the above exercise by fitting an ordinal logistic regression model. Make sure that you order the responses so that $0 < 2 < 1$. Do you get better predictions?

7.14 **(Program choices by high-school students)** Entering high-school students make a choice from three programs: academic, general, and vocational. The file `program.csv` contains data on the program choices of 200 students along with the following possible predictors: reading, writing, math and science test scores, gender, type of school (public or private), and socio-economic status (seslow, sesmiddle, and seshigh). The data are taken from https://stats.idre.ucla.edu/sas/dae/multinomiallogistic-regression/.

(a) Fit a nominal logistic regression model and use it to calculate the probabilities of three program choices for a male student from high ses and a private school with median scores on four tests: reading $= 50$, writing $= 54$, math $= 52$, and science $= 53$ (note that some of these predictors may not be in the final model). Which choice is this student likely to make?

(b) Repeat the above for the ordinal logistic regression model. Compare the results for the two models, in particular, with respect to the predictors in the final model and their interpretations.

(c) Compute the classification matrices for the two models and the CCRs. Which model gives a higher CCR?

Chapter 8

Discriminant analysis

Discriminant analysis is an alternative to logistic regression as a method of classification. However, there is a fundamental conceptual difference between the two. In logistic regression, we condition on predictors $x_1, \ldots, x_p$ and regard the binary outcome y as random with a Bernoulli distribution. We then model the logit of the success probability of this distribution as a function of the x's. In discriminant analysis, we condition on the outcome $y = 0$ or 1 and the x's as random with different distributions depending on the value of y. Because of this difference, in discriminant analysis we often refer to y as a grouping variable (and not as a response or outcome variable) with values of y as group labels. As an example, different groups of customers (e.g. high volume buyers, low volume buyers, and non-buyers) may be characterized by certain attributes (predictors), and these attributes have a different distribution depending on the group that a customer belongs to.

In discriminant analysis, the predictors are ideally required to be numerical. Generally, we assume that the distribution of predictors is multivariate normal (MVN) (the normality assumption) with different mean vectors for different groups but a common covariance matrix for all groups (the homoscedasticity assumption). Under these assumptions, if there are two groups of observations, then we can derive a linear function of the predictors which best discriminates between the two groups. If there are $m > 2$ groups, then we need more linear functions. This methodology is referred to as **linear discriminant analysis (LDA)**. If the homoscedasticity assumption is dropped, then the discriminating functions are quadratic in the x's, and the methodology is referred to as **quadratic discriminant analysis (QDA)**. We will not cover QDA in this book.

Predictive Analytics: Parametric Models for Regression and Classification Using R,
First Edition. Ajit C. Tamhane.
© 2021 John Wiley & Sons, Inc. Published 2021 by John Wiley & Sons, Inc.
Companion website: www.wiley.com/go/tamhane/predictiveanalytics

8.1 Linear discriminant analysis based on Mahalnobis distance

8.1.1 Mahalnobis distance

The linear discriminant function (LDF) of the predictor variables can be derived in several different ways. The simplest way is to use the minimum distance rule. Suppose we have p numerical predictors and we have a sample of n_i observations $\boldsymbol{x}_{ik} = (x_{i1k}, \ldots, x_{ipk})'$ from the ith group $(i = 1, \ldots, m, k = 1, \ldots, n_i)$. Let $\overline{x}_{ij}$ denote the sample mean of the jth predictor from the ith group. Then we can form the sample mean vectors $\overline{\boldsymbol{x}}_i = (\overline{x}_{i1}, \ldots, \overline{x}_{ip})'$ for different groups, $i = 1, \ldots, m$.

Consider a new observation $\boldsymbol{x} = (x_1, \ldots, x_p)'$. The squared **Euclidean distance** between $\boldsymbol{x}$ and $\overline{\boldsymbol{x}}_i$ can be computed as

$$d_i^2 = (x_1 - \overline{x}_{i1})^2 + \cdots + (x_p - \overline{x}_{ip})^2 = (\boldsymbol{x} - \overline{\boldsymbol{x}}_i)'(\boldsymbol{x} - \overline{\boldsymbol{x}}_i) \quad (i = 1, \ldots, m). \quad (8.1)$$

The minimum Euclidean distance rule assigns $\boldsymbol{x}$ to that Group i for which d_i^2 is smallest, i.e. $\boldsymbol{x}$ is nearest to the mean $\overline{\boldsymbol{x}}_i$.

There are three problems with the Euclidean distance function.

1. It combines variables with different units and scales of measurement, e.g. blood pressure and cholesterol level.

2. It does not take into account the different variances of the variables. For example, the variables with large variances should be weighted less and vice versa.

3. It also does not take into account the correlations among the variables. If two variables are highly correlated, then both should not be highly weighted.

The first two problems can be addressed by using standardized variables to calculate the Euclidean distances. In other words,

$$d_i^2 = \left(\frac{x_1 - \overline{x}_{i1}}{s_1}\right)^2 + \cdots + \left(\frac{x_p - \overline{x}_{ip}}{s_p}\right)^2 \quad (i = 1, \ldots, m), \quad (8.2)$$

where s_j^2 is the pooled sample variance of the jth variable estimated from all the groups.

To address the problem of correlations among the variables, we use the sample covariance matrix. Let $\boldsymbol{S} = \{s_{jk}\}$ denote the pooled sample covariance matrix from the m groups. (Calculation of $\boldsymbol{S}$ is explained in Section 8.4.) The diagonal entries of $\boldsymbol{S}$ are the pooled sample variances s_j^2 and the off-diagonal entries are the pooled sample covariances s_{jk} for $j \neq k$. We can then generalize (8.2) to

$$d_i^2 = (\boldsymbol{x} - \overline{\boldsymbol{x}}_i)' \boldsymbol{S}^{-1} (\boldsymbol{x} - \overline{\boldsymbol{x}}_i) \quad (i = 1, \ldots, m). \tag{8.3}$$

This is called the squared **Mahalnobis distance** between the new observation $\boldsymbol{x}$ and the sample mean vector $\overline{\boldsymbol{x}}_i$ of the ith group.

The minimum Mahalnobis distance rule classifies the observation $\boldsymbol{x}$ to that group, which has the smallest d_i^2. The new observation $\boldsymbol{x}$ is closer to the ith group mean $\overline{\boldsymbol{x}}_i$ than to the jth group mean $\overline{\boldsymbol{x}}_j$ in terms of the Mahalnobis distance if

$$
\begin{aligned}
d_i^2 < d_j^2 &\iff (\boldsymbol{x} - \overline{\boldsymbol{x}}_i)' \boldsymbol{S}^{-1} (\boldsymbol{x} - \overline{\boldsymbol{x}}_i) < (\boldsymbol{x} - \overline{\boldsymbol{x}}_j)' \boldsymbol{S}^{-1} (\boldsymbol{x} - \overline{\boldsymbol{x}}_j) \\
&\iff \boldsymbol{x}' \boldsymbol{S}^{-1} \boldsymbol{x} - 2\overline{\boldsymbol{x}}_i' \boldsymbol{S}^{-1} \boldsymbol{x} + \overline{\boldsymbol{x}}_i' \boldsymbol{S}^{-1} \overline{\boldsymbol{x}}_i < \boldsymbol{x}' \boldsymbol{S}^{-1} \boldsymbol{x} - 2\overline{\boldsymbol{x}}_j' \boldsymbol{S}^{-1} \boldsymbol{x} + \overline{\boldsymbol{x}}_j' \boldsymbol{S}^{-1} \overline{\boldsymbol{x}}_j \\
&\iff \overline{\boldsymbol{x}}_i' \boldsymbol{S}^{-1} \boldsymbol{x} - \frac{1}{2}\overline{\boldsymbol{x}}_i' \boldsymbol{S}^{-1} \overline{\boldsymbol{x}}_i > \overline{\boldsymbol{x}}_j' \boldsymbol{S}^{-1} \boldsymbol{x} - \frac{1}{2}\overline{\boldsymbol{x}}_j' \boldsymbol{S}^{-1} \overline{\boldsymbol{x}}_j \\
&\iff L_i > L_j, \tag{8.4}
\end{aligned}
$$

where

$$L_i = -\frac{1}{2}\overline{\boldsymbol{x}}_i' \boldsymbol{S}^{-1} \overline{\boldsymbol{x}}_i + \overline{\boldsymbol{x}}_i' \boldsymbol{S}^{-1} \boldsymbol{x} = -C_i + \overline{\boldsymbol{x}}_i' \boldsymbol{S}^{-1} \boldsymbol{x} \quad (i = 1, \ldots, m), \tag{8.5}$$

and $C_i = -\frac{1}{2}\overline{\boldsymbol{x}}_i' \boldsymbol{S}^{-1} \overline{\boldsymbol{x}}_i$ is a constant depending on the mean $\overline{\boldsymbol{x}}_i$ of the Group i. L_i is called the **linear discriminant function (LDF)** for the ith group since it is linear in the new observation vector $\boldsymbol{x}$. The quadratic term, $\boldsymbol{x}' \boldsymbol{S}^{-1} \boldsymbol{x}$, which is common to both d_i^2 and d_j^2, gets canceled from both sides of the above inequality.

If the homoscedasticity assumption is dropped, then we cannot pool the sample covariance matrices $\boldsymbol{S}_i$ from the different groups. In that case, the inequality $d_i^2 < d_j^2$ becomes $Q_i > Q_j$, where

$$Q_i = -\frac{1}{2}\boldsymbol{x}' \boldsymbol{S}_i^{-1} \boldsymbol{x} + \overline{\boldsymbol{x}}_i' \boldsymbol{S}_i^{-1} \boldsymbol{x} - \frac{1}{2}\overline{\boldsymbol{x}}_i' \boldsymbol{S}_i^{-1} \overline{\boldsymbol{x}}_i \quad (i = 1, \ldots, m). \tag{8.6}$$

Q_i is called the **quadratic discriminant function (QDF)** since it is a quadratic function of the observation $\boldsymbol{x}$.

8.1.2 Bayesian classification

Suppose that some prior information is available about the prevalence of the m groups and this information can be quantified in terms of **prior probabilities** $\pi_i = P(y = i)$ where $\sum_{i=1}^{m} \pi_i = 1$. The **posterior probability** that a new observation $\boldsymbol{x}$ belongs to Group i, $P(y = i|\boldsymbol{x})$, can be computed by combining the prior probability π_i with the probability mass or density function of $\boldsymbol{x}$ assuming that it comes from Group i, namely $f(\boldsymbol{x}|y = i) = f_i(\boldsymbol{x})$, by applying the **Bayes rule** (see Section 8.4.3):

$$
\begin{aligned}
\widehat{p}_i(\boldsymbol{x}) &= P(y = i|\boldsymbol{x}) \\
&= \frac{P(y = i)f(\boldsymbol{x}|y = i)}{f(\boldsymbol{x})} \\
&= \frac{\pi_i f_i(\boldsymbol{x})}{\sum_{j=1}^{m} \pi_j f_j(\boldsymbol{x})} \quad (i = 1, \ldots, m).
\end{aligned}
\tag{8.7}
$$

The **Bayes classifier** assigns the observation $\boldsymbol{x}$ to that Group i for which the posterior probability $\widehat{p}_i(\boldsymbol{x})$ is maximum.

If the π_i are equal, then they cancel out from the numerator and the denominator of (8.7), and we get $\widehat{p}_i \propto f_i(\boldsymbol{x})$. In this case, $\widehat{p}_i > \widehat{p}_j \iff f_i(\boldsymbol{x}) > f_j(\boldsymbol{x})$. This is sometimes referred to as **likelihood classification rule**, which ignores prior probabilities. This rule can be simplified if we assume that $f_i(\boldsymbol{x})$ is MVN with mean vector $\boldsymbol{\mu}_i = (\mu_{i1}, \ldots, \mu_{ip})'$ and a common covariance matrix $\boldsymbol{\Sigma}$ whose diagonal elements are $\sigma_j^2 = \text{Var}(x_j)$ and the off-diagonal elements are $\sigma_{jk} = \text{Cov}(x_j, x_k)$ $(1 \leq j < k \leq p)$. This MVN p.d.f. has the formula:

$$
f_i(\boldsymbol{x}) = \frac{1}{(2\pi)^{p/2}|\boldsymbol{\Sigma}|^{1/2}} \exp\left\{ -\frac{1}{2}(\boldsymbol{x} - \boldsymbol{\mu}_i)'\boldsymbol{\Sigma}^{-1}(\boldsymbol{x} - \boldsymbol{\mu}_i) \right\} \quad (i = 1, \ldots, m), \tag{8.8}
$$

where $|\boldsymbol{\Sigma}|$ is the determinant of $\boldsymbol{\Sigma}$. This formula generalizes the well-known formula for the univariate normal p.d.f. Note that $(\boldsymbol{x} - \boldsymbol{\mu}_i)'\boldsymbol{\Sigma}^{-1}(\boldsymbol{x} - \boldsymbol{\mu}_i)$ is the squared *population* Mahalnobis distance, whereas d_i^2 defined earlier is squared *sample* Mahalnobis distance. For notational convenience, we continue to denote this squared population Mahalnobis distance by d_i^2. Thus

$$
f_i(\boldsymbol{x}) \propto \exp\left(-\frac{1}{2}d_i^2 \right).
$$

The constant of proportionality is the same for all groups i under the homoscedasticity assumption. So the likelihood classification rule simplifies to

$$
f_i(\boldsymbol{x}) > f_j(\boldsymbol{x}) \iff \exp\left(-\frac{1}{2}d_i^2 \right) > \exp\left(-\frac{1}{2}d_j^2 \right) \iff d_i^2 < d_j^2.
$$

Thus, the likelihood classification rule is equivalent to the minimum Mahalnobis distance rule for classification (after replacing the population quantities $\boldsymbol{\mu}_i$ and $\boldsymbol{\Sigma}$ by $\overline{\boldsymbol{x}}_i$ and $\boldsymbol{S}$, respectively).

We can further simplify both these classification rules as follows. Substitute the MVN p.d.f. in (8.7) and cancel the common terms from the numerator and denominator, to obtain

$$\widehat{p}_i(\boldsymbol{x}) = \frac{\pi_i \exp\{\boldsymbol{\mu}_i'\boldsymbol{\Sigma}^{-1}\boldsymbol{x} - (1/2)\boldsymbol{\mu}_i'\boldsymbol{\Sigma}^{-1}\boldsymbol{\mu}_i\}}{\sum\limits_{j=1}^{m} \pi_j \exp\{\boldsymbol{\mu}_j'\boldsymbol{\Sigma}^{-1}\boldsymbol{x} - (1/2)\boldsymbol{\mu}_j'\boldsymbol{\Sigma}^{-1}\boldsymbol{\mu}_j\}} \quad (i = 1, \ldots, m). \tag{8.9}$$

After replacing $\boldsymbol{\mu}_i$ by the sample mean vector $\overline{\boldsymbol{x}}_i$ and $\boldsymbol{\Sigma}$ by the pooled sample covariance matrix $\boldsymbol{S}$, we see that the above posterior probabilities simplify to

$$\widehat{p}_i(\boldsymbol{x}) = \frac{\pi_i \exp(L_i)}{\sum\limits_{j=1}^{m} \pi_j \exp(L_j)} \quad (i = 1, \ldots, m), \tag{8.10}$$

where the L_i are the LDFs defined in (8.5). We assign the new observation $\boldsymbol{x}$ to that group i, which has the highest posterior probability $\widehat{p}_i(\boldsymbol{x})$. If the prior probabilities are equal, $\pi_i = 1/m$, then (8.10) simplifies to

$$\widehat{p}_i(\boldsymbol{x}) = \frac{\exp(L_i)}{\sum\limits_{j=1}^{m} \exp(L_j)}. \tag{8.11}$$

Example 8.1 *(MBA Admissions: LDA Using Mahalnobis Distance and Bayes Classification)* The LDFs L_1, L_2, L_3 computed using `Minitab` are shown below.

Linear Discriminant Function for Groups

	1	2	3
Constant	-240.37	-177.32	-133.90
GPA	106.25	92.67	78.09
GMAT	0.21	0.17	0.17

To classify a new observation GPA $= 3.20$ and GMAT $= 450$, we calculate

$$
\begin{aligned}
L_1 &= -240.37 + 106.25(3.20) + 0.21(450) &= 194.13, \\
L_2 &= -177.32 + 92.67(3.20) + 0.17(450) &= 195.72, \\
L_3 &= -133.90 + 78.09(3.20) + 0.17(450) &= 192.49.
\end{aligned}
$$

Since L_2 is the largest, this applicant is classified to the "wait-list" category.

Next, we calculate posterior probabilities for the three admission decisions. To simplify this calculation, we subtract a common number, say 190, from all three L_i's since the common factor $\exp(190)$ would cancel from both the numerator and denominator of each L_i. If the prior probabilities are equal, then we get

$$\widehat{p}_1 = \frac{\exp(4.13)}{\exp(4.13) + \exp(5.72) + \exp(2.49)} = 0.164,$$

$$\widehat{p}_2 = \frac{\exp(5.72)}{\exp(4.13) + \exp(5.72) + \exp(2.49)} = 0.804,$$

$$\widehat{p}_3 = \frac{\exp(2.49)}{\exp(4.13) + \exp(5.72) + \exp(2.49)} = 0.032.$$

These probabilities are in agreement with the respective L_i with $\widehat{p}_2$ being the highest. In Example 7.12, using ordinal logistic regression we obtained the probability of admit to be 33% and the probability of wait-list to be 67%, so the probabilities predicted by the two methods are different but are similarly ordered.

If the prior probabilities are unequal, say $\pi_1 = 0.3, \pi_2 = 0.2$, and $\pi_3 = 0.5$, then the posterior probabilities are calculated as follows.

$$\widehat{p}_1 = \frac{0.3 \exp(4.13)}{0.3 \exp(4.13) + 0.2 \exp(5.72) + 0.5 \exp(2.49)} = 0.218,$$

$$\widehat{p}_2 = \frac{0.2 \exp(5.72)}{0.3 \exp(4.13) + 0.2 \exp(5.72) + 0.5 \exp(2.49)} = 0.712,$$

$$\widehat{p}_3 = \frac{0.5 \exp(2.49)}{0.3 \exp(4.13) + 0.2 \exp(5.72) + 0.5 \exp(2.49)} = 0.070.$$

These probabilities are somewhat different from the previous ones but the order is unchanged.

If we apply these discriminant functions to the MBA admissions data from which they are estimated, then we get the following confusion matrix:

```
                    True Group
Put into Group      1      2      3
1                   27     1      0
2                   4      25     2
3                   0      0      26
Total N             31     26     28
N correct           27     25     26
```

So the correct classification rate is $(27 + 25 + 26)/85 = 91.8\%$. $\square$

8.2 Fisher's linear discriminant function

8.2.1 Two groups

Fisher's linear discriminant (LD) function is an alternative but equivalent way to the Mahalnobis distance function classification rule. First consider the $m = 2$ groups case. Since the classification rule (8.4) in this case depends only on a single difference $L_1 - L_2$, it can be reformulated in terms of a single linear function of $\boldsymbol{x}$ given by:

$$\text{LD} = \text{LD}(\boldsymbol{x}) = (\overline{\boldsymbol{x}}_1 - \overline{\boldsymbol{x}}_2)' \boldsymbol{S}^{-1} \boldsymbol{x}, \qquad (8.12)$$

called Fisher's linear discriminant (LD) function. As shown in Figure 8.1, $\text{LD} = \text{LD}(\boldsymbol{x})$ is the projection of the point $\boldsymbol{x}$ on the straight line defined by this function. Let $\text{LD}_1 = \text{LD}(\overline{\boldsymbol{x}}_1)$ and $\text{LD}_2 = \text{LD}(\overline{\boldsymbol{x}}_2)$ denote the projections of the group means $\overline{\boldsymbol{x}}_1$ and $\overline{\boldsymbol{x}}_2$ on this line and let $\overline{\text{LD}} = (1/2)(\text{LD}_1 + \text{LD}_2) = C_1 - C_2$, where the C_i are defined in (8.5). Then (8.4) is equivalent to the following classification rule:

Classify $\boldsymbol{x}$ to Group 1 if $\text{LD} > \overline{\text{LD}}$, otherwise classify $\boldsymbol{x}$ to Group 2. $\qquad$ (8.13)

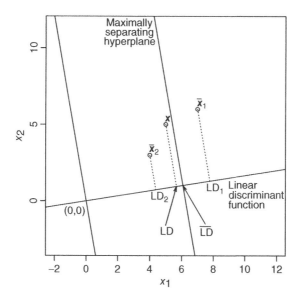

Figure 8.1. *Linear discriminant function for two groups*

Fisher derived this discriminant function using a geometric approach. He aimed to find a linear transformation of the observation vectors $\boldsymbol{x}_{ik}$ to univariate values $y_{ik} = \boldsymbol{c}' \boldsymbol{x}_{ik} = c_1 x_{i1k} + \cdots + c_p x_{ipk}$ $(i = 1, 2; k = 1, \ldots, n_i)$ such that the transformed values y_{1k} from Group 1 and y_{2k} from Group 2 are maximally separated in terms of the squared t-statistic (or equivalently the F-statistic) for comparing their means:

$$t^2(\boldsymbol{c}) = \frac{(\bar{y}_1 - \bar{y}_2)^2}{s_y^2(1/n_1 + 1/n_2)}.$$

We can ignore the term $(1/n_1 + 1/n_2)$ in maximization. Here $\bar{y}_1$ and $\bar{y}_2$ are the sample means of y_{1k} and y_{2k}, respectively, and s_y^2 is the pooled sample variance of the y_{ik} $(i = 1, 2)$ given by

$$s_y^2 = \frac{1}{n_1 + n_2 - 2} \left[\sum_{k=1}^{n_1} (y_{1k} - \bar{y}_1)^2 + \sum_{k=1}^{n_2} (y_{2k} - \bar{y}_2)^2 \right].$$

Now note that $\bar{y}_1 = \boldsymbol{c}'\bar{\boldsymbol{x}}_1, \bar{y}_2 = \boldsymbol{c}'\bar{\boldsymbol{x}}_2$, and $s_y^2 = \boldsymbol{c}'\boldsymbol{S}\boldsymbol{c}$. Hence,

$$t^2(\boldsymbol{c}) \propto \frac{[\boldsymbol{c}'(\bar{\boldsymbol{x}}_1 - \bar{\boldsymbol{x}}_2)]^2}{\boldsymbol{c}'\boldsymbol{S}\boldsymbol{c}}.$$

Exercise 8.1 asks you to show that $t^2(\boldsymbol{c})$ is maximized by $\boldsymbol{c} = \boldsymbol{S}^{-1}(\bar{\boldsymbol{x}}_1 - \bar{\boldsymbol{x}}_2)$, which results in Fisher's LD given by (8.12). Section 8.4 extends this to multiple groups.

We give a small numerical example to illustrate the calculation of Fisher's linear discriminating functions and classification.

Example 8.2 (Calculation of Fisher's LD for Two Predictors) Consider two-group discriminant analysis with two predictors, x_1 and x_2. Suppose the mean vectors of the two groups are $\bar{\boldsymbol{x}}_1 = (7, 6)'$ and $\bar{\boldsymbol{x}}_2 = (4, 3)'$, and the sample covariance matrix of x_1 and x_2 is a diagonal matrix (i.e. x_1 and x_2 are uncorrelated):

$$\boldsymbol{S} = \begin{bmatrix} 4 & 0 \\ 0 & 9 \end{bmatrix}.$$

Its inverse is

$$\boldsymbol{S}^{-1} = \begin{bmatrix} 1/4 & 0 \\ 0 & 1/9 \end{bmatrix}.$$

So the LDFs are

$$L_1 = (7, 6) \begin{bmatrix} 1/4 & 0 \\ 0 & 1/9 \end{bmatrix} \begin{bmatrix} x_1 \\ x_2 \end{bmatrix} - \frac{1}{2}(7, 6) \begin{bmatrix} 1/4 & 0 \\ 0 & 1/9 \end{bmatrix} \begin{bmatrix} 7 \\ 6 \end{bmatrix} = \frac{7}{4}x_1 + \frac{2}{3}x_2 - \frac{65}{8}$$

and

$$L_2 = (4, 3) \begin{bmatrix} 1/4 & 0 \\ 0 & 1/9 \end{bmatrix} \begin{bmatrix} x_1 \\ x_2 \end{bmatrix} - \frac{1}{2}(4, 3) \begin{bmatrix} 1/4 & 0 \\ 0 & 1/9 \end{bmatrix} \begin{bmatrix} 4 \\ 3 \end{bmatrix} = x_1 + \frac{1}{3}x_2 - \frac{5}{2}.$$

Consider a new observation $\boldsymbol{x} = (x_1, x_2)' = (5,5)'$. Then, we have $L_1 = (7/4)5 + (2/3)5 - 65/8 = 3.958$ and $L_2 = 5 + (1/3)5 - 5/2 = 4.167$. Since $L_2 > L_1$, this observation is classified to Group 2.

Fisher's LD function is given by

$$\mathrm{LD} = (7 - 4, 6 - 3) \begin{bmatrix} 1/4 & 0 \\ 0 & 1/9 \end{bmatrix} \begin{bmatrix} x_1 \\ x_2 \end{bmatrix} = \frac{3}{4}x_1 + \frac{3}{9}x_2.$$

For the new observation $\boldsymbol{x} = (x_1, x_2)' = (5,5)'$, we have $\mathrm{LD} = (3/4)5 + (3/9)5 = 5.147$. Furthermore, $\mathrm{LD}_1 = (3/4)7 + (3/9)6 = 7.250$ and $\mathrm{LD}_2 = (3/4)4 + (3/9)3 = 4.000$, so $\overline{\mathrm{LD}} = 5.625$. Since $\mathrm{LD} < \overline{\mathrm{LD}}$, the observation is classified to Group 2. $\qquad\square$

8.2.2 Multiple groups*

In case of multiple groups, Fisher's approach to LDA is more complicated; its derivation is given in Section 8.4. Whereas in the case of two groups we have a single discriminant function LD, in the case of $m > 2$ groups we have $q = \min(m - 1, p)$ discriminant functions, denoted by $\mathrm{LD}_1, \ldots, \mathrm{LD}_q$. These discriminant functions are ordered in terms of the extent to which they separate the m groups; LD_1 separates the groups maximally, LD_2 is next, and so on. They are uncorrelated with each other. The fraction of separation captured by any discriminant function is measured by the proportion of the trace of a certain matrix (defined in Section 8.4.2). We may decide to use only the first $r < q$ discriminant functions if they capture most of the separation. The basic idea is similar to that of principal components analysis discussed in Section 5.3.1.

To classify a new observation $\boldsymbol{x}$, we project it onto the first $r \leq q$ discriminant function axes giving $\mathrm{LD}_1(\boldsymbol{x}), \ldots, \mathrm{LD}_r(\boldsymbol{x})$, referred to as **discriminant scores**, where $\mathrm{LD}_k(\boldsymbol{x}) = \boldsymbol{c}_k' \boldsymbol{x}$ and $\boldsymbol{c}_k = (c_{k1}, \ldots, c_{kp})'$ is the coefficient vector of LD_k ($k = 1, \ldots, r$). We can similarly compute the r discriminant scores for the group means $\overline{\boldsymbol{x}}_1, \ldots, \overline{\boldsymbol{x}}_m$. Denote them by $\mathrm{LD}_1(\overline{\boldsymbol{x}}_i), \ldots, \mathrm{LD}_r(\overline{\boldsymbol{x}}_i)$ ($i = 1, \ldots, m$). Then we compute the distances between the vector of discriminant scores of $\boldsymbol{x}$ from the vector of discriminant scores of each group mean $\overline{\boldsymbol{x}}_i$ and assign $\boldsymbol{x}$ to that group for which the distance is the least. Since the discriminant scores are uncorrelated, we can use the Euclidean distance. Thus define the squared distance between the vector of discriminant scores of $\boldsymbol{x}$ from the vector of discriminant scores of $\overline{\boldsymbol{x}}_i$ by

$$d_i^2 = [\mathrm{LD}_1(\boldsymbol{x}) - \mathrm{LD}_1(\overline{\boldsymbol{x}}_i)]^2 + \cdots + [\mathrm{LD}_r(\boldsymbol{x}) - \mathrm{LD}_r(\overline{\boldsymbol{x}}_i)]^2, \quad i = 1, \ldots, m.$$

We classify $\boldsymbol{x}$ to that group for which d_i^2 is smallest.

Example 8.3 *(MBA Admissions: Discriminant Analysis Using Fisher's Method)* The MASS package in R calculates Fisher's LDs. We run the following script to calculate these functions and to estimate the probabilities of the three classifications.

```
> library(MASS)
> fit=lda(admit ~ GPA+GMAT,data=MBA,prior=c(1,1,1)/3)
> fit
> predict(fit,newdata=data.frame(GPA=3.20,GMAT=450))
> prob = fitted(fit, outcome= FALSE)
```

The R output is as follows.

```
Prior probabilities of groups:
        1         2         3
0.3333333 0.3333333 0.3333333

Group means:
       GPA      GMAT
1 3.403871 561.2258
2 2.992692 446.2308
3 2.482500 447.0714

Coefficients of linear discriminants:
           LD1         LD2
GPA   5.017202736  1.85401003
GMAT 0.008503148 -0.01448967

Proportion of trace:
   LD1    LD2
0.9644 0.0356
$'class'
[1] 2
Levels: 1 2 3

$posterior
          1         2           3
1 0.1119752 0.8870063 0.001018516
```

Thus predicted chances of different outcomes for this applicant are 11.2% admit, 88.7% wait-list, and 0.1% deny admission. These predicted chances are in line with those obtained in Examples 7.11 and 7.12.

We see that LD_1 captures 96.44% of the separation. However, we shall use both LD_1 and LD_2 for illustration purposes in this example. The discriminant scores for the new observation (GPA= 3.20, GMAT= 450) are obtained by the following matrix product:

$$\begin{bmatrix} 5.0172 & 0.0085 \\ 1.8540 & -0.0145 \end{bmatrix} \begin{bmatrix} 3.20 \\ 450 \end{bmatrix} = \begin{bmatrix} 19.8800 \\ -0.5922 \end{bmatrix}.$$

A similar calculation gives the discriminant scores for the three group means:

$$\begin{bmatrix} 5.0172 & 0.0085 \\ 1.8540 & -0.0145 \end{bmatrix} \begin{bmatrix} 3.4039 & 2.9927 & 2.4825 \\ 561.23 & 446.23 & 447.07 \end{bmatrix} = \begin{bmatrix} 21.8485 & 18.8079 & 16.2553 \\ -1.8270 & -0.9219 & -1.8800 \end{bmatrix}.$$

Then, we calculate the squared Euclidean distances between the discriminant scores for the new observation and the three group means as follows:

$$\begin{aligned} d_1^2 &= (19.8800 - 21.8485)^2 + (-0.5922 + 1.8270)^2 &= 5.400, \\ d_2^2 &= (19.8800 - 18.8079)^2 + (-0.5922 + 0.9219)^2 &= 1.258, \\ d_3^2 &= (19.8800 - 16.2553)^2 + (-0.5922 + 1.8800)^2 &= 14.797. \end{aligned}$$

Since d_2^2 is the smallest of the three, we assign this student to the "wait-list" category. This result is in agreement with the previous result obtained using the Mahalnobis distance method, as it should be, since the two methods are equivalent. □

8.3 Naive Bayes

In many applications, the number of predictors, p, is very large and can be even much greater than the sample size, n. In genomics data, for example, the number of genes that are measured to discriminate between disease patients and non-disease patients can be in thousands, but the number of blood samples from the two groups of patients is relatively small. Another example is text analytics where the features are thousands of different words which are used to discriminate between different text samples. The data consist of counts of words in the samples. Some examples are determination of authorship of an anonymous document based on the writing samples of the possible authors; another example is sentiment analysis of tweets. In these examples, the joint distribution of the features is difficult to estimate. In particular, the covariance matrix S, which has $p(p+1)/2$ terms, is impossible to estimate if $p > n$. Even if $p < n$ and still quite large, S does not give a good estimate of Σ and is difficult to invert. The **naive Bayes** method obviates these difficulties by treating the features as independent. Thus in (8.7) we write

$$f_i(\boldsymbol{x}) = \prod_{j=1}^{p} f_{ij}(x_j),$$

where $\boldsymbol{x} = (x_1, \dots, x_p)'$ and $f_{ij}(x_j)$ is the marginal distribution of x_j conditioned on $y = i$, i.e. if the observation $\boldsymbol{x}$ comes from Group i. If the joint distribution of $\boldsymbol{x}$ from Group i is MVN with mean vector $\boldsymbol{\mu}_i = (\mu_{i1}, \dots, \mu_{ip})'$ and covariance matrix $\boldsymbol{\Sigma}$, then the marginal distributions $f_{ik}(x_k)$ are univariate normal with means μ_{ik} and variances σ_k^2 (the diagonal elements of $\boldsymbol{\Sigma}$). Therefore,

$$f_i(\boldsymbol{x}) = \frac{1}{(2\pi)^{p/2} \prod\limits_{j=1}^{p} \sigma_j} \exp\left\{ -\frac{1}{2} \sum_{j=1}^{p} [(x_j - \mu_{ij})^2 / \sigma_j^2] \right\}.$$

Here, the variance σ_j^2 for the jth predictor is taken to be the same for all groups i according to the homoscedasticity assumption. Then, (8.9) simplifies to

$$\widehat{p}_i = \frac{\pi_i \exp\left\{ -\frac{1}{2} \sum\limits_{j=1}^{p} [(x_j - \mu_{ij})^2 / \sigma_j^2] \right\}}{\sum\limits_{k=1}^{m} \pi_k \exp\left\{ -\frac{1}{2} \sum\limits_{j=1}^{p} [(x_j - \mu_{kj})^2 / \sigma_j^2] \right\}} \quad (i = 1, \dots, m).$$

Finally, to calculate $\widehat{p}_i$ we replace μ_{ij} and σ_j^2 by the sample means $\overline{x}_{ij}$ and the pooled sample variances s_j^2, respectively.

If the predictors are Bernoulli random variables (occurrence or nonoccurrence of a certain trait), then their joint multivariate Bernoulli distribution is difficult to model and estimate. In that case, the naive Bayes method ignores the correlations between them and uses the product of Bernoulli probabilities as an approximation.

8.4   Technical notes*

8.4.1   Calculation of pooled sample covariance matrix

Given n_i observations $\boldsymbol{x}_{ik} = (x_{i1k}, \dots, x_{ipk})'$ $(k = 1, \dots, n_i)$ from the ith group, we first calculate the sample means $\overline{x}_{ij}$:

$$\overline{x}_{ij} = \frac{1}{n_i} \sum_{k=1}^{n_i} x_{ijk} \ (i = 1, \dots, m, j = 1, \dots, p).$$

Then, we form the sample mean vectors $\overline{\boldsymbol{x}}_i = (\overline{x}_{i1}, \dots, \overline{x}_{ip})'$.

The diagonal entries of the pooled sample covariance matrix $\boldsymbol{S}$ are the pooled sample variances s_j^2, and the off-diagonal entries are the pooled sample covariances s_{jk} for $j \neq k$. The s_j^2 are given by

$$s_j^2 = \frac{1}{n-m} \sum_{i=1}^{m} (n_i - 1) s_{ij}^2,$$

where $n = \sum_{i=1}^{m} n_i$ is the total sample size and s_{ij}^2 is the sample variance of the jth predictor variable computed for the ith group:

$$s_{ij}^2 = \frac{1}{n_i - 1} \sum_{k=1}^{n_i} (x_{ijk} - \overline{x}_{ij})^2 \quad (i = 1, \ldots, m, j = 1, \ldots, p).$$

The pooled sample covariances are calculated in the same way:

$$s_{jk} = \frac{1}{n-m} \sum_{i=1}^{m} (n_i - 1) s_{ijk} \ (1 \leq j < k \leq p),$$

where

$$s_{ijk} = \frac{1}{n_i - 1} \left[\sum_{\ell=1}^{n_i} (x_{ij\ell} - \overline{x}_{ij})(x_{ik\ell} - \overline{x}_{ik}) \right] \ (i = 1, \ldots, m, 1 \leq j < k \leq p).$$

8.4.2 Derivation of Fisher's linear discriminant functions

First, consider one-way univariate analysis of variance (ANOVA) with data x_{ik} $(i = 1, \ldots, m, k = 1, \ldots, n_i)$ from $m \geq 2$ groups. Let $\overline{x}_{i.}$ denote the mean of the ith group and $\overline{x}_{..}$ denote the grand mean of all x_{ik}. Then, the total sum of squares is defined as $T = \sum_{i=1}^{m} \sum_{k=1}^{n_i} (x_{ij} - \overline{x}_{..})^2$, which can be partitioned into the between sum of squares, $B = \sum_{i=1}^{m} n_i (\overline{x}_{i.} - \overline{x}_{..})^2$ and the within sum of squares, $W = \sum_{i=1}^{m} \sum_{k=1}^{n_i} (x_{ik} - \overline{x}_{i.})^2$, so that $T = B + W$. We aim to maximally separate the groups by maximizing the between groups variation relative to the within groups variation. This is done by maximizing the ratio B/W, which is proportional to the ANOVA F-statistic.

In one-way multivariate analysis of variance (MANOVA), we have multivariate data vectors $\boldsymbol{x}_{ik} = (x_{i1k}, \ldots, x_{ipk})'$ $(i = 1, \ldots, m, k = 1, \ldots, n_i)$ from $m \geq 2$ groups. Analogous to B, W, and T, we have matrices $\boldsymbol{B}$, $\boldsymbol{W}$, and $\boldsymbol{T}$, referred to as between, within and total sums of squares and cross-product (SSCP) matrices defined as follows:

$$\boldsymbol{B} = \sum_{i=1}^{m} n_i (\overline{\boldsymbol{x}}_{i.} - \overline{\boldsymbol{x}}_{..})(\overline{\boldsymbol{x}}_{i.} - \overline{\boldsymbol{x}}_{..})',$$

$$W = \sum_{i=1}^{m} \sum_{k=1}^{n_i} (\boldsymbol{x}_{ik} - \overline{\boldsymbol{x}}_{i\cdot})(\boldsymbol{x}_{ik} - \overline{\boldsymbol{x}}_{i\cdot})',$$

and

$$T = \sum_{i=1}^{m} \sum_{k=1}^{n_i} (\boldsymbol{x}_{ik} - \overline{\boldsymbol{x}}_{\cdot\cdot})(\boldsymbol{x}_{ik} - \overline{\boldsymbol{x}}_{\cdot\cdot})',$$

where $\overline{\boldsymbol{x}}_{i\cdot}$ and $\overline{\boldsymbol{x}}_{\cdot\cdot}$ are vector analogs $\overline{x}_{i\cdot}$ and $\overline{x}_{\cdot\cdot}$, respectively. They satisfy the MANOVA identity $T = B + W$. As in Section 8.2, this multivariate problem can be reduced to the univariate ANOVA problem by taking linear combinations $y_{ik} = \boldsymbol{c}'\boldsymbol{x}_{ik} = c_1 x_{i1k} + \cdots + c_p x_{ipk}$. Then it is not difficult to show that the between sum of squares for the y_{ik}'s equals $B = \boldsymbol{c}'\boldsymbol{B}\boldsymbol{c}$ and the within sum of squares equals $W = \boldsymbol{c}'\boldsymbol{W}\boldsymbol{c}$. Thus we want to find $\boldsymbol{c}$ to maximize

$$\frac{\boldsymbol{c}'\boldsymbol{B}\boldsymbol{c}}{\boldsymbol{c}'\boldsymbol{W}\boldsymbol{c}}.$$

Since this ratio is invariant to any scalar multiple of $\boldsymbol{c}$, we put the constraint $\boldsymbol{c}'\boldsymbol{W}\boldsymbol{c} = 1$. Then, by the Lagrangian multiplier method, we maximize

$$f(\boldsymbol{c}, \lambda) = \boldsymbol{c}'\boldsymbol{B}\boldsymbol{c} - \lambda(\boldsymbol{c}'\boldsymbol{W}\boldsymbol{c} - 1),$$

where λ is the Lagrangian multiplier. Write

$$\frac{\partial f(\boldsymbol{c}, \lambda)}{\partial \boldsymbol{c}} = 2\boldsymbol{B}\boldsymbol{c} - 2\lambda\boldsymbol{W}\boldsymbol{c} = \boldsymbol{0} \implies (\boldsymbol{B} - \lambda\boldsymbol{W})\boldsymbol{c} = \boldsymbol{0} \implies (\boldsymbol{W}^{-1}\boldsymbol{B} - \lambda\boldsymbol{I})\boldsymbol{c} = \boldsymbol{0},$$

where $\boldsymbol{0}$ is a null vector of dimension p. This is an eigenvalue problem. The solution $\boldsymbol{c} = \boldsymbol{c}_1$ to this equation is the eigenvector of $\boldsymbol{W}^{-1}\boldsymbol{B}$ corresponding to the largest eigenvalue $\lambda = \lambda_1$, which is the solution to the determinantal equation $|\boldsymbol{W}^{-1}\boldsymbol{B} - \lambda\boldsymbol{I}| = 0$. It can be shown that λ_1 is the maximum of the ratio $(\boldsymbol{c}'\boldsymbol{B}\boldsymbol{c})/(\boldsymbol{c}'\boldsymbol{W}\boldsymbol{c})$. The first LD function LD_1 equals $\boldsymbol{c}_1'\boldsymbol{x}$. Note that $\boldsymbol{W}^{-1}\boldsymbol{B}$ is a matrix analog of B/W.

$\boldsymbol{W}^{-1}\boldsymbol{B}$ is of rank $q = \min(m - 1, p)$ and so has q nonzero eigenvalues $\lambda_1 > \cdots > \lambda_q$ and associated eigenvectors $\boldsymbol{c}_1, \ldots, \boldsymbol{c}_q$. These eigenvectors are mutually orthogonal w.r.t. $\boldsymbol{W}$, i.e. $\boldsymbol{c}_i'\boldsymbol{W}\boldsymbol{c}_j = 0$ for all $i \neq j$. Thus, we have q LD functions, $\mathrm{LD}_1, \ldots, \mathrm{LD}_q$. Their associated eigenvalues quantify the proportion of trace of $\boldsymbol{W}^{-1}\boldsymbol{B}$ (which is the sum of the eigenvalues, $\sum_{i=1}^{q} \lambda_i$, of $\boldsymbol{W}^{-1}\boldsymbol{B}$) that they explain with LD_1 explaining the largest fraction $\lambda_1/\sum_{i=1}^{q} \lambda_i$ of the trace.

8.4.3 Bayes rule

The Bayes formula is an extension of the conditional probability formula: If A and B are two random events, then the conditional probability of B conditioned on A is given by

$$P(B|A) = \frac{P(A \cap B)}{P(A)} = \frac{P(A|B)P(B)}{P(A)}.$$

Bayes' extension involves expressing the event B as the union of $m \geq 2$ disjoint events $B = B_1 \cup B_2 \cup \cdots \cup B_m$. Then, the Bayes formula gives the conditional probability $P(B_i|A)$ for $i = 1, \ldots, m$ as follows:

$$
\begin{aligned}
P(B_i|A) &= \frac{P(A \cap B_i)}{P(A)} \\
&= \frac{P(A \cap B_i)}{P(A \cap B_1) + \cdots + P(A \cap B_m)} \\
&= \frac{P(A|B_i)P(B_i)}{P(A|B_1)P(B_1) + \cdots + P(A|B_m)P(B_m)} \\
&= \frac{P(A|B_i)P(B_i)}{\sum_{j=1}^{m} P(A|B_j)P(B_j)} \quad (i = 1, 2, \ldots, m).
\end{aligned}
$$

We can think of A as an observed outcome and $B_1, B_2, \ldots, B_m$ as the possible causes of A; also, $P(B_i)$ as the prior probability of cause B_i. Then, the posterior probability that B_i is the cause of event A is given by the above formula.

Exercises

Theoretical Exercises

8.1 **(Fisher's linear discriminant function for two groups)** For two groups show that the coefficient vector for Fisher's LD function is given by c, where c is proportional to $S^{-1}(\bar{x}_1 - \bar{x}_2)$ subject to the normalizing constraint $c'Sc = 1$.

8.2 **(Equivalence between LDF and Fisher's LD classification rules)** Show that the decision rule (8.4) in terms of LDFs is equivalent to the decision rule (8.13) in terms of Fisher's LD.

Applied Exercises

8.3 (Calculation of LDFs for correlated x_1 and x_2) Assume the same variances as in Example 8.2 but suppose that the sample $\text{Corr}(x_1, x_2) = 0.5$ so that $\text{Cov}(x_1, x_2) = 0.5\sqrt{4 \times 9} = 3$. Thus, the sample covariance matrix is

$$S = \begin{bmatrix} 4 & 3 \\ 3 & 9 \end{bmatrix}.$$

(a) Calculate the linear discriminant functions L_1 and L_2. To which group will a new observation $\boldsymbol{x} = (5, 5)'$ be classified to?

(b) Calculate Fisher's linear discriminant function LD. Classify a new observation $\boldsymbol{x} = (5, 5)'$ using LD and check that it is classified to the same group as in Part (a).

8.4 (Coronary heart disease data) Baseline measurements were made on 832 white males free of coronary heart disease (CHD) on three risk factors: age, diastolic blood pressure (DBP), and cholesterol level (CHL). By the end of the study period, 71 subjects had developed CHD while 761 did not (NCHD). The following linear discriminant functions were computed.

Predictor	L_{NCHD}	L_{CHD}
Const.	-23.561	-28.726
Age	0.027	0.072
DBP	0.338	0.360
CHL	0.075	0.079

(a) Based on the coefficients of these linear discriminant functions explain why the probability of CHD increases with increases in each of the three risk factors?

(b) For a person of 50 years of age, 95 mm of Hg diastolic blood pressure and cholesterol level of $210\,\text{mg dL}^{-1}$, calculate the posterior probabilities of the CHD and NCHD assuming equal prior probabilities.

(c) The mean vector for the NCHD group is $(44.81, 86.99, 201.27)'$ and that for the CHD group is $(56.86, 95.62, 221.51)'$. Calculate the Euclidean distances d_1 and d_2 of a new observation vector $(50, 95, 210)'$ from the NCHD and CHD group means. To which group does the Euclidean distance rule classify this person to? Why does this classification differ from that obtained in Part (b)?

8.5 (Fisher's iris data) Fisher (1936) used this data set to introduce discriminant analysis. The data consist of measurements on the sepal and petal dimensions of 50 samples of three species of iris flowers (iris setosa, iris virginica, and iris versicolor). The data are in file `Iris.csv`.

(a) Do a discriminant analysis of the data. Give Fisher's linear discriminant functions.

(b) For an iris flower with the following dimensions: sepal length $= 5.5\,\mathrm{mm}$, sepal width $= 3.0\,\mathrm{mm}$, petal length $= 4.0\,\mathrm{mm}$, petal width $= 1.5\,\mathrm{mm}$, calculate the posterior probabilities of the three species assuming equal prior probabilities.

Chapter 9

Generalized linear models

In Chapters 3–7, we studied mainly two classes of models: multiple regression
and logistic regression. In multiple regression the response variable was assumed
to be normally distributed, while in logistic regression, the response variable was
assumed to be Bernoulli distributed. In practice, many other types of response
variable distributions are encountered, e.g. in survival and reliability studies,
lifetimes may be assumed to be exponentially or gamma distributed. In other
applications, count type of response variables are encountered, which may be
assumed to be Poisson distributed. **Generalized linear models (GLMs)** introduced by Nelder and Wedderburn (1972) deal with these and many other response
distributions which belong to a class called the **exponential family**.

GLMs use another key concept, called the **link function**. In multiple regression, a linear model is postulated on $E(y) = \mu$. In logistic regression, a linear
model is postulated on the logistic transform of $E(y) = P(y = 1) = p$. A link
function generalizes this idea. It is a function of $E(y)$ on which a linear model is
postulated. GLMs employ these two generalizations to provide a class of useful
predictive models. For more details see McCullagh and Nelder (1989).

9.1 Exponential family and link function

9.1.1 Exponential family

A distribution belonging to the **exponential family** has the following general
form for its p.d.f. or p.m.f.:

$$f(y; \theta, \phi) = \exp\left\{ \frac{y\theta - b(\theta)}{a(\phi)} + c(y; \phi) \right\} \tag{9.1}$$

Predictive Analytics: Parametric Models for Regression and Classification Using R,
First Edition. Ajit C. Tamhane.
© 2021 John Wiley & Sons, Inc. Published 2021 by John Wiley & Sons, Inc.
Companion website: www.wiley.com/go/tamhane/predictiveanalytics

for some functions $a(\cdot), b(\cdot)$, and $c(\cdot)$. This form of the distribution is known as the **canonical form** and θ is referred to as the **natural parameter**, which is the main parameter of interest. On the other hand, ϕ is a nuisance parameter (not the main parameter of interest), which is a measure of the variability of the distribution. Hence, it is referred to as the **dispersion parameter**. In many examples, $a(\phi) = \phi = 1$ and so it may be dropped from (9.1) and the expression can be simplified.

For any exponential family distribution, it can be shown that $E(y)$ and $\text{Var}(y)$ are given by (see Section 9.6 for the derivation)

$$E(y) = \mu = b'(\theta) \quad \text{and} \quad \text{Var}(y) = a(\phi)b''(\theta), \tag{9.2}$$

where $b'(\theta)$ and $b''(\theta)$ are the first and second derivatives of $b(\theta)$. $\text{Var}(y)$ is generally a function of μ, so it is often denoted by $V(\mu)$, called the **variance function**.

Here are four common examples of distributions belonging to the exponential family.

Normal distribution

The normal distribution with mean μ and variance σ^2 can be expressed in the form (9.1) as follows:

$$f(y; \mu, \sigma^2) = \frac{1}{\sqrt{2\pi\sigma^2}} \exp\left\{-\frac{(y-\mu)^2}{2\sigma^2}\right\}$$

$$= \exp\left\{\frac{y\mu - \mu^2/2}{\sigma^2} - \frac{1}{2}\left[\frac{y^2}{\sigma^2} + \ln(2\pi\sigma^2)\right]\right\}$$

for $-\infty < y < \infty$. Hence, we have

$$a(\phi) = \sigma^2, \quad \theta = \mu, \quad b(\theta) = \frac{\mu^2}{2} \quad \text{and} \quad c(y; \phi) = -\frac{1}{2}\left[\frac{y^2}{\sigma^2} + \ln(2\pi\sigma^2)\right].$$

Note that μ is the natural parameter, which is the mean of the distribution as follows from $b'(\theta) = (d/d\mu)(\mu^2/2) = \mu$ and σ^2 is the dispersion parameter, which is the variance of the distribution as follows from $a(\phi)b''(\theta) = \sigma^2(d^2/d\mu^2)$ $(\mu^2/2) = \sigma^2$.

Binomial distribution

The binomial distribution is the distribution of the number of successes in n i.i.d. Bernoulli trials each with success probability p. It can be expressed in the form (9.1) as follows:

$$f(y; n, p) = \binom{n}{y} p^y (1-p)^{n-y}$$

$$= \binom{n}{y} \left(\frac{p}{1-p} \right)^y (1-p)^n$$

$$= \exp \left[y \ln \left(\frac{p}{1-p} \right) + n \ln(1-p) + \ln \binom{n}{y} \right]$$

for $y = 0, \ldots, n$. Hence, we have

$$a(\phi) = 1, \quad \theta = \ln \left(\frac{p}{1-p} \right), \quad b(\theta) = -n \ln(1-p), \quad c(y; \phi) = \ln \binom{n}{y}.$$

Thus the logistic transform $\ln[p/(1-p)]$ is the natural parameter.

We can check that the mean of the binomial distribution is

$$E(y) = b'(\theta) = \frac{db(\theta)}{dp} \frac{dp}{d\theta} = \frac{db(\theta)}{dp} \left(\frac{d\theta}{dp} \right)^{-1} = \frac{n}{1-p} \left(\frac{1}{p(1-p)} \right)^{-1} = np$$

and the variance is

$$\mathrm{Var}(y) = a(\phi) b''(\theta) = \frac{d(np)}{dp} \frac{dp}{d\theta} = n \left(\frac{d\theta}{dp} \right)^{-1} = n \left(\frac{1}{p(1-p)} \right)^{-1} = np(1-p).$$

Of course, these formulae are well-known and can be derived more directly.

Poisson distribution

The Poisson distribution can be expressed in the form (9.1) as follows:

$$f(y; \mu) = \frac{e^{-\mu} \mu^y}{y!} = \exp(y \ln \mu - \mu - \ln y!)$$

for $y = 0, 1, 2, \ldots$. Hence, we have

$$a(\phi) = 1, \quad \theta = \ln \mu, \quad b(\theta) = \mu \quad \text{and} \quad c(y; \phi) = -\ln y!.$$

Thus $\ln \mu$ is the natural parameter.

We can check that the mean of the Poisson distribution is

$$E(y) = b'(\theta) = \frac{d\mu}{d(\ln \mu)} = \left(\frac{d \ln \mu}{d\mu} \right)^{-1} = \mu$$

and the variance is

$$\mathrm{Var}(y) = a(\phi)b''(\theta) = \frac{d\mu}{d(\ln \mu)} = \left(\frac{d \ln \mu}{d\mu}\right)^{-1} = \mu.$$

Gamma distribution

The gamma distribution can be expressed in the form (9.1) as follows:

$$f(y; \lambda, \alpha) = \frac{1}{\Gamma(\alpha)} \lambda^\alpha e^{-\lambda y} y^{\alpha-1} = \exp\{-\lambda y + \alpha \ln \lambda + (\alpha - 1) \ln y - \ln \Gamma(\alpha)\}$$

$$(9.3)$$

for $y \geq 0$, where $\Gamma(\alpha)$ is the **gamma function**.[1] Hence, we have

$$\phi = \alpha, \quad a(\phi) = 1, \quad \theta = -\lambda, \quad b(\theta) = -\alpha \ln \lambda, \quad c(y, \phi) = (\alpha - 1) \ln y - \ln \Gamma(\alpha).$$

Here, α is the dispersion (shape) parameter and λ is known as the scale parameter. For $\alpha = 1$ we have $\Gamma(\alpha) = 1$, and we get the exponential distribution.
We can check that

$$E(y) = b'(\theta) = \frac{d}{d(-\lambda)}(-\alpha \ln \lambda) = \frac{\alpha}{\lambda} \quad \text{and}$$

$$\mathrm{Var}(y) = a(\phi)b''(\theta) = \frac{d}{d(-\lambda)}\left(\frac{\alpha}{\lambda}\right) = \frac{\alpha}{\lambda^2}.$$

9.1.2   Link function

A link function is a monotone differentiable function $g(\mu)$ of $\mu = E(y)$. If $g(\mu)$ is chosen to be equal to the natural parameter θ, then it is known as the **canonical link function**. In this case, the inverse of the link function, $g^{-1}(\theta)$, equals $\mu = b'(\theta)$.
   In general, $g(\mu)$ can be arbitrary and we denote it by η, called the **linear predictor** because a linear model is postulated on it:

$$g(\mu) = \eta = \beta_0 + \beta_1 x_1 + \cdots + \beta_p x_p = \boldsymbol{x}'\boldsymbol{\beta}. \qquad (9.4)$$

If $\eta = \theta$, i.e. if we use the canonical link function as the linear predictor then the estimation of the regression parameters is simplified. For the gamma distribution, the natural parameter $\lambda = \alpha/\mu$, but the inverse function $g(\mu) = 1/\mu$ is generally used as the canonical link function. In some cases, it is more convenient to use a link function different from the canonical link function. For instance, for the

---

[1]For all α except for negative integers, $\Gamma(\alpha) = \int_0^\infty x^{\alpha-1} e^{-x}\, dx$ and $\Gamma(0) = \Gamma(1) = 1$. $\Gamma(\alpha)$ satisfies the recursive relation $\Gamma(\alpha) = (\alpha - 1)\Gamma(\alpha - 1)$. If α is a positive integer, then we get $\Gamma(\alpha) = (\alpha - 1)!$.

gamma distribution, $\ln \mu$ may be used as the link function instead of $1/\mu$ to ensure non-negativity constraint on μ.

It is important to remember that the fitting algorithm applies the link function to the mean of y. That transformation is not applied to y itself. Thus, if we use the log-link function, y itself is not log-transformed.

9.2   Estimation of parameters of GLM

9.2.1   Maximum likelihood estimation

Let $y_1, \ldots, y_n$ denote independent observations, each y_i having the same exponential family distribution (9.1) but with a different θ_i, which depends on the predictors $\boldsymbol{x}_i = (1, x_{i1}, \ldots, x_{ip})'$ $(i = 1, \ldots, n)$. Assuming the canonical link function, we have

$$g(\mu_i) = \eta_i = \theta_i = \beta_0 + \beta_1 x_{i1} + \cdots + \beta_p x_{ip} = \boldsymbol{x}_i' \boldsymbol{\beta} \quad (i = 1, \ldots, n). \qquad (9.5)$$

We use the MLE to estimate the unknown regression parameters β_j's. The equation for finding the MLE of $\boldsymbol{\beta}$ has the general form (see Section 9.6 for a derivation):

$$\boldsymbol{X}' \boldsymbol{\mu} = \boldsymbol{X}' \boldsymbol{y}. \qquad (9.6)$$

Here, $\boldsymbol{X}$ is the $n \times (p+1)$ model matrix with row vectors $\boldsymbol{x}_i' = (1, x_{i1}, \ldots, x_{ip})$ $(i = 1, \ldots, n)$, $\boldsymbol{y} = (y_1, \ldots, y_n)'$, and $\boldsymbol{\mu} = (\mu_1, \ldots, \mu_n)'$, where $\mu_i = E(y_i)$. Note that

$$\boldsymbol{X}' \boldsymbol{\mu} = \begin{bmatrix} \boldsymbol{x}_1 & \boldsymbol{x}_2 & \cdots & \boldsymbol{x}_n \end{bmatrix} \begin{bmatrix} \mu_1 \\ \mu_2 \\ \vdots \\ \mu_n \end{bmatrix} = \sum_{i=1}^{n} \mu_i \boldsymbol{x}_i.$$

Similarly, $\boldsymbol{X}' \boldsymbol{y} = \sum_{i=1}^{n} y_i \boldsymbol{x}_i$. Thus, (9.6) can be written as a column vector of equations:

$$\sum_{i=1}^{n} \mu_i \boldsymbol{x}_i = \sum_{i=1}^{n} y_i \boldsymbol{x}_i \quad \text{or} \quad E \sum_{i=1}^{n} y_i \boldsymbol{x}_i = \sum_{i=1}^{n} y_i \boldsymbol{x}_i. \qquad (9.7)$$

This is a system of $p+1$ equations in $p+1$ unknowns, $\beta_0, \beta_1, \ldots, \beta_p$. Notice that this system equates certain linear combinations (defined by the predictor vectors $\boldsymbol{x}_i$'s) of the $E(y_i)$'s to the same linear combinations of the observed y_i's.

Once the MLE $\widehat{\boldsymbol{\beta}}$ of $\boldsymbol{\beta}$ is obtained, the fitted or predicted values of the y_i can be calculated from $\widehat{y}_i = \widehat{\mu}_i = g^{-1}(\boldsymbol{x}_i' \widehat{\boldsymbol{\beta}})$. For example, in Poisson regression, $g(\mu_i) = \ln \mu_i$ and so $\widehat{y}_i = \exp(\boldsymbol{x}_i' \widehat{\boldsymbol{\beta}})$. In gamma regression, $g(\mu_i) = 1/\mu_i$ and so $\widehat{y}_i = 1/\boldsymbol{x}_i' \widehat{\boldsymbol{\beta}}$.

As an illustration of (9.6), consider the MLE of $\boldsymbol{\beta}$ for multiple regression. Putting $\boldsymbol{\mu} = \boldsymbol{X\beta}$ in (9.6), we see that it reduces to $\boldsymbol{X'X\beta} = \boldsymbol{X'y}$, which is the same equation (3.6) for the LS estimator of $\boldsymbol{\beta}$.

For another example, consider the simple logistic regression model (7.1). In this case

$$
\boldsymbol{X} = \begin{bmatrix} 1 & x_1 \\ 1 & x_2 \\ \vdots & \vdots \\ 1 & x_n \end{bmatrix}, \quad \boldsymbol{y} = \begin{bmatrix} y_1 \\ y_2 \\ \vdots \\ y_n \end{bmatrix} \quad \text{and} \quad \boldsymbol{\mu} = \begin{bmatrix} p_1 \\ p_2 \\ \vdots \\ p_n \end{bmatrix}.
$$

Simplifying the matrix equation, we get

$$
p_1 \begin{pmatrix} 1 \\ x_1 \end{pmatrix} + \cdots + p_n \begin{pmatrix} 1 \\ x_n \end{pmatrix} = y_1 \begin{pmatrix} 1 \\ x_1 \end{pmatrix} + \cdots + y_n \begin{pmatrix} 1 \\ x_n \end{pmatrix}.
$$

This is equivalent to

$$
\sum_{i=1}^{n} p_i = \sum_{i=1}^{n} y_i \quad \text{and} \quad \sum_{i=1}^{n} x_i p_i = \sum_{i=1}^{n} x_i y_i,
$$

which is Equation (7.4).

Although (9.6) looks simple, it is not easy to solve since in general $\mu_i = E(y_i)$ is a nonlinear function of $\boldsymbol{\beta}$, e.g. in logistic regression $p_i = \exp(\boldsymbol{x}_i'\boldsymbol{\beta})/[1 + \exp(\boldsymbol{x}_i'\boldsymbol{\beta})]$. Therefore, an iterative algorithm described below is used to solve it. Only in the case of multiple regression, (9.6) is a linear equation in $\boldsymbol{\beta}$, which has a closed form solution (3.7).

9.2.2   Iteratively reweighted least squares (IRWLS) Algorithm*

In the above, we discussed the MLE approach to fit the general linear model $g(\mu_i) = \boldsymbol{x}_i'\boldsymbol{\beta}$ $(1 \leq i \leq n)$. In this section, we give a weighted least squares (WLS) approach to fit this model. In Section 9.6, we show that the WLS approach is asymptotically equivalent to the MLE approach.

For the WLS approach, we define a new response variable z_i by expanding $g(y_i)$ around $g(\mu_i)$ using the first order Taylor series:

$$
g(y_i) \approx g(\mu_i) + (y_i - \mu_i)g'(\mu_i) = z_i \quad (i = 1, \ldots, n). \tag{9.8}
$$

It follows that

$$
\text{Var}(z_i) = \text{Var}(y_i - \mu_i)[g'(\mu_i)]^2 = V(\mu_i)[g'(\mu_i)]^2 \quad (i = 1, \ldots, n). \tag{9.9}
$$

Let $\boldsymbol{z} = (z_1, \ldots, z_n)'$ denote the response vector and consider fitting the linear model $E(\boldsymbol{z}) = \boldsymbol{X\beta}$. Since, in general, the variances of the z_i's are unequal we use the WLS estimator of $\boldsymbol{\beta}$ given by (3.27): $\widehat{\boldsymbol{\beta}}_{\mathrm{WLS}} = (\boldsymbol{X}'\,\boldsymbol{W}\,\boldsymbol{X})^{-1}\boldsymbol{X}'\,\boldsymbol{W}\,\boldsymbol{z}$, where $\boldsymbol{W} = \mathrm{diag}\{w_1, \ldots, w_n\}$ and the weights $w_i = 1/\mathrm{Var}(z_i)$ (see Section 3.4.1). As can be seen from (9.8) and (9.9), these weights as well as the z_i's are functions of the μ_i's and hence of the unknown $\boldsymbol{\beta}$ since $g(\mu_i) = \boldsymbol{x}'_i\boldsymbol{\beta}$. So they need to be iteratively estimated. At the rth iteration, the next estimate of $\boldsymbol{\beta}$ is given by

$$\widehat{\boldsymbol{\beta}}_{\mathrm{WLS}}^{(r+1)} = \left(\boldsymbol{X}'\,\widehat{\boldsymbol{W}}^{(r)}\,\boldsymbol{X}\right)^{-1}\boldsymbol{X}'\,\widehat{\boldsymbol{W}}^{(r)}\widehat{\boldsymbol{z}}^{(r)}, \tag{9.10}$$

where

$$\widehat{z}_i^{(r)} = g\left(\widehat{\mu}_i^{(r)}\right) + \left(y_i - \widehat{\mu}_i^{(r)}\right)g'\left(\widehat{\mu}_i^{(r)}\right). \tag{9.11}$$

The estimated weight matrix $\widehat{\boldsymbol{W}}^{(r)}$ is a diagonal matrix with diagonal elements

$$\widehat{w}_i^{(r)} = \left[V\left(\widehat{\mu}_i^{(r)}\right)\left\{g'\left(\widehat{\mu}_i^{(r)}\right)\right\}^2\right]^{-1},$$

where

$$\widehat{\mu}_i^{(r)} = g^{-1}\left(\boldsymbol{x}'_i\widehat{\boldsymbol{\beta}}^{(r)}\right).$$

Having updated $\widehat{\boldsymbol{\beta}}_{\mathrm{WLS}}^{(r)}$ to $\widehat{\boldsymbol{\beta}}_{\mathrm{WLS}}^{(r+1)}$, we set $r \leftarrow r + 1$ and iterate. The starting value $\widehat{\boldsymbol{\beta}}^{(0)}$ may be taken to be the LS estimate $\widehat{\boldsymbol{\beta}}^{(0)} = (\boldsymbol{X}'\boldsymbol{X})^{-1}\boldsymbol{X}'\boldsymbol{y}$ from which $\widehat{\mu}_i^{(0)}$, $\widehat{\boldsymbol{W}}^{(0)}$ and $\widehat{\boldsymbol{z}}^{(0)}$ can be computed.

Example 9.1 *(IRWLS Algorithm for Logistic Regression)* Suppose that we have binary observations y_i, where $y_i = 1$ with probability p_i and $y_i = 0$ with probability $1 - p_i$. We want to fit the logistic regression model:

$$\ln\left(\frac{p_i}{1 - p_i}\right) = \boldsymbol{x}'_i\boldsymbol{\beta} \quad (i = 1, \ldots, n)$$

to these data. In this case, $E(y_i) = \mu_i = p_i$ and $\mathrm{Var}(y_i) = V(p_i) = p_i(1 - p_i)$. Further,

$$g(p_i) = \ln\left(\frac{p_i}{1 - p_i}\right) \quad \text{and} \quad g'(p_i) = \frac{1}{p_i(1 - p_i)}.$$

Hence,

$$z_i = \ln\left(\frac{p_i}{1 - p_i}\right) + \frac{y_i - p_i}{p_i(1 - p_i)}$$

and

$$w_i = [V(p_i)\{g'(p_i)\}^2]^{-1} = \left[p_i(1 - p_i)\left(\frac{1}{p_i(1 - p_i)}\right)^2\right]^{-1} = p_i(1 - p_i).$$

At the rth iteration, the estimate of p_i is given by

$$\widehat{p}_i^{(r)} = \frac{\exp(\boldsymbol{x}_i'\widehat{\boldsymbol{\beta}}^{(r)})}{1 + \exp(\boldsymbol{x}_i'\widehat{\boldsymbol{\beta}}^{(r)})}.$$

This estimate is substituted in the above expressions for z_i and w_i to yield $\widehat{z}_i^{(r)}$ and $\widehat{w}_i^{(r)}$. The initial estimate $\widehat{\boldsymbol{\beta}}^{(0)}$ of $\boldsymbol{\beta}$ may be obtained by running LS regression of the $\boldsymbol{x}_i$'s on the 0–1 responses y_i's. $\square$

9.3   Deviance and AIC

Deviance as a measure of goodness of fit for a given model M in comparison to the saturated model (SM) was defined for the logistic regression model in Equation (7.10) as

$$D^2 = -2[\ln L_{\max}(M) - \ln L_{\max}(SM)]. \tag{9.12}$$

This definition applies generally to any exponential family distribution. Here $L_{\max}(M)$ is the maximum of the likelihood function under the model M and $L_{\max}(SM)$ is obtained by substituting $\widehat{\mu}_i = y_i$, i.e. by estimating the unknown means of the y_i's by the observed y_i's (thus having as many parameters in the model as the number of observations).

The Akaike information criterion (AIC) of a given model M with $p + 1$ parameters is defined as

$$\text{AIC} = -2\ln L_{\max}(M) + 2(p + 1) = D^2 - 2\ln L_{\max}(SM) + 2(p + 1). \tag{9.13}$$

For the logistic regression model, $\ln L_{\max}(SM) = 0$ and so AIC simplifies to $D^2 + 2(p + 1)$. However, this is not the case in general for other regression models as illustrated in the following examples.

Example 9.2 *(Deviance and AIC for Normal Distribution)* For the normal distribution, the likelihood function equals

$$L = \prod_{i=1}^{n} \left[\frac{1}{\sigma\sqrt{2\pi}} e^{-\frac{1}{2\sigma^2}(y_i - \mu_i)^2} \right] = \left(\frac{1}{2\pi\sigma^2} \right)^{n/2} e^{-\frac{1}{2\sigma^2}\sum_{i=1}^{n}(y_i-\mu_i)^2}.$$

The log-likelihood function equals

$$\ln L = -\frac{n}{2}\ln(2\pi\sigma^2) - \frac{1}{2\sigma^2}\sum_{i=1}^{n}(y_i - \mu_i)^2.$$

If σ^2 is a known parameter then

$$\ln L_{\max}(M) = -\frac{n}{2}\ln(2\pi\sigma^2) - \frac{1}{2\sigma^2}\sum_{i=1}^{n}(y_i - \widehat{\mu}_i)^2,$$

where $\widehat{\mu}_i = \boldsymbol{x}_i'\widehat{\boldsymbol{\beta}}$ and

$$\ln L_{\max}(SM) = -\frac{n}{2}\ln(2\pi\sigma^2) - \frac{1}{2\sigma^2}\sum_{i=1}^{n}(y_i - y_i)^2 = -\frac{n}{2}\ln(2\pi\sigma^2).$$

Therefore,

$$D^2 = -2[\ln L_{\max}(M) - \ln L_{\max}(SM)] = \frac{1}{\sigma^2}\sum_{i=1}^{n}(y_i - \widehat{\mu}_i)^2 = \frac{SSE}{\sigma^2}.$$

Note that $D^2 \sim \chi^2_{n-(p+1)}$. The `glm` function in R reports the SSE as the residual deviance and the SST as the null deviance ignoring the scaling factor $1/\sigma^2$.

For calculation of AIC, `glm` regards σ^2 as an unknown parameter and uses its MLE $\widehat{\sigma}^2 = SSE/n$. Also, it counts σ^2 among the number of estimated parameters. Then,

$$-2\ln L_{\max}(M) = n\ln 2\pi + n\ln\left(\frac{SSE}{n}\right) + \left(\frac{n}{SSE}\right)SSE$$

$$= n\ln 2\pi + n\ln(SSE/n) + n.$$

Hence,

$$AIC = -2\ln L_{\max}(M) + 2(p+1)$$

$$= n\ln 2\pi + n\ln\left(\frac{SSE}{n}\right) + n + 2(p+1). \tag{9.14}$$

Frequently, the constant term $n\ln 2\pi + n$ is omitted as is done in the AIC formula given in Chapter 6 since only the differences in AICs matter when comparing different models.

These formulae are illustrated by refitting the multiple regression model of Example 3.3 to the GPA versus college entrance test scores data using the `glm` function in the following example. □

Example 9.3 (College GPA and Entrance Test Scores: Deviance and AIC) We fit a multiple regression model $GPA = \beta_0 + \beta_1 Verbal + \beta_2 Math + \varepsilon$ using the `glm` function resulting in the following output.

```
glm(formula = GPA ~ Verbal + Math, family = gaussian, data = gpa)

Coefficients:
            Estimate Std. Error t value Pr(>|t|)
(Intercept) -1.570537   0.493749  -3.181  0.00297 **
Verbal       0.025732   0.004024   6.395 1.83e-07 ***
Math         0.033615   0.004928   6.822 4.90e-08 ***
---
Signif. codes:  0 1'***' 0.001 '**' 0.01 '*' 0.05 '.' 0.1 ' ' 1

(Dispersion parameter for gaussian family taken to be 0.1618257)

    Null deviance: 18.7735  on 39  degrees of freedom
Residual deviance:  5.9876  on 37  degrees of freedom
AIC: 45.547

Number of Fisher Scoring iterations: 2
```

Note that the null deviance $= \text{SST} = 18.7735$ and the residual deviance $=$ SSE $= 5.9876$ by comparing them with the ANOVA in Table 3.4. Next the AIC can be calculated using (9.14) as follows:

$$\text{AIC} = 40 \ln 5.9876 + 40 - 40 \ln 40 + 40 \ln 2\pi + 2(3+1) = 45.547.$$

Although $p = 2$ in this example, $\texttt{glm}$ uses $p = 3$ by counting σ^2 as an additional unknown parameter. $\square$

Analogous to the deviance $D^2 = \text{SSE} = \sum_{i=1}^{n} e_i^2$ for the normal distribution, we can express the deviance for any distribution in the exponential family as the sum of squared residuals: $D^2 = \sum_{i=1}^{n} d_i^2$. So we may define d_i as the **deviance residual** with its sign being that of $y_i - \widehat{\mu}_i$. We will see examples of this definition below. These residuals can be used to detect outliers in a similar manner as in the normal distribution case.

Example 9.4 (*Deviance and AIC for Poisson Distribution*) The likelihood function for the Poisson distribution is

$$L = \prod_{i=1}^{n} \left[\frac{e^{-\mu_i} \mu_i^{y_i}}{y_i!} \right]$$

and the log-likelihood function is

$$\ln L = \sum_{i=1}^{n} [-\mu_i + y_i \ln \mu_i - \ln y_i!].$$

Hence, the maximum of the log-likelihood function under the given model is

$$\ln L_{\max}(\text{M}) = \sum_{i=1}^{n} [-\widehat{\mu}_i + y_i \ln \widehat{\mu}_i - \ln y_i!],$$

where $\widehat{\mu}_i = \exp(\boldsymbol{x}_i'\widehat{\boldsymbol{\beta}})$. The log-likelihood function under SM is obtained by setting $\widehat{\mu}_i = y_i$ yielding

$$\ln L_{\max}(\text{SM}) = \sum_{i=1}^{n} [-y_i + y_i \ln y_i - \ln y_i!].$$

Hence, we get

$$D^2 = -2[\ln L_{\max}(\text{M}) - \ln L_{\max}(\text{SM})] = 2 \sum_{i=1}^{n} \left[y_i \ln \left(\frac{y_i}{\widehat{\mu}_i} \right) - (y_i - \widehat{\mu}_i) \right]. \quad (9.15)$$

The deviance residuals are given by

$$d_i = \text{sign}(y_i - \widehat{\mu}_i) \sqrt{2 \left[y_i \ln \left(\frac{y_i}{\widehat{\mu}_i} \right) - (y_i - \widehat{\mu}_i) \right]} \quad (i = 1, \dots, n). \quad (9.16)$$

The AIC equals

$$\text{AIC} = -2 \ln L_{\max}(\text{M}) + 2(p + 1)$$

$$= -2 \sum_{i=1}^{n} [-\widehat{\mu}_i + y_i \ln \widehat{\mu}_i - \ln y_i!] + 2(p + 1). \qquad \square$$

Example 9.5 (*Deviance and AIC for Gamma Distribution*) Assume that the response variables y_i follow the gamma distributions:

$$f(y_i) = \frac{1}{\Gamma(\alpha)} \lambda_i^{\alpha} y_i^{\alpha-1} e^{-\lambda_i y_i},$$

with a common dispersion parameter α (the homoscedasticity assumption). Note $E(y_i) = \mu_i = \alpha/\lambda_i$ or $\lambda_i = \alpha/\mu_i$. Although λ_i is the natural parameter, `glm` uses the inverse link function $g(\mu_i) = 1/\mu_i$ by regarding α as a known parameter. The likelihood function is

$$L = \prod_{i=1}^{n} \left[\frac{1}{\Gamma(\alpha)} \lambda_i^{\alpha} y_i^{\alpha-1} e^{-\lambda_i y_i} \right]$$

and the log-likelihood function is

$$\ln L = \sum_{i=1}^{n} [-\ln \Gamma(\alpha) + \alpha \ln \lambda_i + (\alpha - 1) \ln y_i - \lambda_i y_i].$$

Hence the maximum of the log-likelihood function under the given model is

$$\ln L_{\max}(\text{M}) = \sum_{i=1}^{n} [-\ln \Gamma(\alpha) + \alpha \ln \widehat{\lambda}_i + (\alpha - 1) \ln y_i - \widehat{\lambda}_i y_i].$$

Substituting $\widehat{\lambda}_i = \alpha / \widehat{\mu}_i$ where $1/\widehat{\mu}_i = \boldsymbol{x}_i' \widehat{\boldsymbol{\beta}}$, we get

$$\ln L_{\max}(\text{M}) = \sum_{i=1}^{n} \left[-\ln \Gamma(\alpha) + \alpha \ln \left(\frac{\alpha}{\widehat{\mu}_i} \right) + (\alpha - 1) \ln y_i - \left(\frac{\alpha}{\widehat{\mu}_i} \right) y_i \right].$$

The log-likelihood function under SM is obtained by setting $\widehat{\mu}_i = y_i$ or equivalently $\widehat{\lambda}_i = \alpha / y_i$, thus yielding

$$\ln L_{\max}(\text{SM}) = \sum_{i=1}^{n} \left[-\ln \Gamma(\alpha) + \alpha \ln \left(\frac{\alpha}{y_i} \right) + (\alpha - 1) \ln y_i - \left(\frac{\alpha}{y_i} \right) y_i \right].$$

Hence, the deviance equals

$$D^2 = -2[\ln L_{\max}(\text{M}) - \ln L_{\max}(\text{SM})]$$
$$= 2\alpha \sum_{i=1}^{n} \left[\left(\frac{y_i}{\widehat{\mu}_i} - 1 \right) - \ln \left(\frac{y_i}{\widehat{\mu}_i} \right) \right]. \tag{9.17}$$

Notice that this D^2 is simply the D^2 for the exponential regression multiplied by the scale parameter α. However, $\text{AIC} = -2 \ln L_{\max}(M) + 2(p+1)$ is a more complex function of α.

The deviance residuals are given by

$$d_i = \text{sign}(y_i - \widehat{\mu}_i) \sqrt{2\alpha \left[\left(\frac{y_i}{\widehat{\mu}_i} - 1 \right) - \ln \left(\frac{y_i}{\widehat{\mu}_i} \right) \right]} \quad (i = 1, \ldots, n). \tag{9.18}$$

The AIC equals

$$\text{AIC} = -2 \ln L_{\max}(\text{M}) + 2(p+1)$$
$$= -2 \sum_{i=1}^{n} \left[-\ln \Gamma(\alpha) + \alpha \ln \left(\frac{\alpha}{\widehat{\mu}_i} \right) + (\alpha - 1) \ln y_i - \left(\frac{\alpha}{\widehat{\mu}_i} \right) y_i \right] + 2(p+1).$$

$\square$

9.4 Poisson regression

Poisson regression of count data is a common application of GLM. **Log-linear models** used to analyze contingency tables (tables of count data cross-classified by several categorical variables) is an example of Poisson regression where all predictors are categorical. More generally, the x's can be any type of covariates. The canonical link is $g(\mu) = \ln \mu$. The model is fitted using the `glm` function by specifying `family=Poisson`.

A limitation of the Poisson distribution is that its variance equals its mean. It does not allow the flexibility of having variance greater or less than the mean (referred to as **overdispersion** or **underdispersion**, respectively). Overdispersion is more common primarily due to omitted predictors. There are several options available to model count data having over or underdispersion. One option is to include a scale parameter ϕ in the model so that the variance function equals $V(\mu) = \phi\mu$. We can estimate ϕ by $\widehat{\phi} = D^2/[n - (p + 1)]$ after the model is fitted.

The estimated regression coefficients $\widehat{\beta}$'s are unaffected by the introduction of the dispersion parameter ϕ but their standard errors are scaled by $\sqrt{\widehat{\phi}}$. Thus, the $\widehat{\beta}$'s are less significant if $\widehat{\phi} > 1$ for overdispersed data and more significant if $\widehat{\phi} < 1$ for underdispersed data. The reason for the $\widehat{\beta}$'s to be unaffected by the introduction of the dispersion parameter is that Equation (9.6) for computing $\widehat{\beta}$'s is independent of the dispersion parameter and depends only on the means μ_i. For the same reason, as we shall see in Section 9.5, exponential regression and gamma regression give the same fitted model since the exponential distribution is a special case of the gamma distribution for the dispersion parameter $\alpha = 1$.

Another option is to use the **negative binomial distribution** to model the data; see Exercise 9.1. This distribution also belongs to the exponential family. For more options see the books by Myers et al. (2010) and Hardin and Hilbe (2012).

Example 9.6 (*Melanoma Cancer Data: Poisson Regression of Cases*) The `melanoma.csv` file contains the data reported by Koch et al. (1986) from the Third National Cancer Survey. This data set contains the number of new melanoma cases (Cases) from 1969 to 1971 among white males in two areas (Area 1, Area 2) of the country for the following age groups: < 35, 35–44, 45–54, 55–64, 65–74, and >74, which are coded as 1, ... , 6 under the variable Age. The variable Population gives the size of the population at risk. The variable Rate is the incidence rate of new melanoma cases per 100,000 population. The data are shown in Table 9.1.

We are mainly interested in the effect of Age on Cases. However, using Age as a numerical predictor would imply that its effect varies linearly. So we will use it as a categorical variable (factor). We will also adjust for Area. In Figure 9.1, we have plotted Cases versus six age groups for the two areas. We see that the

Table 9.1. Melanoma cases and rates (per 100,000) data

No.	Area	Age	Population	Cases	Rate
1	1	1	2,880,262	61	2.118
2	1	2	564,535	76	13.462
3	1	3	592,983	98	16.527
4	1	4	450,740	104	23.073
5	1	5	270,908	63	23.255
6	1	6	161,850	80	49.428
7	2	1	1,074,246	64	5.958
8	2	2	220,407	75	34.028
9	2	3	198,119	68	34.323
10	2	4	134,084	63	46.985
11	2	5	70,708	45	63.642
12	2	6	34,233	27	78.871

Source: Koch et al. (1986).

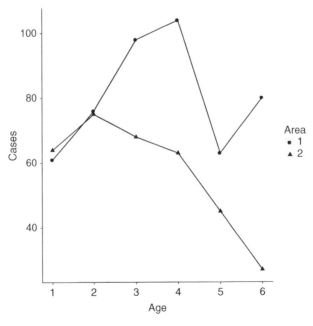

Figure 9.1. *Melanoma cases for two areas versus age groups.*

two areas differ significantly with Area 1 having much higher numbers of cases than Area 2 for all age groups except one. The R output for Poisson regression is shown below.

```
> fit1=glm(Cases~Area+factor(Age), data=melanoma, family=poisson(log))
> summary(fit1)

Call:
glm(formula = Cases ~ Area + factor(Age), family = poisson,
    data = melanoma)

Coefficients:
            Estimate Std. Error z value Pr(>|z|)
(Intercept)   4.6352     0.1342  34.540  < 2e-16 ***
Area         -0.3431     0.0707  -4.853 1.21e-06 ***
factor(Age)2  0.1890     0.1209   1.563  0.1181
factor(Age)3  0.2837     0.1184   2.395  0.0166 *
factor(Age)4  0.2897     0.1183   2.449  0.0143 *
factor(Age)5 -0.1462     0.1314  -1.113  0.2658
factor(Age)6 -0.1555     0.1317  -1.181  0.2378
---
Signif. codes:  0 '***' 0.001 '**' 0.01 '*' 0.05 '.' 0.1 ' ' 1

(Dispersion parameter for poisson family taken to be 1)

    Null deviance: 74.240  on 11  degrees of freedom
Residual deviance: 22.259  on  5  degrees of freedom
AIC: 108.48

Number of Fisher Scoring iterations: 4

> SSE1 = sum((melanoma$Cases-fit1$fitted)^2) # SSE
> SSE1
[1] 1285.294
```

As seen in Figure 9.1, Area is highly significant with the coefficient of Area being negative since Area 2 has a lower incidence of melanoma cases than Area 1, which R uses as reference. We also see a clear nonlinear effect of age with incidence of melanoma cases increasing compared to the youngest group until the age of 64 and then slightly decreasing (although the decline is not statistically significant). The SSE for this model is 1285.294.

To illustrate the calculation of a fitted value, consider Area $= 2$ and Age $= 3$. Then $\ln \widehat{\mu} = 4.6352 - 0.3431 + 0.2837 = 4.5758$ and so $\widehat{y} = \widehat{\mu} = e^{4.5758} = 97.106$. The observed number of cases for this combination is 68. So the residual is

$68 - 97.106 = -29.106$. The deviance residual is

$$\text{sign}(68 - 97.106)\sqrt{2\left[68\ln\left(\frac{68}{97.106}\right) - (68 - 97.106)\right]} = -3.123.$$

Note that this residual is on the log-scale.

For comparison sake, we performed multiple regression using the square-root transformation of Cases as the response variable. The results are shown below.

```
> fit2=lm(sqrt(Cases)~Area+factor(Age), data=melanoma)
> summary(fit2)

Call: lm(formula = sqrt(Cases) ~ Area + factor(Age), data =
melanoma)

Coefficients:
            Estimate Std. Error t value Pr(>|t|)
(Intercept)  10.0949     1.1541   8.747 0.000324 ***
Area         -1.4598     0.5960  -2.449 0.057976 .
factor(Age)2  0.7839     1.0323   0.759 0.481867
factor(Age)3  1.1677     1.0323   1.131 0.309286
factor(Age)4  1.1625     1.0323   1.126 0.311224
factor(Age)5 -0.5824     1.0323  -0.564 0.596996
factor(Age)6 -0.8349     1.0323  -0.809 0.455374
---
Signif. codes:  0 '***' 0.001 '**' 0.01 '*' 0.05 '.' 0.1 ' ' 1

Residual standard error: 1.032 on 5 degrees of freedom
Multiple R-squared:  0.7267,    Adjusted R-squared:  0.3987
F-statistic: 2.216 on 6 and 5 DF,  p-value: 0.2002

> SSE2 = sum((melanoma$Cases-(fit2$fitted)^2)^2) # SSE
> SSE2
[1] 1297.955
```

We see that the effects of all predictors are similar to those obtained from Poisson regression, but they are all nonsignificant except the effect of Area, which is close to being significant. Surprisingly, the SSE for this model is only slightly higher than that for the Poisson regression model.   □

9.4.1   Poisson regression for rates

Poisson regression is used to model count data but often the count of events depends on the amount of exposure. In Example 9.6, the number of cases depends on the population size. In epidemiological studies, the number of cases of a disease depends on the amount or time of exposure of some pollutant. In traffic studies the number of car accidents depends on the number of cars on the road or the traffic density. In such applications, it is more appropriate to model the event rates rather than the counts of events while maintaining the Poisson nature of the counts.

For the ith observation, denote the number of cases by y_i, the covariate vector by $\boldsymbol{x}_i$ and the size of the exposure by N_i. Assume that y_i is Poisson distributed with mean μ_i. The mean event rate for the ith observation is then μ_i/N_i. We fit a linear model $\ln(\mu_i/N_i) = \boldsymbol{x}_i'\boldsymbol{\beta}$. This model can be written as $\ln\mu_i = \ln N_i + \boldsymbol{x}_i'\boldsymbol{\beta}$, where $\ln N_i$ can be regarded as an additional covariate except that its β coefficient is fixed at 1. Such a term is called an **offset**.

This model can be fitted using `glm` as follows. The observed rates $r_i = y_i/N_i$ are treated as responses. The r_i are not in general integers as required by the Poisson distribution. This problem can be handled by specifying `family=quasipoisson`, which is an option available in the `MASS` library. Furthermore, the N_i are used as weights. These steps are illustrated in the example below.

Example 9.7 (Melanoma Cancer Data: Poisson Regression for Rates) In Figure 9.2, we have plotted Rate versus Age for the two areas. Notice the systematic increasing trends in Rate with Age for both areas. Cases showed decreasing trends with Age in Figure 9.1, but that was an artifact of decreasing population sizes with Age. Also observe that the number of melanoma cases are significantly higher for Area 1, but the incidence rate is

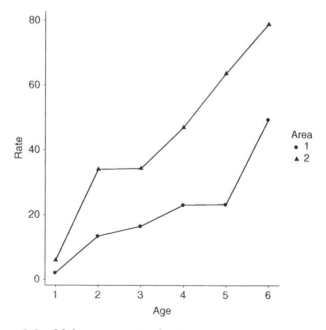

Figure 9.2. *Melanoma rates for two areas versus age group.*

significantly lower. Again, this is the result of substantially smaller population sizes in Area 2 for each Age.

In the following, we fit the Poisson regression model to Rate using the log link. If we use Cases/Population as the response variable then only the intercept term will be different; all other regression coefficients will be the same. Since Rate is defined as the number of Cases per 100,000 of Population, the intercept term for Cases/Population will be the intercept term for Rate $- \ln 10^5 = 0.0352 - 11.5129 = -11.4777$.

The R output is shown below.

```
Call:
glm(formula = Rate ~ Area + factor(Age), family = quasipoisson,
    data = melanoma, weights = Population)

Coefficients:
             Estimate Std. Error t value Pr(>|t|)
(Intercept)   0.03519    0.15135   0.232 0.825383
Area          0.81949    0.07855  10.432 0.000139 ***
factor(Age)2  1.79730    0.13373  13.439 4.08e-05 ***
factor(Age)3  1.91304    0.13098  14.605 2.72e-05 **
factor(Age)4  2.24173    0.13087  17.130 1.24e-05 ***
factor(Age)5  2.36566    0.14545  16.265 1.60e-05 ***
factor(Age)6  2.94461    0.14603  20.164 5.55e-06 ***
---
Signif. codes:  0 '***' 0.001 '**' 0.01 '*' 0.05 '.'0.1 ' ' 1

(Dispersion parameter for quasipoisson family taken to be 122308)

    Null deviance: 89579678  on 11   degrees of freedom
Residual deviance:   621520  on  5   degrees of freedom
AIC: NA

Number of Fisher Scoring iterations: 4

> SSE3 = sum((melanoma$Cases-(fit4$fitted)*(melanoma$Population)
> /100000)^2) # SSE
> SSE3
[1] 350.7735
```

Notice that the SSE for this model is nearly one-fourth of the SSEs for the two models fitted for Cases. (SSE in both examples is the sum of the squared differences between the observed number of Cases and the predicted number of Cases.) The reason is that the Rate is a much more well-behaved variable and hence can be more accurately modeled.

An alternative approach to modeling the Rate is to treat Rate/10^5 = Cases/Population as a binomial proportion and use the logistic regression model with Population sizes as weights. The R output for this analysis is shown below.

```
Call:
glm(formula = Cases/Population ~ Area + factor(Age), family = binomial,
    data = melanoma, weights = Population)

Coefficients:
              Estimate Std. Error z value Pr(>|z|)
(Intercept)  -11.47811    0.13688  -83.86  <2e-16 ***
Area           0.81973    0.07104   11.54  <2e-16 ***
factor(Age)2   1.79756    0.12093   14.86  <2e-16 ***
factor(Age)3   1.91330    0.11845   16.15  <2e-16 ***
factor(Age)4   2.24211    0.11835   18.95  <2e-16 ***
factor(Age)5   2.36608    0.13153   17.99  <2e-16 ***
factor(Age)6   2.94531    0.13207   22.30  <2e-16 ***
---
Signif. codes:  0 '***' 0.001 '**' 0.01 '*' 0.05 '.' 0.1 ' ' 1

(Dispersion parameter for binomial family taken to be 1)

    Null deviance: 895.9558  on 11  degrees of freedom
Residual deviance:   6.2102  on  5  degrees of freedom
AIC: 92.431

Number of Fisher Scoring iterations: 4

> SSE4 = sum((melanoma$Cases-(fit4$fitted)*(melanoma$Population))^2) # SSE
> SSE4
[1] 350.347
```

Notice that the fitted model is almost identical to the Poisson regression model for Rate and the SSE is also almost exactly the same. The reason is that the proportion p_i of people having melanoma in each (Area, Age) category is very small and so $\ln\left(\frac{p_i}{1-p_i}\right) \approx \ln p_i$. Thus, the link function is approximately $\ln p_i$, the same as for the Poisson regression model for Rate.   □

9.5   Gamma regression*

As noted before, in gamma regression the canonical link is the inverse function $g(\mu) = 1/\mu$. To fit the model, we use the glm function and specify family=Gamma. As in the case of Poisson regression, the dispersion parameter α is estimated after the specified linear model is fitted. The estimated value of α is printed by leaving out the value of the dispersion parameter in the summary(fit)function. A particular value of α can be specified in the summary(fit) function. For example, if we want to fit the exponential regression model then we use the function summary(fit,dispersion=1). The estimates of the regression coefficients are the same in both cases but the standard errors are multiplied by a factor $\sqrt{\widehat{\alpha}}$ in the former case. This factor follows from the fact that $Var(y) \propto \alpha$. Thus, if $\widehat{\alpha} < 1$ then the estimated regression coefficients are more significant and vice versa.

Example 9.8 *(Leukemia Survival and WBC Count)* Table 9.2 gives data on the survival times of 17 leukemia patients and their $\log_{10}$(WBC) counts (WBC = white blood cell). It is known that the WBC count and survival are negatively correlated. This is demonstrated in the plot of Survival versus $\log_{10}$(WBC) in Figure 9.3.

Table 9.2. Leukemia survival (in days) data

No.	$\log_{10}$(WBC)	Survival	No.	$\log_{10}$(WBC)	Survival
1	3.36	65	10	3.85	143
2	2.88	156	11	3.97	56
3	3.63	100	12	4.51	26
4	3.41	134	13	4.54	22
5	3.78	16	14	5.00	1
6	4.02	108	15	5.00	1
7	4.00	121	16	4.72	5
8	4.23	4	17	5.00	65
9	3.73	39			

Source: Unknown

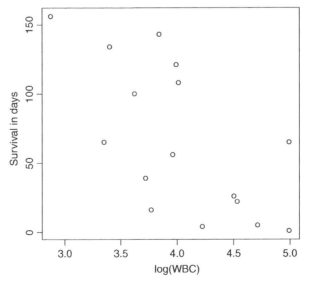

Figure 9.3. *Survival in days versus* $\log_{10}$*(WBC) for leukemia patients.*

We first fit the exponential regression model to these data as shown in the following R output.

```
Call: glm(formula = Survival ~ logWBC, family = Gamma, data =
leukemia)

Coefficients:
             Estimate Std. Error z value Pr(>|z|)
(Intercept) -0.034657   0.018626  -1.861   0.0628 .
   logWBC    0.013528   0.005519   2.451   0.0142 *
---
Signif. codes:  0 '***' 0.001 '**' 0.01 '*' 0.05 '.'0.1 ' ' 1

(Dispersion parameter for Gamma family taken to be 1)

    Null deviance: 26.282  on 16  degrees of freedom
Residual deviance: 20.956  on 15  degrees of freedom
AIC: 175.46

Number of Fisher Scoring iterations: 6
```

The gamma regression model is obtained by simply printing the summary of the fit without specifying the dispersion parameter. The estimated dispersion parameter is $\widehat{\alpha} = 0.7813$. We get the same regression coefficients as for exponential regression but their standard errors are multiplied by $\sqrt{0.7813} = 0.884$ as shown in the R output below. So the t-values are scaled up by $1/0.884 = 1.131$ and hence are more significant. However, the deviance residuals are not scaled by $\sqrt{\widehat{\alpha}}$ as they should be according to the formula (9.18). Similarly, the deviance is not scaled by $\widehat{\alpha}$ as it should be. Thus, the residual deviance for the gamma regression should be $0.7813 \times 20.956 = 16.374$ instead of 20.956, which is for the exponential regression.

```
Call:
glm(formula = Survival ~ logWBC, family = Gamma,
data = leukemia)

Coefficients:
             Estimate Std. Error t value Pr(>|t|)
(Intercept) -0.034657   0.016465  -2.105   0.0526 .
logWBC       0.013528   0.004879   2.773   0.0142 *
---
Signif. codes:  0 '***' 0.001 '**' 0.01 '*' 0.05 '.' 0.1 ' ' 1

(Dispersion parameter for Gamma family taken to be 0.7813441)
```

```
    Null deviance: 26.282  on 16  degrees of freedom
Residual deviance: 20.956  on 15  degrees of freedom
AIC: 175.46

Number of Fisher Scoring iterations: 6
```

The fitted values for this model are given by $\widehat{y}_i = 1/(-0.0347 + 0.0135x_i)$, and the residuals are given by $e_i = y_i - \widehat{y}_i$. The SSE $= \sum_{i=1}^{n} e_i^2 = 33{,}574.12$. The deviance residuals are calculated using (9.18). These calculations are readily done in a spreadsheet and are shown in the Table 9.3.

Table 9.3. Fitted values and residuals using exponential regression model for leukemia data

log(WBC)	Survival	$\widehat{y}_i$	e_i	d_i
3.36	65	92.618	-27.618	-0.3344
2.88	156	232.361	-76.361	-0.3736
3.63	100	69.206	30.794	0.3921
3.41	134	87.158	46.842	0.4633
3.78	16	60.684	-44.684	-1.0925
4.02	108	50.696	57.304	0.8650
4	121	51.401	69.599	0.9979
4.23	4	44.314	-40.314	-1.7293
3.73	39	63.281	-24.281	-0.4479
3.85	143	57.386	85.614	1.0760
3.97	56	52.496	3.504	0.0653
4.51	26	37.945	-11.945	-0.3556
4.54	22	37.369	-15.369	-0.4869
5	1	30.319	-29.319	-2.2112
5	1	30.319	-29.319	-2.2112
4.72	5	34.252	-29.252	-1.4631
5	65	30.319	34.681	0.8732

We see that many of the fitted values deviate significantly from the observed Survival values, which means that the predictions made by this model are not very accurate. The main reason for the prediction inaccuracies is survival times fluctuate widely.

An alternative approach to fitting a model with log-transformation is to use the gamma family with the log link instead of the inverse link. The corresponding R output is shown below.

```
Call:
glm(formula = Survival ~ logWBC, family = Gamma(log),
    data = leukemia)

Coefficients:
            Estimate Std. Error t value Pr(>|t|)
(Intercept)   8.4775     1.6034   5.287 9.13e-05 ***
logWBC       -1.1093     0.3872  -2.865   0.0118 *
---
Signif. codes:  0 '***' 0.001 '**' 0.01 '*' 0.05 '.' 0.1 ' ' 1

(Dispersion parameter for Gamma family taken to be 0.9388638)

    Null deviance: 26.282  on 16  degrees of freedom
Residual deviance: 19.457  on 15  degrees of freedom
AIC: 173.97

Number of Fisher Scoring iterations: 8
```

The fitted values for this model are given by $\widehat{y}_i = \exp(8.4775 - 1.1093 x_i)$ and the residuals are given by $e_i = y_i - \widehat{y}_i$. The SSE $= \sum_{i=1}^{n} e_i^2 = 27{,}211.34$, which is about 19% smaller than that obtained using the inverse link. This is an example where non-canonical link (namely the log link) may be preferred to the canonical link (namely the inverse link). □

9.6 Technical notes*

9.6.1 Mean and variance of the exponential family of distributions

In this section we derive the formulae (9.2). We use the following two relationships from (B.7):

$$E\left[\frac{d \ln L}{d\theta}\right] = 0 \quad \text{and} \quad E\left[\left(\frac{d \ln L}{d\theta}\right)^2\right] = -E\left[\frac{d^2 \ln L}{d\theta^2}\right].$$

Substituting (9.1) for L we get

$$E\left[\frac{d\ln L}{d\theta}\right] = \frac{1}{a(\phi)}E[y - b'(\theta)] = 0 \quad \Rightarrow \quad E(y) = \mu = b'(\theta).$$

Next

$$-E\left[\frac{d^2\ln L}{d\theta^2}\right] = \frac{1}{a(\phi)}b''(\theta)$$

and

$$E\left[\left(\frac{d\ln L}{d\theta}\right)^2\right] = \frac{1}{a^2(\phi)}E(y - b'(\theta))^2 = \frac{1}{a^2(\phi)}\mathrm{Var}(y).$$

Equating these two expressions, we get $\mathrm{Var}(y) = V(\mu) = a(\phi)b''(\theta)$.

9.6.2 MLE of β and its evaluation using the IRWLS algorithm

The log-likelihood function for GLM equals

$$\ln L(\boldsymbol{\beta}) = \sum_{i=1}^{n}\left[\frac{1}{a(\phi)}[y_i\theta_i - b(\theta_i)] + c(y_i, \phi)\right].$$

To find the MLE of $\boldsymbol{\beta}$, we treat ϕ as a known parameter. It is estimated separately from the deviance residuals after estimating $\boldsymbol{\beta}$, just as how σ^2 is estimated in multiple regression from residuals by $\mathrm{MSE} = \mathrm{SSE}/[n - (p + 1)]$ after estimating $\boldsymbol{\beta}$.

We assume that $a(\phi)$ is known and constant. If we use the canonical link function $\theta_i = \boldsymbol{x}_i'\boldsymbol{\beta}$, then we get

$$\frac{d\ln L(\boldsymbol{\beta})}{d\boldsymbol{\beta}} = \sum_{i=1}^{n}\frac{d\ln L(\boldsymbol{\beta})}{d\theta_i}\frac{d\theta_i}{d\boldsymbol{\beta}}$$

$$= \frac{1}{a(\phi)}\sum_{i=1}^{n}(y_i - b'(\theta_i))\boldsymbol{x}_i \quad (\text{since } \theta_i = \boldsymbol{x}_i'\boldsymbol{\beta}, \text{so } d\theta_i/d\boldsymbol{\beta} = \boldsymbol{x}_i)$$

$$= \frac{1}{a(\phi)}\sum_{i=1}^{n}(y_i - \mu_i)\boldsymbol{x}_i.$$

Setting this derivative equal to $\mathbf{0}$ (the null vector) and canceling the constant factor $1/a(\phi)$ results in Equation (9.6) by expressing it in matrix notation.

Next we show that solving this equation is asymptotically equivalent to solving Equation (9.10) with the $z_i = g(\mu_i) + (y_i - \mu_i)g'(\mu_i)$ as the response variables and $w_i = 1/\text{Var}(z_i)$ as the weights. The jth element of the vector $d\ln L(\boldsymbol{\beta})/d\boldsymbol{\beta}$ can be written as

$$\frac{\partial \ln L(\boldsymbol{\beta})}{\partial \beta_j} = \sum_{i=1}^{n} \frac{\partial \ell_i}{\partial \beta_j},$$

where

$$\ell_i = \ln L_i = \frac{1}{a(\phi)}(y_i - b(\theta_i)) + c(y_i, \phi) \quad (i = 1, \dots, n).$$

By the chain rule, we have

$$\frac{\partial \ell_i}{\partial \beta_j} = \frac{d\ell_i}{d\theta_i} \frac{d\theta_i}{d\mu_i} \frac{d\mu_i}{d\eta_i} \frac{\partial \eta_i}{\partial \beta_j}. \tag{9.19}$$

These derivatives can be written as follows:

$$\frac{d\ell_i}{d\theta_i} = \frac{y_i - b'(\theta_i)}{a(\phi)} = \frac{y_i - \mu_i}{a(\phi)},$$

$$\frac{d\theta_i}{d\mu_i} = \left[\frac{d\mu_i}{d\theta_i}\right]^{-1} = [b''(\theta_i)]^{-1} = \frac{a(\phi)}{V(\mu_i)} \quad \text{(using (9.2))},$$

$$\frac{\partial \eta_i}{\partial \beta_j} = \frac{\partial(\boldsymbol{x}_i'\boldsymbol{\beta})}{\partial \beta_j} = x_{ij}.$$

The second equation above can be re-expressed in terms of the weights w_i using (9.9) as follows:

$$\frac{d\theta_i}{d\mu_i} = \frac{a(\phi)}{V(\mu_i)(d\eta_i/d\mu_i)^2}\left(\frac{d\eta_i}{d\mu_i}\right)^2 = \frac{a(\phi)}{\text{Var}(z_i)}\left(\frac{d\eta_i}{d\mu_i}\right)^2 = w_i a(\phi)\left(\frac{d\eta_i}{d\mu_i}\right)^2.$$

Substituting in (9.19), we get

$$\frac{d\ell_i}{d\beta_j} = \frac{y_i - \mu_i}{a(\phi)}w_i a(\phi)\left(\frac{d\eta_i}{d\mu_i}\right)^2\frac{d\mu_i}{d\eta_i}x_{ij} = w_i x_{ij}(y_i - \mu_i)\frac{d\eta_i}{d\mu_i}.$$

Thus we have to solve

$$\frac{\partial \ln L(\boldsymbol{\beta})}{\partial \beta_j} = \sum_{i=1}^{n} w_i x_{ij}(y_i - \mu_i)\frac{d\eta_i}{d\mu_i} = 0 \quad (j = 0, 1, \dots, p).$$

We use the Fisher scoring algorithm from Section B.3 to solve this equation. To simplify the notation, we denote the current estimate $\widehat{\boldsymbol{\beta}}^{(r)}$ by $\widehat{\boldsymbol{\beta}}$ and the new estimate $\widehat{\boldsymbol{\beta}}^{(r+1)}$ by $\widehat{\boldsymbol{\beta}}^*$. Then the recursion equation of the Fisher scoring algorithm (see (B.8)) becomes

$$\widehat{\boldsymbol{\beta}}^* = \widehat{\boldsymbol{\beta}} + \boldsymbol{\mathcal{I}}^{-1}(\widehat{\boldsymbol{\beta}}) \frac{d \ln L(\widehat{\boldsymbol{\beta}})}{d\boldsymbol{\beta}},$$

where $\boldsymbol{\mathcal{I}}(\widehat{\boldsymbol{\beta}})$ is the expected information matrix and $d \ln L(\widehat{\boldsymbol{\beta}})/d\boldsymbol{\beta}$ is the derivative $d \ln L(\boldsymbol{\beta})/d\boldsymbol{\beta}$ both evaluated at $\boldsymbol{\beta} = \widehat{\boldsymbol{\beta}}$. Multiplying both sides of the above equation by $\boldsymbol{\mathcal{I}}(\widehat{\boldsymbol{\beta}})$ we get

$$\boldsymbol{\mathcal{I}}(\widehat{\boldsymbol{\beta}})\widehat{\boldsymbol{\beta}}^* = \boldsymbol{\mathcal{I}}(\widehat{\boldsymbol{\beta}})\widehat{\boldsymbol{\beta}} + \frac{d \ln L(\widehat{\boldsymbol{\beta}})}{d\boldsymbol{\beta}}. \tag{9.20}$$

To derive an expression for the (j,k)th element of $\boldsymbol{\mathcal{I}}(\widehat{\boldsymbol{\beta}})$, we compute

$$E\left[-\left(\frac{\partial^2 \ln L(\boldsymbol{\beta})}{\partial \beta_j \partial \beta_k}\right)\right]$$
$$= E \sum_{i=1}^{n} -\left[(y_i - \mu_i)x_{ij}\frac{\partial\{w_i(d\eta_i/d\mu_i)\}}{\partial \beta_k} - w_i x_{ij}\frac{\partial(y_i - \mu_i)}{\partial \beta_k}\frac{d\eta_i}{d\mu_i}\right].$$

The first term has expectation zero since $E(y_i - \mu_i) = 0$. The second term equals

$$\sum_{i=1}^{n} w_i x_{ij}\frac{d\eta_i}{d\mu_i}\frac{\partial \mu_i}{\partial \beta_k} = \sum_{i=1}^{n} w_i x_{ij}\frac{\partial \eta_i}{\partial \beta_k} = \sum_{i=1}^{n} w_i x_{ij} x_{ik}.$$

Thus
$$\boldsymbol{\mathcal{I}}(\boldsymbol{\beta}) = \boldsymbol{X}' \boldsymbol{W} \boldsymbol{X} \text{ and } \boldsymbol{\mathcal{I}}(\widehat{\boldsymbol{\beta}}) = \boldsymbol{X}' \widehat{\boldsymbol{W}} \boldsymbol{X}.$$

where $\widehat{\boldsymbol{W}}$ is the estimated diagonal weight matrix evaluated at $\boldsymbol{\beta} = \widehat{\boldsymbol{\beta}}$.

The jth element of the R.H.S. of (9.20) equals

$$(\boldsymbol{\mathcal{I}}(\widehat{\boldsymbol{\beta}})\widehat{\boldsymbol{\beta}})_j + \left(\frac{d \ln L(\widehat{\boldsymbol{\beta}})}{d\boldsymbol{\beta}}\right)_j = \sum_{k=0}^{p} \boldsymbol{\mathcal{I}}_{jk}(\widehat{\boldsymbol{\beta}})\widehat{\beta}_k + \sum_{i=1}^{n} \widehat{w}_i x_{ij}(y_i - \widehat{\mu}_i)\frac{d\widehat{\eta}_i}{d\mu_i}$$
$$= \sum_{k=0}^{p}\sum_{i=1}^{n} \widehat{w}_i x_{ij} x_{ik}\widehat{\beta}_k + \sum_{i=1}^{n} \widehat{w}_i x_{ij} x_{ij}(y_i - \widehat{\mu}_i)\frac{d\widehat{\eta}_i}{d\mu_i}$$
$$= \sum_{i=1}^{n} \widehat{w}_i x_{ij}\sum_{k=0}^{p} x_{ik}\widehat{\beta}_k + \sum_{i=1}^{n} \widehat{w}_i x_{ij} x_{ij}(y_i - \widehat{\mu}_i)\frac{d\widehat{\eta}_i}{d\mu_i}.$$

Now note that
$$\sum_{k=0}^{p} x_{ik}\widehat{\beta}_k = \widehat{\eta}_i.$$

Substituting in the above and combining the first and second terms we get the above equal to

$$\sum_{i=1}^{n} \widehat{w}_i x_{ij} \left\{ \widehat{\eta}_i + (y_i - \widehat{\mu}_i) \frac{d\widehat{\eta}_i}{d\mu_i} \right\} = \sum_{i=1}^{n} \widehat{w}_i x_{ij} \widehat{z}_i,$$

where $\widehat{z}_i$ is given by (9.11). In vector notation, the R.H.S. of (9.20) equals $\boldsymbol{X} \widehat{\boldsymbol{W}} \boldsymbol{z}$. Thus we get the final equation (9.10) for the IRWLS algorithm.

Exercises

Theoretical Exercises

9.1 (Negative binomial distribution) The negative binomial distribution is the probability distribution of the number of trials y needed to get a fixed number $r \geq 1$ of successes in a sequence of i.i.d. Bernoulli trials, each with success probability p. It is given by

$$f(y; r, p) = \binom{y-1}{r-1} p^r (1-p)^{y-r},$$

where p is the unknown parameter of interest and r is specified. For $r = 1$, it reduces to the geometric distribution.

(a) Express the negative binomial distribution in the exponential family form. Give the functions $a(\phi), b(\theta)$, and $c(y, \phi)$ for it.

(b) Find the mean μ and variance σ^2 of this distribution.

(c) Show that the distribution is overdispersed $(\sigma^2 > \mu)$ if $p < 1/2$ and underdispersed $(\sigma^2 < \mu)$ if $p > 1/2$.

9.2 (MLE equations for exponential and Poisson distributions) Derive the equations for finding the MLE of the regression parameter vector $\boldsymbol{\beta}$ if the response variable distribution is exponential or Poisson, and show that they have the form (9.6).

Applied Exercises

9.3 (Airline injury incidents) The file `Airline-Injury.csv` contains data from Chatterjee and Hadi (2012) on the number of injury incidents and the proportion of the total number of flights out of New York for nine airlines.

Fit three different models to predict the number of injury incidents (y) as a function of the proportion of the total number of flights (x): (i) simple linear regression without any transformation of y, (ii) simple linear regression with square-root transformation of y, and (iii) Poisson regression. Calculate the SSE for each model. Which model do you prefer and why?

9.4 (Automobile traffic accidents) The file `crashdata2014.csv` contains data (provided by Professor Hani Mahmassani, Director of the Transportation Center, Northwestern University) on 77 variables for 82,744 automobile crashes in Illinois towns in 2014. The variables include location of the accident (county and township), day and time of the accident, driving conditions, presence of traffic control device, weather conditions, etc. Most of these variables are not relevant as predictor variables for predicting the number of accidents. Furthermore, each variable has many categories, for example, there are 102 counties, time of the day has 24 categories (hours) and so on. An R code was written using the library `dplyr` to create new categorical variables from selected variables as shown in Table 9.4.

Table 9.4. New categorical variables created from existing variables from raw data for exercise

Current variable (values)	New variable	Recoded values
Day of Week Code (1–7)	Day	Weekday (1–5), Weekend (6–7)
Crash Hour (1–24)	Time	Morning (6–10), Midday (11–15), Evening (16–20), Night 21–5)
Traffic Control Device Code (1–14, 99)	Traffic-Control	No control (1), Control (2–14), Unknown (99)
Road Surface Condition Code (1–6, 9)	Road	Dry (1), Wet (2, 3, 4), Other (5, 6, 9)
Light Condition Code (1–5, 9)	Light	Daylight (1), Dawn/Dusk (2, 3), Dark (4, 5), Unknown (9)
Weather Code (1–9)	Weather	Clear (1), Rain/Snow (2, 3, 5), Poor Visibility (4, 8), Other (7, 9)

In addition, a new variable called Weight was created which takes a value of 5 for Weekday and 2 for Weekend, as a measure of the exposure time, to take into account that the number of accidents are likely to be proportional to the number of days in each category, keeping all other conditions fixed. It can be used to perform weighted Poisson regression. Finally, these variables were grouped to create a Count variable, which is the number of cases (accidents) in each group. The resulting data are saved in a data file `crashdata2014-summary.csv`. The marginal counts for each new variable are as follows (the numbers are the counts of crashes in each category).

Variable	Counts
Day	Weekday: 59,949; Weekend: 22,795
Time	Night: 16,865; Morning: 18,892; Midday: 23,399; Evening: 23,588
Traffic Control	No Control: 41,651; Control: 38,072; Unknown: 3021
Road	Dry: 58,089; Wet: 18,676; Other: 5979
Light	Daylight: 52,178; Dawn/Dusk: 3076, Dark: 23,538; Unknown: 3952
Weather	Clear: 64,058; Rain/Snow: 12,117; Poor Visibility: 1753, Other: 4816

Note that there are a total of $2 \times 4 \times 3 \times 3 \times 4 \times 4 = 1152$ grouping combinations.

(a) Do unweighted Poisson regression with Count as the response variable. Comment on the significance of each variable and whether the sign of each coefficient is as expected.

(b) Do weighted Poisson regression with Count as the response variable. Check if the results change much.

9.5 **(Auto insurance claims)** The data file `claims.csv` contains data from a Canadian insurance company for policy years 1956 and 1957. The variables defined in Table 9.5 describe the different categories of insured drivers, the premiums paid by them, the number of claims they filed and the cost to the company of paying the claims. The response variable is Cost.

Table 9.5. Description of the variables in the automobile insurance claims data

Variable	Description
Merit	3 licensed and accident free $\geq$ 3 years
	2 licensed and accident free 2 years
	1 licensed and accident free 1 year
	0 all others
Class	1 pleasure, no male operator < 25
	2 pleasure, non-principal male operator < 25
	3 business use
	4 unmarried owner or principal operator < 25
	5 married owner or principal operator < 25
Insured	Earned car years
Premium	Earned premium in 1000's (adjusted to what the premium would have been had all cars been written at 01 rates)
Claims	Number of claims
Cost	Total cost of the claim in 1000's of dollars

Source: https://www.statistics.ma.tum.de/fileadmin/w00bdb/www/czado/lec8.pdf.

   (a) Fit a gamma regression model relating Cost of insurance to the attributes of the drivers, the premiums paid out and the claims filed by them. Which variables are significant? What is the estimate of the dispersion parameter?
   (b) Perform the best subsets regression to find the model with the minimum AIC.

Chapter 10

Survival analysis

Origins of survival analysis lie in actuarial science of lifetimes and death rates of people. As a discipline, survival analysis has grown from many applications in biomedical fields. However, its methods are applicable in any area where the variable of interest is time to an event. In actuarial and biomedical applications, the event of interest is typically death or recurrence of an adverse outcome such as tumor or hospitalization. In engineering reliability studies, the event of interest is failure of an item. In marketing applications, the event of interest is the next purchase by a customer. Generally, the goal is to model the lifetime as a function of some predictor variables, e.g. prognostic variables of a patient in a biomedical study or load conditions in a reliability study or demographic and socioeconomic attributes and past purchase history of a customer in a marketing study. For convenience, we will use the biomedical terminology of the death of a patient as the event of interest.

A unique feature of the time to an event data is that often they are censored; in other words, the event is not observed within the time frame of the study in all observations. Thus, all we know is that the lifetimes of these patients are greater than their censoring times. For example, the death or failure may not occur before the study is terminated or a patient may withdraw or drop out of the study before his/her outcome is known or a customer may not place a new order before data collection stops. In such cases, the data are said to be **right censored**.

In some applications, the data are **left censored** as happens when the start of the lifetime is not observed. For example, in the study of incubation time from HIV infection to onset of AIDS, some patients may enter the study already infected by HIV virus. So the start of the incubation time is unknown. We will not consider left censoring in this chapter.

Predictive Analytics: Parametric Models for Regression and Classification Using R.
First Edition. Ajit C. Tamhane.
© 2021 John Wiley & Sons, Inc. Published 2021 by John Wiley & Sons, Inc.
Companion website: www.wiley.com/go/tamhane/predictiveanalytics

The challenge in analyzing censored data is that the information is incomplete. Methods discussed in this chapter such as life tables, Kaplan–Meier survival curves and Cox's proportional hazards regression model take into account censoring of the lifetimes. These methods are different from other methods discussed in this book in that the first two methods are nonparametric while the third method is semiparametric, i.e. one part of the model is unspecified and so is nonparametric while the other part, is specified and is parametric.

10.1   Hazard rate and survival distribution

Suppose that the lifetime is a continuous r.v. denoted by T. Let $f(t)$ denote the p.d.f. and $F(t) = P(T \le t)$ denote the c.d.f. of T. Then, $S(t) = 1 - F(t) = P(T > t)$ is called the **survival distribution** of T. The **hazard rate**, denoted by $\lambda(t)$, is defined as

$$\lambda(t) = \lim_{\Delta t \to 0} \left[\frac{P(t < T \le t + \Delta t | T > t)}{\Delta t} \right].$$

In words, $\lambda(t)$ is the instantaneous **failure rate** at time t given that the patient has survived until that time. An explicit expression for $\lambda(t)$ can be obtained as follows:

$$\lambda(t) = \lim_{\Delta t \to 0} \left[\frac{1}{\Delta t} \frac{P(t < T \le t + \Delta t)}{P(T > t)} \right] = \lim_{\Delta t \to 0} \frac{1}{\Delta t} \left[\frac{f(t)\Delta t}{1 - F(t)} \right] = \frac{f(t)}{S(t)}. \quad (10.1)$$

Note that

$$\lambda(t) = -\frac{d}{dt}(\ln(S(t))),$$

so

$$S(t) = \exp\left(-\int_0^t \lambda(u)du \right) = \exp(-\Lambda(t)), \quad (10.2)$$

where $\Lambda(t) = \int_0^t \lambda(u)du$ is called the **cumulative hazard**.

Example 10.1 *(Hazard Rate for the Exponential Distribution)*   The exponential distribution is the simplest continuous lifetime distribution. Its p.d.f. equals $f(t) = \lambda \exp(-\lambda t)$ and its c.d.f. equals $F(t) = 1 - \exp(-\lambda t)$; thus, its survival distribution equals $S(t) = \exp(-\lambda t)$. Hence, its hazard rate equals

$$\lambda(t) = \frac{f(t)}{S(t)} = \frac{\lambda e^{-\lambda t}}{e^{-\lambda t}} = \lambda,$$

which is constant with respect to time.

The **memoryless property** of the exponential distribution derives from its constant hazard rate. This property says that the conditional probability that a patient having survived until time t, will survive for another u time units is independent of t. This can be checked as follows:

$$P(T > t + u | T > t) = \frac{P(T > t + u)}{P(T > t)} = \frac{e^{-\lambda(t+u)}}{e^{-\lambda t}} = e^{-\lambda u},$$

which is independent of t. $\qquad\qquad\square$

The constant hazard rate for the exponential distribution limits its use in practice. Many real-life phenomena exhibit increasing hazard rates, e.g. machine parts become more failure-prone as they age due to wear and tear; the same is true of living beings. The **Weibull distribution** (see Exercise 10.2) generalizes the exponential distribution and allows modeling of increasing or decreasing hazard rates. The gamma distribution introduced in Chapter 9 is another generalization of the exponential distribution.

10.2   Kaplan–Meier estimator

The **Kaplan–Meier (KM) estimator** gives a nonparametric estimate of the survival distribution. The idea of the estimator is very simple and yet powerful. Consider a sample of n patients and let $t_1 < t_2 < \cdots < t_m$ denote their event times (death or censored) where $m \leq n$ since more than one event may take place at any time t_i. Let c_i be the number of censored events and d_i be the number of death events at time t_i. We form $m + 1$ intervals $[0, t_1), [t_1, t_2), \ldots, [t_m, \infty)$ such that the events (whether death or censored) are assumed to occur just after the beginning of each time interval. Let n_i be the number of patients at risk at time t_i (i.e. those who are still alive). Then the number of patients at risk at time t_{i+1} equal $n_{i+1} = n_i - c_i - d_i$ ($i = 1, \ldots, m - 1$) where $n_1 = n$. The estimated hazard rate at time t_i is

$$\widehat{\lambda}(t_i) = \frac{d_i}{n_i}.$$

To find the KM estimator of the survival function, note that $S(t_i) = P(T > t_i) = P(T > t_{i-1})P(T \neq t_i) = S(t_{i-1})[1 - \lambda(t_i)]$ for $i \geq 1$. Applying this formula recursively, we get

$$S(t_i) = \prod_{j=1}^{i}[1 - \lambda(t_j)].$$

Now $\lambda(t_j)$ can be estimated by $\widehat{\lambda}(t_j)$ given above. Further noting that $\widehat{S}(t)$ is constant for $t_i \leq t < t_{i+1}$, leads to the following KM estimator of the survival function:

$$\widehat{S}(t) = \prod_{t_j \leq t} [1 - \widehat{\lambda}(t_j)] = \prod_{t_j \leq t} \left[1 - \frac{d_j}{n_j} \right]. \tag{10.3}$$

These calculations can be presented in the form of a **lifetable**. KM survival curves can be plotted from lifetable calculations as illustrated in the Example 10.2.

If the death times are not tied, i.e. if there is at most one death at any given time, then it is easy to see that the KM estimator for $t_i \leq t < t_{i+1}$ reduces to

$$\widehat{S}(t) = \prod_{j \leq i} \left(1 - \frac{1}{n - j + 1} \right)^{\delta_j} = \prod_{j \leq i} \left(\frac{n - j}{n - j + 1} \right)^{\delta_j},$$

where δ_j is an indicator variable for censoring with $\delta_j = 0$ if t_j is censored and $\delta_j = 1$ if t_j is not censored.

Example 10.2 *(AML Data: Kaplan–Meier Survival Curves)* Tableman and Kim (2003) have given data on the times to relapse for acute myelogenous leukemia (AML) patients in a clinical trial under two treatment arms: extended treatment ("Maintained") and non-extended treatment ("Nonmaintained"). The data for the 11 Maintained group of patients are 9, 13, 13+, 18, 23, 28+, 31, 34, 45+, 48, and 161+ weeks, where the + sign indicates a censored observation. The data for the 12 Nonmaintained group of patients are 5, 5, 8, 8, 12, 16+, 23, 27, 30, 33, 43, 45. These data are available under the file `aml` as part of the `survival` library in R.

Note that the estimated survival function remains constant from one uncensored event to the next regardless of any censored events in between (e.g. $\widehat{S}(t) = 0.614$ from $t = 23$ to $t = 31^-$ in the Maintained treatment group regardless of the censored observation at time $t = 28$ in between). This is clear from the fact that at a censored time if there is no death then $\widehat{\lambda}(t_i) = 0$ and so $1 - \widehat{\lambda}(t_i) = 1$.

The survival curves for the two treatment arms are shown in Figure 10.1. We see that the survival curve for the Maintained treatment lies uniformly above that for the Nonmaintained treatment, thus showing that the Maintained treatment improves survival, i.e. lengthens time to relapse. These survival curves are obtained using the following R code:

```
> library(survival)
> fit<- survfit(Surv(time,status) ~ x,data=aml)
> plot(fit,col=1:2)
> legend("topright", paste(" ",c("Maintained","Nonmaintained")),
  col=1:2, lty=c(1,2))
```

□

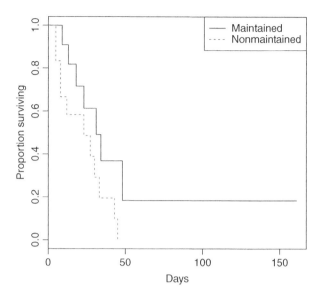

Figure 10.1. *Kaplan–Meier survival curves for Maintained and Nonmaintained treatment arms.*

Since $\widehat{S}(t)$ is an estimate of the true survival function $S(t)$ at some fixed time t, we can assess its precision through its variance. The **Greenwood formula** gives an estimate of the variance of $\widehat{S}(t)$:

$$\widehat{\mathrm{Var}}(\widehat{S}(t)) = [\widehat{S}(t)]^2 \sum_{t_i \leq t} \frac{d_i}{n_i(n_i - d_i)}. \tag{10.4}$$

The following example illustrates this calculation.

Example 10.3 (AML Data: Standard Error Calculation for Kaplan–Meier Survival Curves)   For illustration purposes, consider $\widehat{S}(t) = 0.716$ for $18 \leq t < 23$ for the Maintained group calculated in Table 10.1. Then, $t_1 = 9$, $t_2 = 13$, and $t_3 = 18$ are $\leq t$ (any censored time $t_i \leq t$ with $d_i = 0$ can be omitted from the variance calculation). Thus, for $18 \leq t < 23$ we have

$$\widehat{\mathrm{Var}}(\widehat{S}(t)) = (0.716)^2 \left[\frac{1}{11 \times 10} + \frac{1}{10 \times 9} + \frac{1}{8 \times 7} \right] = 0.0195.$$

Hence, $\mathrm{SE}(\widehat{S}(t)) = \sqrt{0.0195} = 0.1397$. The standard errors for all values of $\widehat{S}(t)$ for both treatment arms calculated using R are given in Table 10.2. These standard errors can be used to calculate pointwise confidence intervals around the estimated survival function. For example, for $18 \leq t < 23$ for the Maintained treatment, the 95% confidence interval is

$$0.716 \pm 1.96 \times 0.1397 = [0.442, 0.990].$$

$\square$

Table 10.1. Calculation of Kaplan–Meier survival curves for AML data (+ indicates a censored observation)

	Maintained							Nonmaintained					
t_i	n_i	c_i	d_i	$\widehat{\lambda}(t_i)$	$1 - \widehat{\lambda}(t_i)$	$\widehat{S}(t_i)$	t_i	n_i	c_i	d_i	$\widehat{\lambda}(t_i)$	$1 - \widehat{\lambda}(t_i)$	$\widehat{S}(t_i)$
0	11	0	0	0.000	1.000	1.000	0	12	0	0	0.000	1.000	1.000
9	11	0	1	0.091	0.909	0.909	5	12	0	2	0.167	0.833	0.833
13+	10	1	1	0.100	0.900	0.818	8	10	0	2	0.200	0.800	0.667
18	8	0	1	0.125	0.875	0.716	12	8	0	1	0.125	0.875	0.583
23	7	0	1	0.143	0.857	0.614	16+	7	1	0	0.000	1.000	0.583
28+	6	1	0	0.000	1.000	0.614	23	6	0	1	0.167	0.833	0.486
31	5	0	1	0.200	0.800	0.491	27	5	0	1	0.200	0.800	0.389
34	4	0	1	0.250	0.750	0.368	30	4	0	1	0.250	0.750	0.292
45+	3	1	0	0.000	1.000	0.368	33	3	0	1	0.333	0.667	0.194
48	2	0	1	0.500	0.500	0.184	43	2	0	1	0.500	0.500	0.097
161+	1	1	0	0.000	1.000	0.184	45	1	0	1	1.000	0.000	0.000

10.3   Logrank test

We saw in Figure 10.1 that the survival distribution for the Maintained group lies uniformly above that of the Nonmaintained group, but is the difference statistically significant? Denoting the two survival distributions by $S_1(t)$ and $S_2(t)$, we would like to test the null hypothesis $H_0 : S_1(t) = S_2(t)$ for all t. The alternative hypothesis may be one-sided $H_1 : S_1(t) \geq S_2(t)$ for all t with a strict inequality for at least some t or two-sided $H_1 : S_1(t) \leq S_2(t)$ for all t with a strict inequality for at least some t.

Logrank test provides a test of the *overall* difference between the two survival distributions by cumulating the differences between $\widehat{S}_1(t)$ and $\widehat{S}_2(t)$ at the observed death times. Let $t_1 < t_2 < \cdots < t_m$ denote the observed death times in the combined data set of the two treatments. The deaths may occur in one treatment arm or both. For each time point t_i, let n_{1i} and n_{2i} denote the numbers at risk and d_{1i} and d_{2i} denote the numbers of deaths from the two treatment arms with $n_i = n_{1i} + n_{2i}$ being the total number at risk at time t_i. This data can be summarized in a 2×2 table shown in Table 10.3.

It is well-known that if we fix the row and column totals (i.e. n_{1i}, n_{2i}, d_i, and $n_i - d_i$) then d_{1i} determines the remaining entries in the table and the

Table 10.2. Calculation of standard errors for Kaplan–Meier survival curves for AML data

	Maintained			Nonmaintained	
t_i	$\widehat{S}_1(t_i)$	$\mathrm{SE}(\widehat{S}_1(t_i))$	t_i	$\widehat{S}_2(t_i)$	$\mathrm{SE}(\widehat{S}_2(t_i))$
0	1.000	0.0000	0	1.000	0.000
9	0.909	0.0867	5	0.833	0.1076
13+	0.818	0.1163	8	0.667	0.1361
18	0.716	0.1397	12	0.583	0.1423
23	0.614	0.1526	16+	0.583	0.1423
28+	0.614	0.1526	23	0.486	0.1481
31	0.491	0.1642	27	0.389	0.1470
34	0.368	0.1627	30	0.292	0.1387
45+	0.368	0.1627	33	0.194	0.1219
48	0.184	0.1535	43	0.097	0.0919
161+	0.184	0.1535	45	0.000	N/A

Table 10.3. Number of deaths, survivals, and at risk patients at time t_i

		Death Yes	Death No	
Treatment	Maintained	d_{1i}	$n_{1i} - d_{1i}$	n_{1i}
	Nonmaintained	d_{2i}	$n_{2i} - d_{2i}$	n_{2i}
		d_i	$n_i - d_i$	n_i

distribution of d_{1i} under H_0 is the **hypergeometric distribution** given by

$$f(d_{1i}) = \frac{\binom{n_{1i}}{d_{1i}} \binom{n_{2i}}{d_{2i}}}{\binom{n_i}{d_i}}.$$

The mean and variance of this distribution are as follows:

$$E(d_{1i}) = \frac{d_i n_{1i}}{n_i} \quad \text{and} \quad \mathrm{Var}(d_{1i}) = \frac{n_{1i} n_{2i} d_i (n_i - d_i)}{n_i^2 (n_i - 1)}.$$

Calculations of the Logrank statistic for the AML data.

Time	d_{1i}	d_{2i}	n_{1i}	n_{2i}	d_i	n_i	$E(d_{1i})$	$\mathrm{Var}(d_{1i})$
5	0	2	11	12	2	23	0.9565	0.4764
8	0	2	11	10	2	21	1.0476	0.4739
9	1	0	11	8	1	19	0.5789	0.2438
12	0	1	10	8	1	18	0.5556	0.2469
13	1	0	10	7	1	17	0.5882	0.2422
18	1	0	8	6	1	14	0.5714	0.2449
23	1	1	7	6	2	13	1.0769	0.4556
27	0	1	6	5	1	11	0.5454	0.2479
30	0	1	5	4	1	9	0.5556	0.2469
31	1	0	5	3	1	8	0.6250	0.2344
33	0	1	4	3	1	7	0.5714	0.2449
34	1	0	4	2	1	6	0.6667	0.2222
43	0	1	3	2	1	5	0.6000	0.2400
45	0	1	3	1	1	4	0.7500	0.1875
48	1	0	2	0	1	2	1	0

The logrank statistic is obtained as the standardized sum of the deviations of the d_{1i} from $E(d_{1i})$ summed over all time points:

$$z = \frac{\sum_{i=1}^{m} d_{1i} - \sum_{i=1}^{m} E(d_{1i})}{\left(\sum_{i=1}^{m} \mathrm{Var}(d_{1i})\right)^{1/2}}. \tag{10.5}$$

Asymptotically (for large n_{1i}, n_{2i}), this statistic has a standard normal distribution under H_0.

Example 10.4 *(AML Data: Logrank Test)* For the AML data shown in Table 10.1, the number of distinct observed event times are $m = 15$. Table 10.4 shows the calculation of the logrank statistic.

We calculate $\sum_{i=1}^{15} d_{1i} = 7, \sum_{i=1}^{15} E(d_{1i}) = 10.689$, and $\sum_{i=1}^{15} \mathrm{Var}(d_{1i}) = 4.008$. Hence, the logrank statistic equals $z = (7 - 10.689)/\sqrt{4.008} = -1.843$ with a one-sided P-value $= 0.0326$. So we can reject $H_0 : S_1(t) = S_2(t)$ for all t at

$\alpha = 0.05$ and conclude that the Maintained group experiences significantly longer survival times than the Nonmaintained group. The R code for performing the logrank test and its output are shown below. Note that R reports a two-sided test with $\chi^2 = (-1.843)^2 = 3.40$ with a two-sided P-value $= 0.0653$, which is twice that of the one-sided P-value of 0.0326.

```
> library(survival)
> survdiff(Surv(time,status) ~ x,data=aml)
Call: survdiff(formula = Surv(time, status) ~ x, data = aml)

                N Observed Expected (O-E)^2/E (O-E)^2/V
x=Maintained   11        7    10.69      1.27       3.4
x=Nonmaintained 12       11     7.31      1.86       3.4

 Chisq= 3.4  on 1 degrees of freedom, p= 0.0653
```

□

10.4 Cox's proportional hazards model

Cox (1972) proposed a novel regression model for censored lifetimes as a function of a set of covariates. This model postulates how the hazard rate depends on the covariates. Let $\lambda(t) = \lambda(t|\boldsymbol{x})$ denote the hazard rate at time t given the covariate vector $\boldsymbol{x} = (x_1, \ldots, x_p)'$. Let $\boldsymbol{\beta} = (\beta_1, \ldots, \beta_p)'$ denote the corresponding unknown regression coefficient vector. Then Cox's model is

$$\lambda(t) = \lambda(t|\boldsymbol{x}) = \lambda_0(t)\exp(\beta_1 x_1 + \cdots + \beta_p x_p) = \lambda_0(t)\exp(\boldsymbol{x}'\boldsymbol{\beta}), \qquad (10.6)$$

where $\lambda_0(t)$, called the **base hazard rate**, is independent of $\boldsymbol{x}$ and is assumed to be completely unspecified. Thus, the model has a nonparametric component $\lambda_0(t)$ and a parametric component $\exp(\boldsymbol{x}'\boldsymbol{\beta})$. Hence, it is called a **semiparametric model**. Note that this model does not have an intercept term β_0 since $\exp(\beta_0)$ can be absorbed in $\lambda_0(t)$.

This model has an interesting property that if there are two individuals, i and j, with covariate vectors $\boldsymbol{x}_i$ and $\boldsymbol{x}_j$ then the ratio of their hazard rates (called the **hazard ratio**) at any time t:

$$\frac{\lambda_i(t)}{\lambda_j(t)} = \frac{\lambda_0(t)\exp(\boldsymbol{x}_i'\boldsymbol{\beta})}{\lambda_0(t)\exp(\boldsymbol{x}_j'\boldsymbol{\beta})} = \exp((\boldsymbol{x}_i - \boldsymbol{x}_j)'\boldsymbol{\beta})$$

is independent of t. Hence, this is called the **proportional hazards (PH) model**. This property is similar to the proportional odds property of the ordinal logistic regression model mentioned in Section 7.5.2.

A special case of interest is when there is a single covariate x_1, which is a binary treatment factor, say treatment $(x_1 = 1)$ and control $(x_1 = 0)$. Then, the hazard ratio is

$$\frac{\lambda(t|x_1 = 1)}{\lambda(t|x_1 = 0)} = \frac{\lambda_0(t)\exp(\beta_1)}{\lambda_0(t)} = \exp(\beta_1),$$

which is constant. If $\beta_1 < 0$ then the hazard rate for the treatment is less than that for the control. Thus, a negative β_1 represents an effective treatment since it reduces the hazard rate. More generally, $\exp(\beta_j)$ represents the ratio of hazard rates if x_j is increased by one unit, keeping all other covariates fixed.

10.4.1   Estimation

Suppose we have data on the lifetimes t_i, covariate vectors $\boldsymbol{x}_i$ and censoring indicators δ_i ($\delta_i = 0$ if t_i is censored and $\delta_i = 1$ if t_i is not censored) for $i = 1, \ldots, n$. Denote by $C = \{i : \delta_i = 0\}$ the set of censored individuals and by $D = \{i : \delta_i = 1\}$ the set of died individuals. Assume that $t_1 < t_2 < \ldots < t_n$ and thus there are no ties among the lifetimes. For convenience, suppose that the individuals are labeled so that the ith individual is associated with the observation at time t_i. Since $\lambda_0(t)$ is arbitrary and unspecified, it is not possible to find the MLE of $\boldsymbol{\beta}$ by maximizing the full likelihood. Cox (1972) proposed to use the likelihood conditioned on the observed lifetimes of individuals.

Consider the event time t_i at which individual $i \in D$ dies. Let $R(t_i)$ denote the **risk set** of all individuals who are alive and still under observation at time t_i^- and hence are at risk of dying at time t_i. Then the conditional probability that out of all the individuals in this risk set, the particular individual i dies at time t_i is given by

$$\frac{\lambda_0(t_i)\exp(\boldsymbol{x}_i'\boldsymbol{\beta})}{\sum_{j \in R(t_i)}\lambda_0(t_i)\exp(\boldsymbol{x}_j'\boldsymbol{\beta})} = \frac{\exp(\boldsymbol{x}_i'\boldsymbol{\beta})}{\sum_{j \in R(t_i)}\exp(\boldsymbol{x}_j'\boldsymbol{\beta})} = \frac{\psi_i}{\sum_{j \in R(t_i)}\psi_j},$$

where $\psi_j = \exp(\boldsymbol{x}_j'\boldsymbol{\beta})$ is referred to as the jth individual's **risk score**. One can view this as an urn model in which one individual is drawn to die out of all the individuals in the risk set $R(t_i)$ and the probability of drawing individual $i \in R(t_i)$ is proportional to ψ_i. In the above, the unknown base hazard function $\lambda_0(t_i)$ cancels from the numerator and denominator in each term and so its unknown form does not affect the MLE of $\boldsymbol{\beta}$.

The so-called **partial likelihood** is obtained by regarding the observed death times as independent resulting in

$$L = \prod_{i \in D}\left[\frac{\psi_i}{\sum_{j \in R(t_i)}\psi_j}\right]. \tag{10.7}$$

The MLE of β maximizes L or equivalently $\ln L$. Equations for obtaining β are derived in Section 10.5.

Note that the survival times t_i enter the partial likelihood L only through the risk sets $R(t_i)$ and hence L is a function only of the ranks of the t_i's. Therefore, any monotone transformation, such as the log transformation, of the t_i's does not affect the partial likelihood and hence the MLE of β.

10.4.2 Examples

First, we give a toy example to illustrate how the partial likelihood is constructed.

Example 10.5 (Toy Example) Suppose there are three patients, aged $x_1 = 60$, 55, and 50, which is the only covariate. The patients are observed at three time points: $t_1 < t_2 < t_3$. Patient 1 dies at t_1, Patient 2 dies at t_2, while Patient 3 survives past t_3 and is censored since the study is terminated at t_3. Then the risk sets are

$$R(t_1) = \{1, 2, 3\}, R(t_2) = \{2, 3\}, R(t_3) = \{3\}.$$

Suppose we want to fit the model

$$\lambda(t) = \lambda_0(t) \exp(\beta_1 x_1).$$

At time t_1, Patient 1 dies out of the risk set $R(t_1)$, at time t_2, Patient 2 dies out of the risk set $R(t_2)$ and at time t_3, no death is observed. Therefore, the partial likelihood equals

$$L = \left[\frac{\exp(60\beta_1)}{\exp(60\beta_1) + \exp(55\beta_1) + \exp(50\beta_1)} \right] \left[\frac{\exp(55\beta_1)}{\exp(55\beta_1) + \exp(50\beta_1)} \right].$$

Note again that only the order $t_1 < t_2 < t_3$ matters, their numerical values do not enter into the partial likelihood. $\square$

In the next example, we apply the PH model to the AML data and perform a two sample test of Maintained versus Nonmaintained groups.

Example 10.6 (AML Data: Proportional Hazards Model) The only covariate in this example is the treatment variable indicating whether the individual belongs to the Maintained group ($x = 0$) or to the Nonmaintained group ($x = 1$). The R code used to perform this analysis and the resulting output are shown below.

```
> library(survival)
> fit1 <- coxph(Surv(time, status) ~ x, data=aml)
> summary(fit1)
```

```
Call: coxph(formula = Surv(time, status) ~ x, data = aml)

  n= 23, number of events= 18

               coef exp(coef) se(coef)      z Pr(>|z|)
xNonmaintained 0.9155    2.4981   0.5119 1.788   0.0737 .
---
Signif. codes:  0 '***' 0.001 '**' 0.01 '*' 0.05 '.' 0.1 ' ' 1

               exp(coef) exp(-coef)  lower .95  upper .95
xNonmaintained     2.498     0.4003     0.9159      6.813

Concordance= 0.619  (se = 0.073 )
Rsquare= 0.137    (max possible= 0.976 )
Likelihood ratio test= 3.38  on 1 df,    p=0.06581
Wald test = 3.2  on 1 df,    p=0.07371
Score (logrank) test = 3.42  on 1 df,    p=0.06454
```

We see that the fitted model is $\lambda(t) = \lambda_0(t) \exp(0.9155x)$. So the hazard for the Nonmaintained group is $\exp(0.9155) = 2.498$ or almost 2.5 times that of the Maintained group. The results of the likelihood ratio (LR) test, Wald test, and logrank test, agree closely. The Wald statistic equals $[\hat{\beta}/\mathrm{SE}(\hat{\beta})]^2 = [0.9155/0.5119]^2 = 1.788^2 = 3.20$ which is a chi-square statistic with 1 d.f. The P-value is approximately 0.07 and so $\hat{\beta}_1$ is not significant at $\alpha = 0.05$.

A large sample 95% CI on the hazard ratio between the nonmaintained group and the maintained group is obtained by first obtaining a 95% CI on β_1 as

$$\hat{\beta}_1 \pm z_{.025}\mathrm{SE}(\hat{\beta}_1) = 0.9155 \pm 1.960 \times 0.5119 = [-0.0878, 1.9188].$$

Hence the corresponding 95% CI on the hazard ratio equals

$$[\exp(-0.0878), \exp(1.9188)] = [0.9159, 6.8128],$$

which is given in the R output above. Note that these CIs are in agreement with the test result above in that the CI for β_1 includes 0 and correspondingly the CI for the hazard ratio includes 1.   □

The final example involves multiple covariates including a treatment factor with two levels.

Example 10.7 (Recidivism Study: Proportional Hazards Model)
Rossi et al. (1980) performed an experimental study of recidivism of 432 male prisoners, who were observed for a year after being released from prison. The goal of the study was to evaluate whether financial aid (the treatment factor) helps these prisoners not get arrested again. Half the prisoners were randomly

assigned to receive financial aid. The following variables were recorded for each prisoner.

1. *Week*: Week of first arrest after release or censoring time.

2. *Arrest*: The event indicator (1 if arrested, 0 if not arrested).

3. *Aid*: Financial aid (yes = 1 or no = 0)

4. *Age*: In years at the time of release.

5. *Race*: Black = 1 or other = 0.

6. *Work*: Yes = 1 if the individual had full-time work experience prior to incarceration, no=0 if he did not.

7. *Married*: Married = 1 or not married = 0.

8. *Parole*: Released on parole = 1 or not = 0.

9. *Prior*: Number of prior convictions.

10. *Education*: An ordinal variable with codes 2 (grade 6 or less), 3 (grades 6–9), 4 (grades 10 and 11), 5 (grade 12), or 6 (some post-secondary).

The data are in file `recid.csv`. The file also contains data on the employment status for each of 52 weeks. This is a time-dependent covariate, which may vary from week to week. In this example, we ignore this variable. Exercise 10.11 asks you to include it in the analysis.

First, we compare the Aid versus No Aid group by plotting their Kaplan–Meier curves and performing the logrank test. This plot ignores other covariates. The Kaplan–Meier curves are shown in Figure 10.2. We see that the survival curves for the two groups overlap until about 20 weeks but later the survival curve for the Aid group lies above that for the No Aid group. Thus in the first 20 weeks, prisoners get arrested at roughly the same rate whether they received financial aid or not. After 20 weeks, prisoners who received financial aid stay out of jail longer. Thus, financial aid does not have a short-term positive effect but has a long-term positive effect.

The logrank test results are shown below. We see that the difference between the two survival distributions is nearly significant at the 0.05 level.

```
> recid=read.csv("c:/data/recid.csv")
> fit <- coxph(Surv(Week, Arrest) ~ Aid, data=recid)
> summary(fit)
Call: coxph(formula = Surv(Week, Arrest) ~ Aid, data = recid)

  n= 432, number of events= 114
```

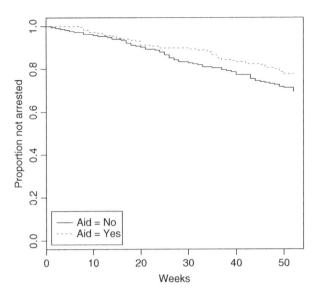

Figure 10.2. *Kaplan–Meier survival curves for Aid and No Aid groups for Recidivism data.*

```
        coef exp(coef) se(coef)       z Pr(>|z|)
Aid -0.3691    0.6914   0.1897 -1.945    0.0517.
---
Signif. codes:  0 '***' 0.001 '**' 0.01 '*' 0.05 '.' 0.1 ' ' 1

    exp(coef) exp(-coef) lower.95 upper.95
Aid    0.6914      1.446    0.4767    1.003

Concordance= 0.546  (se = 0.024 )
Rsquare= 0.009    (max possible=0.956 )
Likelihood ratio test= 3.84  on 1 df,    p=0.05013
Wald test            = 3.78  on 1 df,    p=0.05174
Score (logrank) test = 3.83  on 1 df,    p=0.05042
```

Next, we fit a full proportional hazards model with all covariates using the following R code.

```
> library(survival)
> recid=read.csv("c:/data/recid.csv")
> fit1 <- coxph(Surv(Week, Arrest) ~., data=recid)
> summary(fit1)
```

The output is as follows.

```
Call:
coxph(formula = Surv(Week, Arrest) ~., data = recid)

  n= 432, number of events= 114
```

```
            coef exp(coef) se(coef)       z Pr(>|z|)
Aid      -0.35963   0.69794  0.19180 -1.875  0.06079 .
Age      -0.05768   0.94395  0.02187 -2.638  0.00835 **
Race      0.34554   1.41276  0.30907  1.118  0.26356
Work     -0.11439   0.89191  0.21311 -0.537  0.59145
Married  -0.42496   0.65380  0.38209 -1.112  0.26605
Parole   -0.08991   0.91401  0.19568 -0.459  0.64589
Prior     0.08469   1.08838  0.02919  2.902  0.00371 **
Education -0.18578  0.83046  0.13153 -1.412  0.15782
---

Concordance= 0.656  (se = 0.027 )
Rsquare= 0.079    (max possible=0.956 )
Likelihood ratio test= 35.35  on 8 df,    p=2.31e-05
Wald test             = 33.74  on 8 df,    p=4.529e-05
Score (logrank) test = 35.1   on 8 df,    p=2.568e-05
```

Note that the effect of Aid is now less significant ($P = 0.0608$). Of the remaining covariates, only Age and Prior are significant. Including only these three variables, another proportional hazards model was fitted whose results are shown below.

```
Call: coxph(formula = Surv(Week, Arrest) ~ Aid + Age + Prior,
          data =recid)

  n= 432, number of events= 114

        coef exp(coef) se(coef)       z Pr(>|z|)
Aid  -0.34695   0.70684  0.19025 -1.824 0.068197 .
Age  -0.06711   0.93510  0.02085 -3.218 0.001289 **
Prior 0.09689   1.10174  0.02725  3.555 0.000378 ***
---

Concordance= 0.63  (se = 0.027 )
Rsquare= 0.065    (max possible=0.956 )
Likelihood ratio test= 29.05  on 3 df,    p=2.189e-06
Wald test             = 27.94  on 3 df,    p=3.741e-06
Score (logrank) test = 29.03  on 3 df,    p=2.203e-06
```

Now Age and Prior are much more significant but Aid is even less significant ($P = 0.0682$). Our overall conclusion is that Aid has a positive effect on the recidivism of the prisoners but not very significant. □

10.4.3 Time-dependent covariates

In the basic Cox model (10.6), we have assumed that the covariates are fixed over time. But in practice, many covariates vary with time. For example, many lab measurements such as cholesterol level or hemoglobin A1c vary with time.

We do not consider the age of a patient as a time-dependent covariate since it increases at the same rate for all patients, so its value at the baseline is used as a fixed covariate. If the time-dependent nature of covariates is ignored in the analysis, then misleading conclusions may result as shown in Example 10.8.

Denote a time-dependent covariate x_j as $x_j(t)$. By letting $x'_j(t) = x_j(t - s)$, we can model lagged effect of a variable where $s \geq 1$ is a specified lag. More generally, denote the covariate vector as $\boldsymbol{x}(t)$, some of whose components are functions of time whereas other components may be fixed in time. Then the Cox model may be written as

$$\lambda(t) = \lambda_0(t) \exp(\boldsymbol{x}'(t)\boldsymbol{\beta}).$$

Note that this model does not have the proportional hazards property since the ratio of hazards for any pair of individuals is no longer fixed but varies with time. The proportional hazards property of the Cox model only holds if the covariates are not time-dependent. The partial likelihood is still the product of the conditional probabilities at different death times t_i; each conditional probability being the probability that the ith patient dies at time t_i conditioned on all the patients in the risk set $R(t_i)$.

To analyze time-dependent covariate data using the `survival` library, the data need to be set up in what is called the **long format**. Essentially, this format consists of multiple rows of entries for each subject for successive time intervals, `tstart` $\leq t <$ `tstop`, such that the values of all covariates remain fixed within each time interval. The values of time-dependent covariates change between different time intervals. Note that each time interval is open on the right except the last interval when either death or censoring occurs at time `tstop`. We set `tstart` for any interval (except the first one) equal to `tstop` for the previous interval. Any changes in the data are assumed to take place at `tstart` of each interval and are included in that interval.

As an example, consider the recidivism data which has a time-varying covariate, employment status (0 if unemployed, 1 if employed) for each of 52 weeks. The first prisoner was arrested in Week 20 and was not employed throughout those 20 weeks. The second prisoner was employed from Week 9 until Week 14 and was arrested in Week 17. So the data for these two prisoners (ignoring other covariates) can be represented in long format as shown below.

Prisoner	`tstart`	`tstop`	Employment status	Arrest
1	0	20	0	1
2	0	9	0	0
2	9	14	1	0
2	14	17	0	1

Example 10.8 *(Stanford Heart Transplant Study)* Crowley and Hu (1977) reported a heart transplant study conducted at Stanford University Medical School between 1 October 1967 and 1 April 1974. There were 103 cardiac patients enrolled. Patients had to wait until a suitable donor heart was available. Of the 103 patients, 30 died before receiving a transplant, whereas 4 patients had still not received a transplant when the study ended. Only 24 of the 69 patients who received a transplant were still alive at termination.

The raw data file `jasa`, which is part of the **survival** library, consists of a number of variables for each patient. Several data manipulations have to be done and certain anomalies in the data have to be fixed in order to obtain the variables in the form ready for analysis. The R code for these data manipulations is taken from a preprint by Therneu et al. (2018) and is shown below. The code creates two data files; **tdata.csv** in which the trt variable is regarded as fixed and so there are 103 rows of data (one for each patient) and **sdata.csv** in which the trt variable is regarded as time-dependent, and there are 170 rows of data (one or more for each patient) in the long format.

```
> jasa$subject <- 1:nrow(jasa) #we need an identifier variable
> tdata <- with(jasa, data.frame(subject = subject,
    futime= pmax(.5, fu.date - accept.dt),
    txtime= ifelse(tx.date== fu.date,
    (tx.date -accept.dt) -.5,
    (tx.date - accept.dt)),
    fustat = fustat
))
> sdata <- tmerge(jasa, tdata, id=subject,
    death = event(futime, fustat),
    trt = tdc(txtime),
    options= list(idname="subject"))
> sdata$age <- sdata$age -48
> sdata$year <- as.numeric(sdata$accept.dt - as.Date("1967-10-01"))
  /365.25
> write.csv(sdata, "c:/data/sdata.csv")
> tdata$year <- as.numeric(jasa$accept.dt - as.Date("1967-10-01"))
  /365.25
> tdata$trt = as.numeric(!is.na(tdata$txtime))
> tdata$survtime=tdata$futime
> tdata$age=jasa$age-48
> tdata$death=jasa$fustat
> tdata$surgey=jasa$surgery
> write.csv(tdata, "c:/data/tdata.csv")
```

Table 10.4. Stanford heart transplant data ignoring the time-dependent nature of the trt variable for the first 10 subjects from `tdata.csv` file

Subject	trt	Year	Age	Death	Surgery	Survtime
1	0	0.123	30.845	1	0	49
2	0	0.255	51.836	1	0	5
3	1	0.266	54.297	1	0	15
4	1	0.490	40.263	1	0	38
5	0	0.608	20.786	1	0	17
6	0	0.701	54.595	1	0	2
7	1	0.780	50.869	1	0	674
8	0	0.835	45.350	1	0	39
9	0	0.857	47.162	1	0	84
10	1	0.862	42.502	1	0	57

The variables used in the analysis are listed below.

trt	=	0 if transplant was not done, 1 if transplant was done
age	=	baseline age of the patient
surgery	=	0 if there was no prior heart surgery, 1 if there was prior heart surgery
year	=	time since the start of the study until the enrollment of the patient

The data on the first 10 subjects from the `tdata.csv` are shown in Table 10.4. The data on the same 10 subjects from the `sdata.csv` are shown in Table 10.5. Notice that there are two rows of data each for subjects 4, 7, and 10. Subject 3 received the transplant on the day of the entry to the study but died on Day 15; thus there is only a single row of data for that subject.

First, we give the results of fitting the Cox model to the `tdata.csv`. We see that the trt variable is highly significant.

```
> library(survival)
> tdata=read.csv("c:/data/tdata.csv")
> tfit<-coxph(Surv(survtime, death) ~ trt+age+surgery+year, data= tdata)
> tfit
```

Table 10.5. Stanford heart transplant data in long format regarding the trt variable as time-dependent for the first 10 subjects from sdata.csv file

Subject	trt	Year	Age	Death	Surgery	tstart	tstop
1	0	0.123	30.845	1	0	0	49
2	0	0.255	51.836	1	0	0	5
3	1	0.266	54.297	1	0	0	15
4	0	0.490	40.263	0	0	0	35
4	1	0.490	40.263	1	0	35	38
5	0	0.608	20.786	1	0	0	17
6	0	0.701	54.595	1	0	0	2
7	0	0.780	50.869	0	0	0	50
7	1	0.780	50.869	1	0	50	674
8	0	0.835	45.350	1	0	0	39
9	0	0.857	47.162	1	0	0	84
10	0	0.862	42.502	0	0	0	11
10	1	0.862	42.502	1	0	11	57

```
Call: coxph(formula = Surv(survtime, death) ~ trt + age + surgery +
            year, data = tdata)

          coef  exp(coef)  se(coef)      z        p
trt     -1.7045    0.1819    0.2826   -6.03  1.6e-09
age      0.0575    1.0592    0.0147    3.92  9.0e-05
surgery -0.3178    0.7278    0.3767   -0.84    0.399
year    -0.1177    0.8890    0.0692   -1.70    0.089

Likelihood ratio test=48.8  on 4 df, p=6.47e-10 n= 103, number of
events= 75
```

However, regarding the trt variable as fixed is not correct since the patients who died early did not have a chance to get a matching heart donor and so their deaths may be wrongly attributed to not getting a transplant which would make the trt variable more significant than it actually is. Therefore, we must take into account the waiting time to get a transplant by treating trt as a time-dependent covariate. Below we give the results of fitting the Cox model to the `sdata`. We see that the trt variable is now highly nonsignificant.

```
> library(survival)
> sdata=read.csv("c:/data/sdata.csv")
> sfit<-coxph(Surv(tstart, tstop, death) ~ trt+age + surgery + year,
    data= sdata)
> sfit

Call: coxph(formula = Surv(tstart, tstop, death) ~ trt + age +
surgery +
    year, data = sdata)

          coef  exp(coef) se(coef)     z     p
trt    -0.0129     0.9872   0.3133 -0.04 0.967
age     0.0272     1.0276   0.0137  1.98 0.047
surgery -0.6371    0.5288   0.3672 -1.73 0.083
year   -0.1464     0.8638   0.0705 -2.08 0.038

Likelihood ratio test=15.1  on 4 df, p=0.00447 n= 170, number of
events= 75
```

□

10.5 Technical notes*

10.5.1 ML estimation of the Cox proportional hazards model

The log-likelihood function is given by taking the log of the partial likelihood function (10.7):

$$\ln L = \sum_{i \in D} \ln \psi_i - \sum_{i \in D} \ln \sum_{j \in R(t_i)} \psi_j = \sum_{i \in D} \boldsymbol{x}_i' \boldsymbol{\beta} - \sum_{i \in D} \ln \sum_{j \in R(t_i)} \psi_j.$$

The MLE of $\boldsymbol{\beta}$ is the solution of the equation obtained by taking the derivative of this function with respect to $\boldsymbol{\beta}$ and setting it equal to zero.

Now,

$$\frac{d(\boldsymbol{x}_i' \boldsymbol{\beta})}{d\boldsymbol{\beta}} = \boldsymbol{x}_i,$$

and

$$\frac{d\psi_j}{d\boldsymbol{\beta}} = \frac{d(\exp(\boldsymbol{x}_j' \boldsymbol{\beta}))}{d\boldsymbol{\beta}} = \exp(\boldsymbol{x}_j' \boldsymbol{\beta})\frac{d(\boldsymbol{x}_j' \boldsymbol{\beta})}{d\boldsymbol{\beta}} = \psi_j \boldsymbol{x}_j.$$

Hence,

$$\frac{d}{d\boldsymbol{\beta}} \sum_{i \in D} \ln \sum_{j \in R(t_i)} \psi_j = \sum_{i \in D} \frac{\frac{d}{d\boldsymbol{\beta}}(\sum_{j \in R(t_i)} \psi_j)}{\sum_{j \in R(t_i)} \psi_j} = \sum_{i \in D} \frac{\sum_{j \in R(t_i)} \psi_j \boldsymbol{x}_j}{\sum_{j \in R(t_i)} \psi_j}.$$

Therefore, the MLE $\widehat{\boldsymbol{\beta}}$ of $\boldsymbol{\beta}$ is the solution to the equation

$$\frac{d \ln L}{d\boldsymbol{\beta}} = \sum_{i \in D} \boldsymbol{x}_j - \sum_{i \in D} \frac{\sum_{j \in R(t_i)} \psi_j \boldsymbol{x}_j}{\sum_{j \in R(t_i)} \psi_j} = 0.$$

Asymptotically, $\widehat{\boldsymbol{\beta}}$ can be shown to be fully efficient and normally distributed, and its asymptotic covariance matrix can be obtained by inverting the Hessian matrix of the second partial derivatives of $\ln L$ in the usual manner.

Exercises

Theoretical Exercises

10.1 (Geometric distribution: Hazard rate) The geometric distribution gives the probability of the first success on the tth trial when making successive independent Bernoulli trials, each with success probability θ. It is given by

$$f(t) = P(T = t) = \theta(1 - \theta)^{t-1}, \quad t = 1, 2, \ldots .$$

and is a discrete analog of the exponential distribution. Using the hazard rate formula for discrete lifetime:

$$\lambda(t) = \frac{P(T = t)}{P(T \geq t)}, \quad t = 1, 2, \ldots ,$$

show that the hazard rate for the geometric distribution is constant and equal to θ.

10.2 (Weibull distribution: Hazard rate) The Weibull distribution is widely used to model failure times in engineering reliability applications. Its p.d.f. is given by

$$f(t; \lambda, \gamma) = \lambda\gamma(\lambda t)^{\gamma-1} \exp(-(\lambda t)^\gamma) \quad \text{for } t \geq 0,$$

where λ is the scale parameter and γ is the shape parameter. For $\gamma = 1$, we get the exponential distribution. Another way to think of the Weibull distribution is that if $U = (\lambda T)^\gamma$ has the unit exponential distribution then T has the Weibull distribution.

(a) Show that the survival function of the Weibull distribution is given by

$$S(t) = \exp(-(\lambda t)^\gamma).$$

(b) Show that the hazard rate of the Weibull distribution is given by

$$\lambda(t) = \lambda\gamma(\lambda t)^{\gamma-1}.$$

(c) Show that the hazard rate of the Weibull distribution is increasing in t if $\gamma > 1$, decreasing in t if $\gamma < 1$ and constant if $\gamma = 1$ (the exponential distribution).

10.3 (Uncensored data: Survival function estimation) Show that if there are no censored observations then $\widehat{S}(t)$ is simply 1 minus the empirical c.d.f. of T, i.e. $\widehat{S}(t) = 1 - \sum_{j=1}^{i}(d_j/n_j)$, which is just the binomial proportion of the patients still surviving at time t for $t_i \leq t < t_{i+1}$.

10.4 (Greenwood formula) Derive the Greenwood formula (10.4). (*Hint*: First find

$$\mathrm{Var}[\ln(\widehat{S}(t))] = \sum_{t_i \leq t} \mathrm{Var}[\ln(1 - \widehat{\lambda}(t_i))]$$

using the delta method (4.1). Since $\widehat{\lambda}(t_i) = d_i/n_i$ is a binomial proportion, use $\mathrm{Var}(\widehat{\lambda}(t_i)) = \lambda(t_i)(1 - \lambda(t_i))/n_i$. Finally, obtain $\mathrm{Var}(\widehat{S}(t))$ from $\mathrm{Var}[\ln(\widehat{S}(t))]$ by re-applying the delta method to the transformation $\widehat{S}(t) = \exp(\ln(\widehat{S}(t)))$.

10.5 (Exponential data: One sample problem) Consider n independent observations $(\boldsymbol{x}_i, t_i, \delta_i)$, where t_i is the survival time, $\boldsymbol{x}_i$ is the covariate vector and δ_i is the censoring indicator ($\delta_i = 1$ if t_i is observed and $\delta_i = 0$ if t_i is censored) of the ith patient ($i = 1, \ldots, n$). Let m denote the number of uncensored observations (deaths) and $n - m$ denote the number of censored observations. Assume that the t_i are exponentially distributed and follow the Cox proportional hazards model with a constant (with respect to time) hazard rate $\lambda_i = \lambda_0 \exp(\boldsymbol{x}_i'\boldsymbol{\beta})$ for the ith patient.

(a) Write the full likelihood function for this model and derive the equations for finding the MLEs of λ_0 and $\boldsymbol{\beta}$.

(b) Why is the partial likelihood approach not necessary nor advisable in this case?

(c) What is the MLE of λ_0 if there are no covariates, i.e. the t_i are identically distributed?

10.6 (Exponential data: Two sample problem) Consider the same set up as in the previous exercise but now assume that the only covariate is the indicator variable for the group, $x_i = 0$ for the placebo group (denoted by P) and $x_i = 1$ for the treatment group (denoted by T). Thus $\lambda_i = \lambda_0$ for $i \in P$ and $\lambda_i = \lambda_0 \exp(\beta)$ for $i \in T$. Further denote the set of censored

observations (from both the placebo group and the treatment group) by C and the set of uncensored observations (deaths) by D. Let n_0 and n_1 be the number of patients in the two groups of whom $m_0 = |P \cap D|$ and $m_1 = |T \cap D|$ are uncensored (deaths), respectively.
(a) Show that the MLEs of λ_0 and β are

$$\widehat{\lambda}_0 = \frac{1}{\overline{t}_0} \quad \text{and} \quad \widehat{\beta} = \ln\left(\frac{\overline{t}_0}{\overline{t}_1}\right).$$

(b) Interpret $\widehat{\beta}$. Explain how $\widehat{\beta} < 0$ implies that the treatment is effective.

10.7 (Survival function for the Cox regression model) Use (10.2) to show that in the Cox regression model, for the ith individual with the covariate vector $\boldsymbol{x}_i$ and the risk score $\psi_i = \exp(\boldsymbol{x}_i'\boldsymbol{\beta})$, the survival function is given by

$$S_i(t) = [S_0(t)]^{\psi_i},$$

where $S_0(t)$ is the survival function corresponding to the baseline hazard rate. Thus, the higher the risk score of an individual, the lower is his/her survival distribution.

Applied Exercises

10.8 (Cellphone data: Kaplan–Meier curves and logrank test) A cellular service provider keeps data on how many months their customers maintained service with the company before switching to another company. If a customer did not switch to another service provider then the service time on the customer is censored. The file `cellphone data.csv` contains 4912 records of customers (88 have missing data on at least one variable) with data on the following variables: Months = the number of months of service, Account_Type = Business (B) or Individual (I), Churn = censor indicator (0 if no, 1 if yes), Line_Count = number of phone lines served.

(a) Make Kaplan–Meier curves for business and individual customers. Which customers maintain their service longer? How long do most customers maintain the service? Does this seem to be related to the standard service contract period of two years that the cellphone providers used to have before this restriction was removed?

(b) Do the logrank test to check if there is a significant difference between the two survival curves.

10.9 (Cellphone data: Cox model) Fit a Cox proportional hazards model with Months as the response variable and the Account_Type as the predictor variable. Fit another Cox proportional hazards model with both the Account_Type and Line_Count as predictor variables. Note that the Account_Type goes from being a highly significant variable in the first model to a highly nonsignificant variable in the second model. What could explain this change? Note that the business customers generally have more lines.

10.10 **(Air Miles Reward Program: Cox model)** Air Miles Reward Program (AMRP) is a loyalty program of an airline for redeeming miles for a reward. Data on 5330 program members are in the file `AMRP.csv`. All members in the data set have redeemed at least once. The goal is to model the time until the next (i.e. the second) redemption, so customers who have redeemed only once are censored. The variables included in the data set are as follows.

1. t: time in days until the second redemption or censoring
2. censored: 1 for censored, 0 for event (redemption)
3. totredeem: total number of miles redeemed in the past
4. prevcat: category of the previous redemption (travel, merchandise, gift certificate, entertainment)
5. prefood: previous miles earned/day in food (grocery)
6. pregas: previous miles earned/day gas
7. prebonus: previous miles earned/day that were bonus miles, e.g. double miles under a certain promotion. This measures if they chase promotions.
8. baselen: how long they have been a member.

   Find the strongest predictors of the redemption time.

10.11 **(Recidivism study: Time-dependent employment status)** In Example 10.7, we analyzed the recidivism data using Aid, Age, Race, Work, Married, Parole, Prior, and Education as covariates with Week as the response variable and Arrest as the censoring indicator. We did not use the time-dependent covariate employment status (0 if not employed, 1 if employed), which changes every week and is denoted by Emplt for the tth week ($t = 1, \ldots, 52$) in the data file `recid.csv`. Note that if the person is arrested in the sth week then Emplt is marked NA for $t > s$.

   Fit the Cox model to the recidivism data with the employment status as a time-dependent covariate in addition to those fixed covariates used in Example 10.7. Comment on any changes in significance of the variables.

Appendix A

Primer on matrix algebra and multivariate distributions

A.1 Review of matrix algebra

We assume the basic knowledge of matrix algebra including arithmetic operations with vectors and matrices. We will review a few advanced concepts that are useful in linear models.

We use bold letters to denote vectors and matrices with lower case letters denoting vectors and upper case letters denoting matrices. Their dimensions are not generally indicated notationally and their elements are denoted by the respective unbolded letters with appropriate subscripts, e.g. a_i for an element of vector $\boldsymbol{a}$ and a_{ij} for an element of matrix $\boldsymbol{A}$. All vectors are assumed to be column vectors and transpose of a vector or a matrix is indicated by putting a prime on its symbol.

A symmetric $m \times m$ matrix $\boldsymbol{A}$ is said to be **positive definite** if for all non-null vectors $\boldsymbol{a} = (a_1, \ldots, a_m)'$, the **quadratic form** $\boldsymbol{a}'\boldsymbol{A}\boldsymbol{a} > 0$. If $\boldsymbol{a}'\boldsymbol{A}\boldsymbol{a} \geq 0$ for all non-null vectors $\boldsymbol{a}$, but $\boldsymbol{a}'\boldsymbol{A}\boldsymbol{a} = 0$ for some non-null vector $\boldsymbol{a}$ then $\boldsymbol{A}$ is said to be **positive semidefinite** or **non-negative definite**. An inverse of a positive definite matrix (denoted by $\boldsymbol{A}^{-1}$) exists and $\boldsymbol{A}$ is said to be **non-singular**. In that case, $\boldsymbol{A}^{-1}$ is also positive definite. If $\boldsymbol{A}$ is positive semidefinite, then $\boldsymbol{A}^{-1}$ does not exist and $\boldsymbol{A}$ is said to be **singular**.

As we saw in Chapter 3, the $(p+1) \times (p+1)$ symmetric matrix $(\boldsymbol{X}'\boldsymbol{X})^{-1}$ arises in multiple regression in the computation of the least squares estimates

Predictive Analytics: Parametric Models for Regression and Classification Using R.
First Edition. Ajit C. Tamhane.
© 2021 John Wiley & Sons, Inc. Published 2021 by John Wiley & Sons, Inc.
Companion website: www.wiley.com/go/tamhane/predictiveanalytics

and their covariance matrix. For this inverse to exist, $\boldsymbol{X}'\boldsymbol{X}$ must be positive definite, which can be checked as follows. Let $\boldsymbol{a} = (a_1, \ldots, a_{p+1})'$ be a non-null vector. Denoting $\boldsymbol{b} = \boldsymbol{X}\boldsymbol{a} = (b_1, \ldots, b_n)'$, we have

$$\boldsymbol{a}'\boldsymbol{X}'\boldsymbol{X}\boldsymbol{a} = \boldsymbol{b}'\boldsymbol{b} = \sum_{i=1}^{n} b_i^2 \geq 0$$

for all non-null vectors $\boldsymbol{a}$. Note that $\boldsymbol{b}$ is a linear combination of the columns of $\boldsymbol{X}$ with the coefficients of the linear combination being $a_1, \ldots, a_{p+1}$. The above inequality will be an equality iff $\boldsymbol{b} = \boldsymbol{X}\boldsymbol{a}$ is a null vector for some vector $\boldsymbol{a}$, which means that the columns of $\boldsymbol{X}$ are linearly dependent. Therefore, for $\boldsymbol{X}'\boldsymbol{X}$ to be positive definite and hence invertible, the columns of $\boldsymbol{X}$ must be linearly independent.

Another useful concept from linear algebra is that of **eigenvalues** and **eigenvectors**. The eigenvalue λ and its associated eigenvector $\boldsymbol{u}$ of an $m \times m$ matrix $\boldsymbol{A}$ are defined by the equation

$$[\boldsymbol{A} - \lambda\boldsymbol{I}]\boldsymbol{u} = \boldsymbol{0},$$

where $\boldsymbol{I}$ is an identity matrix and $\boldsymbol{0}$ is a null vector. This is a linear system of equations in unknowns $\boldsymbol{u} = (u_1, \ldots, u_m)'$. By Cramer's rule, this system has a nontrivial solution iff the matrix $\boldsymbol{A} - \lambda\boldsymbol{I}$ is singular, i.e. iff its determinant vanishes:

$$\det(\boldsymbol{A} - \lambda\boldsymbol{I}) = \det \begin{bmatrix} a_{11} - \lambda & a_{12} & \cdots & a_{1m} \\ a_{21} & a_{22} - \lambda & \cdots & a_{2m} \\ \vdots & \vdots & \ddots & \vdots \\ a_{m1} & a_{m2} & \cdots & a_{mm} - \lambda \end{bmatrix} = 0.$$

This is a polynomial equation of degree m and thus has m solutions, $\lambda_1, \ldots, \lambda_m$, not necessarily distinct. If they are distinct then there exist associated m eigenvectors, $\boldsymbol{u}_1, \ldots, \boldsymbol{u}_m$, which are mutually orthogonal, i.e. $\boldsymbol{u}_i'\boldsymbol{u}_j = 0$ for all $i \neq j$. Furthermore, if all eigenvectors are scaled to be of unit length then the matrix $\boldsymbol{U}$ whose columns are the eigenvectors $\boldsymbol{u}_1, \ldots, \boldsymbol{u}_m$ is an orthogonal matrix, i.e. $\boldsymbol{U}\boldsymbol{U}' = \boldsymbol{U}'\boldsymbol{U} = \boldsymbol{I}$.

Let $\boldsymbol{\Lambda} = \mathrm{diag}\{\lambda_1, \ldots, \lambda_m\}$ be the diagonal matrix with eigenvalues of $\boldsymbol{A}$ as its entries. Then the **spectral decomposition theorem** states that

$$\boldsymbol{U}\boldsymbol{A}\boldsymbol{U}' = \boldsymbol{\Lambda} \quad \text{or} \quad \boldsymbol{U}'\boldsymbol{\Lambda}\boldsymbol{U} = \boldsymbol{A}. \tag{A.1}$$

From this decomposition it follows that

$$\mathrm{tr}(\boldsymbol{A}) = \mathrm{tr}(\boldsymbol{\Lambda}) = \sum_{i=1}^{m} \lambda_i \quad \text{and} \quad \det(\boldsymbol{A}) = \det(\boldsymbol{\Lambda}) = \prod_{i=1}^{m} \lambda_i.$$

In this sense, the essential information about A is contained in its eigenvalues and eigenvectors.

Singular value decomposition (SVD) is a generalization of spectral decomposition. Let A be an $m \times p$ matrix of rank $r \leq p$. Then A can be written as

$$A = U \Lambda V',\tag{A.2}$$

where $\Lambda = \operatorname{diag}\{\lambda_1, \ldots, \lambda_r\}$, $U'U = V'V = I_r$. The column vectors of U are the eigenvectors of AA' and the column vectors of V are the eigenvectors of $A'A$. The eigenvalues of both $A'A$ and AA' are $\lambda_1^2, \ldots, \lambda_r^2$, the remaining eigenvalues being 0. SVD has many applications; in the regression context, it is used in the computation of the LS estimates and in the computation of principal component scores.

A.2   Review of multivariate distributions

Let $\boldsymbol{x} = (x_1, x_2, \ldots, x_m)'$ be a random vector where $x_1, x_2, \ldots, x_m$ are jointly distributed r.v.s with means $E(x_i) = \mu_i$, variances

$$\operatorname{Var}(x_i) = E[(x_i - \mu_i)^2] = \sigma_{ii} = \sigma_i^2$$

and covariances

$$\operatorname{Cov}(x_i, x_j) = E[(x_i - \mu_i)(x_j - \mu_j)] = \sigma_{ij}.$$

The **mean vector** of $\boldsymbol{x}$ equals

$$\boldsymbol{\mu} = E(\boldsymbol{x}) = \begin{bmatrix} E(x_1) \\ E(x_2) \\ \vdots \\ E(x_m) \end{bmatrix} = \begin{bmatrix} \mu_1 \\ \mu_2 \\ \vdots \\ \mu_m \end{bmatrix}.$$

The **covariance matrix** of $\boldsymbol{x}$ (denoted by $\boldsymbol{\Sigma} = \operatorname{Cov}(\boldsymbol{x})$)[1] is an $m \times m$ matrix with diagonal elements $\operatorname{Var}(x_i) = \sigma_{ii} = \sigma_i^2$ and off-diagonal elements $\operatorname{Cov}(x_i, x_j) = \sigma_{ij}$:

---

[1]We use the same notation $\operatorname{Cov}(\cdot)$ to denote different types of covariances, e.g. $\operatorname{Cov}(\boldsymbol{x})$ denotes the covariance matrix of a random vector $\boldsymbol{x}$, $\operatorname{Cov}(x_i, x_j)$ denotes the covariance of two scalar r.v.s x_i and x_j, and $\operatorname{Cov}(\boldsymbol{x}, \boldsymbol{y})$ denotes the covariance matrix between two random vectors $\boldsymbol{x}$ and $\boldsymbol{y}$, i.e. the matrix of covariances between the r.v.s x_i and the r.v.s y_j.

$$\boldsymbol{\Sigma} = \begin{bmatrix} \sigma_{11} & \sigma_{12} & \cdots & \sigma_{1m} \\ \sigma_{21} & \sigma_{22} & \cdots & \sigma_{2m} \\ \vdots & \vdots & \ddots & \vdots \\ \sigma_{m1} & \sigma_{m2} & \cdots & \sigma_{mm} \end{bmatrix}.$$

It is easy to see that $\boldsymbol{\Sigma}$ can be expressed as

$$\boldsymbol{\Sigma} = E\left(\begin{bmatrix} x_1 - \mu_1 \\ x_2 - \mu_2 \\ \vdots \\ x_m - \mu_m \end{bmatrix} \begin{bmatrix} x_1 - \mu_1, & x_2 - \mu_2, & \cdots, & x_m - \mu_m \end{bmatrix} \right)$$

$$= E[(\boldsymbol{x} - \boldsymbol{\mu})(\boldsymbol{x} - \boldsymbol{\mu})']. \tag{A.3}$$

Since $\mathrm{Cov}(x_i, x_j) = \mathrm{Cov}(x_j, x_i)$, we have $\sigma_{ij} = \sigma_{ji}$ for all $i \neq j$; therefore, $\boldsymbol{\Sigma}$ is a symmetric matrix. Furthermore, $\boldsymbol{\Sigma}$ is a **positive semidefinite matrix**, i.e. for all non-null vectors $\boldsymbol{a} = (a_1, a_2, \ldots, a_m)'$, we have $\boldsymbol{a}'\boldsymbol{\Sigma}\boldsymbol{a} \geq 0$. In fact, if the r.v.s x_i's are linearly independent, i.e. if there is no vector $\boldsymbol{a} \neq \boldsymbol{0}$ such that $\boldsymbol{a}'\boldsymbol{x} = \sum a_i x_i$ equals a constant, then $\boldsymbol{\Sigma}$ is a **positive definite matrix**. These results follow from (A.5) below.

We know from basic probability that if $a, b, c,$ and d are constants $(a, b \neq 0)$, x and y are r.v.s, and $u = ax + c$ and $v = by + d$ then

$$\mathrm{Cov}(u, v) = ab\mathrm{Cov}(x, y) \quad \text{and} \quad \mathrm{Corr}(u, v) = \pm\mathrm{Corr}(x, y),$$

where the sign is $+$ if $ab > 0$ and the sign is $-$ if $ab < 0$. Thus, if the r.v.s. x and y are linearly transformed, then the additive constants c and d have no effect on the covariance; furthermore, the multiplicative constants a and b have no effect on the correlation except through their signs.

More generally, let $\boldsymbol{x} = (x_1, x_2, \ldots, x_m)'$ and $\boldsymbol{y} = (y_1, y_2, \ldots, y_n)'$ be two random vectors and $\boldsymbol{a} = (a_1, a_2, \ldots, a_m)'$ and $\boldsymbol{b} = (b_1, b_2, \ldots, b_n)'$ be two vectors of constants. Let

$$u = \sum_{i=1}^{m} a_i x_i = \boldsymbol{a}'\boldsymbol{x} \quad \text{and} \quad v = \sum_{j=1}^{n} b_j y_j = \boldsymbol{b}'\boldsymbol{y}.$$

Further let $\boldsymbol{\Omega} = \mathrm{Cov}(\boldsymbol{x}, \boldsymbol{y})$ denote an $m \times n$ covariance matrix between $\boldsymbol{x}$ and $\boldsymbol{y}$ whose elements are $\omega_{ij} = \mathrm{Cov}(x_i, y_j)$ $(1 \leq i \leq m, 1 \leq j \leq n)$. Note that $\boldsymbol{\Omega}$ is not a symmetric matrix if $\boldsymbol{x} \neq \boldsymbol{y}$ even if $m = n$. Then

$$\mathrm{Cov}(u, v) = \mathrm{Cov}(\boldsymbol{a}'\boldsymbol{x}, \boldsymbol{b}'\boldsymbol{y}) = \sum_{i=1}^{m} \sum_{j=1}^{n} a_i b_j \mathrm{Cov}(x_i, y_j) = \boldsymbol{a}'\boldsymbol{\Omega}\boldsymbol{b}. \tag{A.4}$$

A special case of the above formula is obtained by putting $u = v$. Then

$$\text{Var}(\boldsymbol{a}'\boldsymbol{x}) = \sum_{i=1}^{m} \sum_{j=1}^{m} a_i a_j \text{Cov}(x_i, x_j) = \boldsymbol{a}'\boldsymbol{\Sigma}\boldsymbol{a}, \tag{A.5}$$

where $\boldsymbol{\Sigma}$ is the covariance matrix of $\boldsymbol{x}$. Note that $\text{Var}(\boldsymbol{a}'\boldsymbol{x}) = \boldsymbol{a}'\boldsymbol{\Sigma}\boldsymbol{a} \geq 0$ for $\boldsymbol{a} \neq \boldsymbol{0}$ and equals 0 iff $\boldsymbol{a}'\boldsymbol{x}$ equals a constant. This shows that $\boldsymbol{\Sigma}$ is positive semidefinite and is in fact positive definite if the r.v.s x_i's are linearly independent.

Suppose that $\boldsymbol{x}$ is an $m \times 1$ random vector and $\boldsymbol{A}$ is a $p \times m$ matrix of constants. Let $\boldsymbol{u} = \boldsymbol{A}\boldsymbol{x}$. It is readily shown that

$$E(\boldsymbol{A}\boldsymbol{x}) = \boldsymbol{A}E(\boldsymbol{x}) = \boldsymbol{A}\boldsymbol{\mu}. \tag{A.6}$$

Then using (A.3), it follows that

$$\begin{aligned}
\text{Cov}(\boldsymbol{u}) &= \text{Cov}(\boldsymbol{A}\boldsymbol{x}) \\
&= E[(\boldsymbol{A}\boldsymbol{x} - \boldsymbol{A}\boldsymbol{\mu})(\boldsymbol{A}\boldsymbol{x} - \boldsymbol{A}\boldsymbol{\mu})'] \\
&= E[\boldsymbol{A}(\boldsymbol{x} - \boldsymbol{\mu})(\boldsymbol{x} - \boldsymbol{\mu})'\boldsymbol{A}'] \\
&= \boldsymbol{A}E[(\boldsymbol{x} - \boldsymbol{\mu})(\boldsymbol{x} - \boldsymbol{\mu})']\boldsymbol{A}' \\
&= \boldsymbol{A}\boldsymbol{\Sigma}\boldsymbol{A}'. \tag{A.7}
\end{aligned}$$

This is known as the **sandwich formula**, which generalizes (A.5).

A.3 Multivariate normal distribution

The random vector $\boldsymbol{x} = (x_1, x_2, \ldots, x_m)'$ has a multivariate normal (MVN) distribution with mean vector $\boldsymbol{\mu}$ and covariance matrix $\boldsymbol{\Sigma}$ (denoted by $\boldsymbol{x} \sim \text{MVN}(\boldsymbol{\mu}, \boldsymbol{\Sigma})$) if the joint p.d.f. of $\boldsymbol{x} = (x_1, x_2, \ldots, x_m)'$ is given by

$$f(\boldsymbol{x}) = \frac{1}{(2\pi)^{m/2}|\boldsymbol{\Sigma}|^{1/2}} \exp\left\{-\frac{1}{2}(\boldsymbol{x} - \boldsymbol{\mu})'\boldsymbol{\Sigma}^{-1}(\boldsymbol{x} - \boldsymbol{\mu})\right\}, \tag{A.8}$$

where $|\boldsymbol{\Sigma}|$ denotes the determinant of $\boldsymbol{\Sigma}$. In the above, it is assumed that $\boldsymbol{\Sigma}$ is invertible or equivalently positive definite. We will only consider this nonsingular case of the MVN distribution.

The marginal p.d.f. of each component r.v. x_i is $N(\mu_i, \sigma_i^2)$. If $\boldsymbol{\Sigma}$ is a diagonal matrix, $\boldsymbol{\Sigma} = \text{diag}(\sigma_1^2, \sigma_2^2, \ldots, \sigma_m^2)$, i.e. if $\text{Cov}(x_i, x_j) = \sigma_{ij} = 0$ for all $i \neq j$, and thus the x_i's are uncorrelated, then the joint p.d.f. (A.8) of $\boldsymbol{x}$ factors into the product of the marginal p.d.f.s of x_i's:

$$f(\boldsymbol{x}) = \prod_{i=1}^{m} \frac{1}{\sqrt{2\pi}\sigma_i} \exp\left\{-\frac{1}{2\sigma_i^2}(x_i - \mu_i)^2\right\}.$$

Therefore, the x_i's are independent and are distributed as $N(\mu_i, \sigma_i^2)$. The converse of this result is immediate. Therefore if $\boldsymbol{x} = (x_1, x_2, \ldots, x_m)'$ is MVN distributed then the x_i's are independent if and only if they are uncorrelated.

The following is a useful property of the MVN distribution: If $\boldsymbol{x} \sim$ MVN $(\boldsymbol{\mu}, \boldsymbol{\Sigma})$ then any fixed (nonrandom) nonsingular linear transformation of $\boldsymbol{x}$ also has an MVN distribution. Specifically, let $\boldsymbol{A}$ be a $p \times m$ non-random matrix with linearly independent rows. Then, $\boldsymbol{u} = \boldsymbol{Ax} = (u_1, u_2, \ldots, u_p)'$ has an MVN distribution of dimension p with the mean vector and covariance matrix given by

$$E(\boldsymbol{u}) = \boldsymbol{A\mu} \quad \text{and} \quad \text{Cov}(\boldsymbol{u}) = \boldsymbol{A\Sigma A}'.$$

Two cases of this result are of particular interest.

1. Since $\boldsymbol{\Sigma}$ is invertible it can be shown that there exists an $m \times m$ symmetric matrix, say $\boldsymbol{P}$, such that $\boldsymbol{P\Sigma P}' = \boldsymbol{I}$ and $\boldsymbol{P}'\boldsymbol{P} = \boldsymbol{\Sigma}^{-1}$. Then

   $$\boldsymbol{z} = \boldsymbol{P}(\boldsymbol{x} - \boldsymbol{\mu}) \tag{A.9}$$

   is MVN with

   $$E(\boldsymbol{z}) = \boldsymbol{0} \quad \text{and} \quad \text{Cov}(\boldsymbol{z}) = \boldsymbol{P\Sigma P}' = \boldsymbol{I},$$

   i.e. $z_1, z_2, \ldots, z_m$ are i.i.d. $N(0,1)$ r.v.s. Thus, (A.9) is a standardizing transformation.

2. Let $\boldsymbol{a} = (a_1, a_2, \ldots, a_m)'$ be a vector of constants. Then, $u = \boldsymbol{a}'\boldsymbol{x} = \sum a_i x_i$ is univariate normal with

   $$E(u) = \boldsymbol{a}'\boldsymbol{\mu} = \sum_{i=1}^{m} a_i \mu_i \quad \text{and} \quad \text{Var}(u) = \boldsymbol{a}'\boldsymbol{\Sigma a} = \sum_{i=1}^{m} \sum_{j=1}^{m} a_i a_j \sigma_{ij}.$$

Appendix B

Primer on maximum likelihood estimation

B.1 Maximum likelihood estimation

The method of maximum likelihood estimation was proposed by Sir R. A. Fisher. Let y be a r.v., either discrete or continuous, with probability mass or density function (both abbreviated as p.d.f.) $f(y|\theta)$. Here θ is an unknown parameter, which we want to estimate from an independent and identically distributed (i.i.d.) random sample $y_1, \dots, y_n$ drawn from this distribution. We can view the joint p.d.f. of $y_1, \dots, y_n$,

$$f(y_1, \dots, y_n|\theta) = \prod_{i=1}^{n} f(y_i|\theta),$$

as being proportional to the probability of their occurrence if the true parameter is θ. (If the y_i are discrete then this is the probability; if the y_i are continuous then $f(y_1, \dots, y_n|\theta)dy_1 \dots dy_n$ is the probability element.) This is called the **likelihood function**, viewed as a function of θ for given $y_1, \dots, y_n$ (whereas the joint p.d.f. is a function of $y_1, \dots, y_n$ for given θ and has the probability interpretation given above). We denote the likelihood function by

$$L(\theta) = L(\theta|y_1, \dots, y_n) = \prod_{i=1}^{n} f(y_i|\theta).$$

We ask the question: what value of θ makes the observed data $y_1, \dots, y_n$ most likely? This value of θ, which maximizes the likelihood function, is called the **maximum likelihood estimator (MLE)** of θ and is denoted by $\widehat{\theta}$.

Predictive Analytics: Parametric Models for Regression and Classification Using R,
First Edition. Ajit C. Tamhane.
© 2021 John Wiley & Sons, Inc. Published 2021 by John Wiley & Sons, Inc.
Companion website: www.wiley.com/go/tamhane/predictiveanalytics

Since the log is a monotone increasing function, maximizing $L(\theta)$ is equivalent to maximizing its log. So we define the **log-likelihood function** as

$$\ln L(\theta) = \sum_{i=1}^{n} \ln f(y_i|\theta). \tag{B.1}$$

To keep things simple, we will restrict to the so-called regular case where the log-likelihood function is differentiable and concave, so that its maximum can be found by setting the derivative of $\ln L(\theta)$ equal to zero and solving the resulting equation:

$$\frac{d \ln L(\hat{\theta})}{d\theta} = \left[\frac{d \ln L(\theta)}{d\theta}\right]_{\theta=\hat{\theta}} = 0. \tag{B.2}$$

We assume that the solution to this equation, which is the MLE $\hat{\theta}$, exists and is unique.

Example B.1 *(MLE of Bernoulli Parameter)* Suppose $f(y|\theta)$ is a Bernoulli distribution with

$$f(y|\theta) = \theta \quad \text{if } y = 1 \quad \text{and} \quad f(y|\theta) = 1 - \theta \quad \text{if } y = 0,$$

where θ is the probability of success on a single Bernoulli trial. This distribution can be written compactly as

$$f(y|\theta) = \theta^y (1 - \theta)^{1-y} \quad \text{for } y = 0, 1.$$

The likelihood function is then given by

$$L(\theta) = \prod_{i=1}^{n}[\theta^{y_i}(1-\theta)^{1-y_i}] = \theta^{\sum_{i=1}^{n} y_i}(1-\theta)^{\sum_{i=1}^{n}(1-y_i)} = \theta^s(1-\theta)^{n-s},$$

where $s = \sum_{i=1}^{n} y_i$ is the number of successes in n Bernoulli trials. The log-likelihood function is

$$\ln L(\theta) = s \ln \theta + (n - s) \ln(1 - \theta).$$

Taking the derivative of $\ln L(\theta)$ w.r.t. θ and setting it equal to zero, we get

$$\frac{d \ln L(\theta)}{d\theta} = \frac{s}{\theta} - \frac{n - s}{1 - \theta} = 0.$$

The solution to this equation is $\hat{\theta} = s/n$, which is the MLE. Note that this is simply the proportion of successes. It is easy to see that the second derivative of $\ln L(\theta)$ at $\theta = \hat{\theta}$ is negative and hence $\hat{\theta}$ indeed gives the maximum.

Instead of working with n i.i.d. Bernoulli outcomes, we can work directly with the distribution of their sum $s = \sum_{i=1}^{n} y_i$ (which is a sufficient statistic for θ). The distribution of s is binomial:

$$f(s|\theta) = \binom{n}{s} \theta^s (1-\theta)^{n-s},$$

which is the likelihood function of θ. Note that this likelihood function differs from the previous one obtained using the Bernoulli distribution only in the multiplication factor $\binom{n}{s}$. But this factor is irrelevant to maximizing the likelihood function w.r.t. θ since it does not involve θ. Therefore, we get the same MLE $\widehat{\theta} = s/n$. □

B.2   Large sample inference on MLEs

Next we discuss inference on the MLE. Fisher showed that, under certain regularity conditions, asymptotically (as $n \to \infty$) the MLE $\widehat{\theta}$ is normally distributed with mean θ and asymptotic variance $[\boldsymbol{\mathcal{I}}(\theta)]^{-1}$, where

$$\boldsymbol{\mathcal{I}}(\theta) = E\left[\left(\frac{d \ln L(\theta)}{d\theta}\right)^2\right] = -E\left[\frac{d^2 \ln L(\theta)}{d\theta^2}\right] \tag{B.3}$$

is the so-called **Fisher information**. The above identity is derived in Section B.4. A plug-in estimate of $\boldsymbol{\mathcal{I}}(\theta)$, denoted by $\boldsymbol{\mathcal{I}}(\widehat{\theta})$, can be obtained by substituting $\widehat{\theta}$ for θ in the expected value. Another approach is to substitute $\widehat{\theta}$ in the second derivative expression of the log-likelihood function without taking the expected value. The former quantity is called the **expected information**, while the latter quantity is called the **observed information**. They are equal under certain regularity conditions. In that case

$$\boldsymbol{\mathcal{I}}(\widehat{\theta}) = -\left[\frac{d^2 \ln L(\theta)}{d\theta^2}\right]_{\theta=\widehat{\theta}} = -\sum_{i=1}^{n}\left[\frac{d^2 \ln f(y_i|\theta)}{d\theta^2}\right]_{\theta=\widehat{\theta}}.$$

A large sample $100(1-\alpha)\%$ CI on θ is given by

$$\widehat{\theta} - z_{\alpha/2}\frac{1}{\sqrt{\boldsymbol{\mathcal{I}}(\widehat{\theta})}} \leq \theta \leq \widehat{\theta} + z_{\alpha/2}\frac{1}{\sqrt{\boldsymbol{\mathcal{I}}(\widehat{\theta})}}.$$

Example B.2 *(Inference on the MLE of Bernoulli Parameter)* We first evaluate $\boldsymbol{\mathcal{I}}(\theta)$. Using the result from Example B.1 we get

$$\frac{d^2 \ln L(\theta)}{d\theta^2} = -\frac{s}{\theta^2} - \frac{n-s}{(1-\theta)^2}.$$

Hence

$$\mathcal{I}(\theta) = \frac{E(s)}{\theta^2} + \frac{E(n-s)}{(1-\theta)^2} = \frac{n\theta}{\theta^2} + \frac{n(1-\theta)}{(1-\theta)^2} = \frac{n}{\theta} + \frac{n}{1-\theta} = \frac{n}{\theta(1-\theta)}.$$

Thus, $[\mathcal{I}(\theta)]^{-1} = \theta(1-\theta)/n$, which is in fact the *exact* variance of the MLE $\widehat{\theta} = s/n$. It is estimated by $\widehat{\theta}(1-\widehat{\theta})/n$.

We can check that we get the same result if we calculate the observed information. Thus

$$\mathcal{I}(\widehat{\theta}) = \frac{s}{\widehat{\theta}^2} + \frac{n-s}{(1-\widehat{\theta})^2} = \frac{n\widehat{\theta}}{\widehat{\theta}^2} + \frac{n(1-\widehat{\theta})}{(1-\widehat{\theta})^2} = \frac{n}{\widehat{\theta}(1-\widehat{\theta})}.$$

Using this result, we get the following well-known formula for the large sample $100(1-\alpha)\%$ CI on θ:

$$\widehat{\theta} - z_{\alpha/2}\sqrt{\frac{\widehat{\theta}(1-\widehat{\theta})}{n}} \leq \theta \leq \widehat{\theta} + z_{\alpha/2}\sqrt{\frac{\widehat{\theta}(1-\widehat{\theta})}{n}}.$$

$\square$

Next we extend these results to multiple parameters. Suppose the distribution of y depends on $p \geq 2$ unknown parameters, $\theta_1, \ldots, \theta_p$, represented as a vector parameter $\boldsymbol{\theta} = (\theta_1, \ldots, \theta_p)$ and denote the distribution of y by $f(y|\boldsymbol{\theta})$. Assume that we have an i.i.d. sample $y_1, \ldots, y_n$ from this distribution. Then, the log-likelihood function equals

$$\ln L(\boldsymbol{\theta}) = \ln\left[\prod_{i=1}^{n} f(y_i|\boldsymbol{\theta})\right] = \sum_{i=1}^{n} \ln f(y_i|\boldsymbol{\theta}).$$

The MLEs of the θ_j's are obtained by solving the p simultaneous equations obtained by setting the partial derivatives of the log-likelihood function w.r.t. the θ_j's equal to zero:

$$\frac{\partial \ln L(\boldsymbol{\theta})}{\partial \theta_j} = \sum_{i=1}^{n} \frac{\partial \ln f(y_i|\boldsymbol{\theta})}{\partial \theta_j} = 0 \quad (1 \leq j \leq p).$$

To make inferences on the $\widehat{\theta}_j$'s, we need to calculate the asymptotic covariance matrix of $\widehat{\boldsymbol{\theta}} = (\widehat{\theta}_1, \ldots, \widehat{\theta}_p)$. The diagonal elements of this matrix are the asymptotic variances of the $\widehat{\theta}_j$'s. This covariance matrix is the inverse of the **information matrix** $\mathcal{I}(\boldsymbol{\theta})$, where

$$\mathcal{I}_{jk}(\boldsymbol{\theta}) = -E \sum_{i=1}^{n} \left[\frac{\partial^2 \ln f(y_i|\boldsymbol{\theta})}{\partial \theta_j \partial \theta_k}\right] \quad (1 \leq j < k \leq p).$$

The estimated information matrix can be obtained in two ways as before. One way is to compute the above expected information matrix and substitute $\boldsymbol{\theta} = \widehat{\boldsymbol{\theta}}$ in it. The other way is to compute the observed information matrix by substituting $\boldsymbol{\theta} = \widehat{\boldsymbol{\theta}}$ in the second derivative expressions of the log-likelihood function.

Example B.3 *(Information Matrix for Simple Logistic Regression)* In this example, we derive the formula (7.6) for the information matrix for simple logistic regression. From the results in Section 7.1.2, we have

$$\ln L = \sum_{i=1}^{n} y_i(\beta_0 + \beta_1 x_i) - \sum_{i=1}^{n} \ln[1 + \exp(\beta_0 + \beta_1 x_i)].$$

Hence,

$$\frac{\partial \ln L}{\partial \beta_0} = \sum_{i=1}^{n} y_i - \sum_{i=1}^{n} \frac{\exp(\beta_0 + \beta_1 x_i)}{[1 + \exp(\beta_0 + \beta_1 x_i)]} = \sum_{i=1}^{n} y_i - \sum_{i=1}^{n} p_i.$$

Similarly,

$$\frac{\partial \ln L}{\partial \beta_1} = \sum_{i=1}^{n} x_i y_i - \sum_{i=1}^{n} x_i \frac{\exp(\beta_0 + \beta_1 x_i)}{[1 + \exp(\beta_0 + \beta_1 x_i)]} = \sum_{i=1}^{n} x_i y_i - \sum_{i=1}^{n} x_i p_i.$$

Hence,

$$\frac{\partial^2 \ln L}{\partial \beta_0^2} = -\sum_{i=1}^{n} \frac{\partial p_i}{\partial \beta_0} = -\sum_{i=1}^{n} \frac{\exp(\beta_0 + \beta_1 x_i)}{[1 + \exp(\beta_0 + \beta_1 x_i)]^2} = -\sum_{i=1}^{n} p_i(1 - p_i),$$

$$\frac{\partial^2 \ln L}{\partial \beta_1^2} = -\sum_{i=1}^{n} x_i \frac{\partial p_i}{\partial \beta_1} = -\sum_{i=1}^{n} x_i^2 \frac{\exp(\beta_0 + \beta_1 x_i)}{[1 + \exp(\beta_0 + \beta_1 x_i)]^2} = -\sum_{i=1}^{n} x_i^2 p_i(1 - p_i),$$

and

$$\frac{\partial^2 \ln L}{\partial \beta_0 \beta_1} = -\sum_{i=1}^{n} \frac{\partial p_i}{\partial \beta_1} = -\sum_{i=1}^{n} x_i \frac{\exp(\beta_0 + \beta_1 x_i)}{[1 + \exp(\beta_0 + \beta_1 x_i)]^2} = -\sum_{i=1}^{n} x_i p_i(1 - p_i).$$

The information matrix $\boldsymbol{\mathcal{I}}$ in (7.6) follows immediately from these second partial derivatives.

Note that these second derivatives do not involve the y_i's, which are the random quantities. Therefore, there are no expected values to be taken. Hence, both methods of calculating the estimated information matrix give the same result. □

B.3 Newton–Raphson and Fisher scoring algorithms

We will present these algorithms for the case of a single unknown parameter θ and then generalize them to the multiparameter case. They provide an iterative method to solving the equation:

$$\frac{d \ln L(\hat{\theta})}{d\theta} \equiv \left. \frac{d \ln L(\theta)}{d\theta} \right|_{\theta=\hat{\theta}} = 0, \tag{B.4}$$

which is referred to as the **MLE score equation**. We begin with an initial guess $\widehat{\theta}_0$ for its solution. By the first-order Taylor series expansion of $d \ln L(\widehat{\theta})/d\theta$ around $\widehat{\theta}_0$, we get

$$\frac{d \ln L(\widehat{\theta})}{d\theta} \approx \frac{d \ln L(\widehat{\theta}_0)}{d\theta} + (\widehat{\theta} - \widehat{\theta}_0)\frac{d^2 \ln L(\widehat{\theta}_0)}{d\theta^2}. \tag{B.5}$$

We set the R.H.S. of the above equation equal to 0 and solve for $\widehat{\theta}$ to obtain the first approximation $\widehat{\theta}_1$ to the solution of the MLE equation:

$$\widehat{\theta}_1 = \widehat{\theta}_0 - \left[\frac{d^2 \ln L(\widehat{\theta}_0)}{d\theta^2}\right]^{-1}\left[\frac{d \ln L(\widehat{\theta}_0)}{d\theta}\right].$$

Repeating this procedure we get the following recursion at the rth iteration:

$$\widehat{\theta}_{r+1} = \widehat{\theta}_r - \left[\frac{d^2 \ln L(\widehat{\theta}_r)}{d\theta^2}\right]^{-1}\left[\frac{d \ln L(\widehat{\theta}_r)}{d\theta}\right].$$

Recall that $-d^2 \ln L(\widehat{\theta}_r)/d\theta^2 = \boldsymbol{I}(\widehat{\theta}_r)$, namely the observed information evaluated at the current estimate $\widehat{\theta}_r$. Therefore, the above iterative equation can be equivalently written as

$$\widehat{\theta}_{r+1} = \widehat{\theta}_r + [\boldsymbol{I}(\widehat{\theta}_r)]^{-1}\left[\frac{d \ln L(\widehat{\theta}_r)}{d\theta}\right]. \tag{B.6}$$

The algorithm converges when a specified convergence criterion is met, e.g. when $|\widehat{\theta}_{r+1} - \widehat{\theta}_r| < \varepsilon$ for some specified tolerance $\varepsilon > 0$. Note that not only does this algorithm give the MLE $\widehat{\theta}$ but it also gives the asymptotic variance $\mathrm{Var}(\widehat{\theta}) = [\boldsymbol{I}(\widehat{\theta})]^{-1}$ at the final step.

Fisher's **score statistic** is defined as

$$U(\theta) = \frac{d \ln L(\theta)}{d\theta}.$$

In Section B.4, we show under certain regularity conditions that its mean and variance are given by

$$E[U(\theta)] = E\left[\frac{d \ln L(\theta)}{d\theta}\right] = 0 \quad \text{and} \quad \mathrm{Var}[U(\theta)] = -E\left[\frac{d^2 \ln L(\theta)}{d\theta^2}\right]. \tag{B.7}$$

Equation (B.4) is equivalent to solving $U(\widehat{\theta}) = 0$. The Fisher scoring algorithm is essentially the same as the Newton–Raphson algorithm except that the expected information evaluated at $\widehat{\theta}_r$, i.e. $E[\boldsymbol{I}(\theta)]_{\theta=\widehat{\theta}_r}$ is used instead of the observed information $\boldsymbol{I}(\widehat{\theta}_r)$ in the recursion (B.6).

In the multiparameter case, let $\boldsymbol{\theta} = (\theta_1, \ldots, \theta_p)'$ denote the unknown parameter vector of interest and let $\boldsymbol{I}(\boldsymbol{\theta})$ denote the associated information matrix.

Then, the above recursion generalizes to the following:

$$\widehat{\boldsymbol{\theta}}_{r+1} = \widehat{\boldsymbol{\theta}}_r + [\boldsymbol{\mathcal{I}}(\widehat{\boldsymbol{\theta}}_r)]^{-1} \left[\frac{d \ln L(\widehat{\boldsymbol{\theta}}_r)}{d\boldsymbol{\theta}} \right]. \tag{B.8}$$

Once again Fisher's scoring algorithm uses the expected value of the information matrix $\boldsymbol{\mathcal{I}}(\boldsymbol{\theta})$ evaluated at $\boldsymbol{\theta} = \widehat{\boldsymbol{\theta}}$.

B.4   Technical notes*

In this section, we derive the identity (B.3) and provide an outline of the proof of the asymptotic normality of the MLE. We denote $f(y|\theta)$ by a simpler notation $f(y, \theta)$. The information from a single observation y with p.d.f. $f(y, \theta)$ is defined as

$$\boldsymbol{\mathcal{I}}_1(\theta) = E \left[\left(\frac{d \ln L(\theta)}{d\theta} \right)^2 \right] = \int \left(\frac{d \ln L(\theta)}{d\theta} \right)^2 f(y, \theta) \, dy.$$

Under appropriate regularity conditions, we can move the derivative inside the integral sign and write

$$E \left(\frac{d \ln L(\theta)}{d\theta} \right) = \int \frac{d \ln f(y, \theta)}{d\theta} f(y, \theta) \, dy$$

$$= \int \frac{1}{f(y, \theta)} \frac{df(y, \theta)}{d\theta} f(y, \theta) \, dy$$

$$= \int \frac{df(y, \theta)}{d\theta} \, dy$$

$$= \frac{d}{d\theta} \int f(y, \theta) \, dy$$

$$= 0 \quad (\text{since } \int f(y, \theta) \, dy = 1).$$

Differentiating under the integral sign in the first equation above we get

$$\int \left[\frac{d^2 \ln f(y, \theta)}{d\theta^2} f(y, \theta) + \frac{d \ln f(y, \theta)}{d\theta} \frac{df(y, \theta)}{d\theta} \right] dy$$

$$= \int \left[\frac{d^2 \ln f(y, \theta)}{d\theta^2} + \frac{d \ln f(y, \theta)}{d\theta} \frac{df(y, \theta)}{d\theta} \frac{1}{f(y, \theta)} \right] f(y, \theta) \, dy$$

$$= \int \left[\frac{d^2 \ln f(y, \theta)}{d\theta^2} + \frac{d \ln f(y, \theta)}{d\theta} \frac{d \ln f(y, \theta)}{d\theta} \right] f(y, \theta) \, dy$$

$$= \int \left[\frac{d^2 \ln f(y, \theta)}{d\theta^2} + \left(\frac{d \ln f(y, \theta)}{d\theta} \right)^2 \right] f(y, \theta) \, dy$$

$$= 0.$$

Hence,

$$\int \left(\frac{d\ln f(y,\theta)}{d\theta}\right)^2 f(y,\theta)\,dy = -\int \frac{d^2\ln f(y,\theta)}{d\theta^2} f(y,\theta)\,dy.$$

By putting $L(\theta) = f(y,\theta)$, namely the likelihood function from a single observation, we can write the above equation as

$$\boldsymbol{\mathcal{I}}_1(\theta) = E\left[\left(\frac{d\ln L(\theta)}{d\theta}\right)^2\right] = -E\left[\frac{d^2\ln L(\theta)}{d\theta^2}\right],$$

which gives an alternative expression for $\boldsymbol{\mathcal{I}}(\theta)$.

The information from n i.i.d. observations $y_1, \ldots, y_n$ with a common p.d.f. $f(y,\theta)$ equals $\boldsymbol{\mathcal{I}}(\theta) = \boldsymbol{\mathcal{I}}_n(\theta) = n\boldsymbol{\mathcal{I}}_1(\theta)$. This follows since from (B.1) we can write

$$\boldsymbol{\mathcal{I}}(\theta) = -E\left[\left(\frac{d^2\ln L}{d\theta^2}\right)\right] = -E\left[\sum_{i=1}^n \frac{d^2 f(y_i,\theta)}{d\theta^2}\right] = \sum_{i=1}^n -E\left[\frac{d^2 f(y_i,\theta)}{d\theta^2}\right] = nI_1(\theta),$$

where we have used the fact that $\boldsymbol{\mathcal{I}}_1(\theta)$ is the same for all n i.i.d. observations and so $\boldsymbol{\mathcal{I}}(\theta) = n\boldsymbol{\mathcal{I}}_1(\theta)$.

Next consider the proof of the asymptotic normality of the MLE $\widehat{\theta}$. Using the same first order Taylor series expansion as in (B.5) we get

$$\widehat{\theta} - \theta \approx \frac{d\ln L(\theta)/d\theta}{-d^2\ln L(\theta)/d\theta^2}$$

$$= \frac{(1/n)\sum_{i=1}^n (d\ln f(y_i,\theta)/d\theta)}{-(1/n)\sum_{i=1}^n (d^2\ln f(y_i,\theta)/d\theta^2)}.$$

The numerator is the mean of n i.i.d. Fisher score statistics $U_i = d\ln f(y_i,\theta)/d\theta$ and so by the central limit theorem approaches a normal distribution with mean equal to $E(U_i) = 0$ and variance equal to $\mathrm{Var}(U_i)/n = \boldsymbol{\mathcal{I}}(\theta)/n$. The denominator is also the mean of n i.i.d. r.v.s $-d^2\ln f(y_i,\theta)/d\theta^2$, each of which has expectation $\boldsymbol{\mathcal{I}}(\theta)$. Hence by the law of large numbers, the denominator approaches $\boldsymbol{\mathcal{I}}(\theta)$ in probability. Hence, $\widehat{\theta} - \theta$ approaches $N(0, \boldsymbol{\mathcal{I}}^{-1}(\theta)/n)$ in distribution.

Appendix C

Projects

A data analysis project involving a large data set should be an integral part of any course based on this book. I generally assign a team project to teams of three to four students (the same project to all teams) at just past the midpoint of the course when multiple regression and logistic regression topics have been covered in the class. The project involves applying these two methodologies to develop predictive models for large marketing data sets. Three sample projects are described in the following. They are quite similar in the types of data and the business questions of interest. So the guidelines given below are common to all three projects. Also follow the guidelines for multiple regression model building in Section 6.3. Any special features of the projects are briefly discussed in the individual descriptions of projects.

The data sets for the projects are provided at the book's website. Each data set is divided into a training set and a test set (either as two separate data files or as a single file with an indicator variable for the training or test set). Thus, all student teams use the same training set for developing predictive models and the same test set for model validation. So the results of different teams can be directly compared. A 15–20 page report is required from all teams.

The response variable in all three projects is the purchase amount by customers in response to a marketing promotion. The predictor variables are their past purchase histories. The response rates are quite small (less than 10%), so a straightforward multiple regression approach does not work since more than 90% of the purchase amounts are zeros, i.e. those customers did not respond to the marketing promotion. In addition, the distributions of the nonzero purchase amounts are highly right-skewed with some unusually large amounts. Therefore, a log transformation of the purchase amount y to $y \leftarrow \ln(y + 1)$ is recommended.

Predictive Analytics: Parametric Models for Regression and Classification Using R,
First Edition. Ajit C. Tamhane.
© 2021 John Wiley & Sons, Inc. Published 2021 by John Wiley & Sons, Inc.
Companion website: www.wiley.com/go/tamhane/predictiveanalytics

Note that this function transforms zero purchase amounts back to zeros. This log-transformed y is used in what follows.

Although this transformation helps to symmetrize the distribution of nonzero purchase amounts and reduce the number of outliers, the problem of too many zero purchase amounts still remains. To deal with this problem, a two-step modeling approach involving logistic regression followed by multiple regression is recommended. The first step is to use logistic regression to model the response probabilities, $P(y > 0)$. The second step is to use multiple regression to model the conditional mean of the purchase amount given that the customer is a responder, i.e. $E(y|y > 0)$. The predictor variables used in the two models are generally different. To model $E(y|y > 0)$, only the subset of responders in the training set must be used, i.e. those with $y > 0$. Then unconditional means of the purchase amounts for the customers in the test set can be estimated using the formula $E(y) = E(y|y > 0)P(y > 0)$. Note that in this approach logistic regression is not used as a hard classifier of customers into responders and non-responders, but as a soft classifier to estimate their response probabilities. Let $\hat{y}$ denote the predicted value of the purchase amount for a customer in the test set using the given multiplication formula. Then, the predicted value of the purchase amount in dollars for that customer is obtained by applying the inverse transformation: $\hat{y} \leftarrow e^{\hat{y}} - 1$.

Predictor variables should also be transformed as necessary to symmetrize their distributions. New predictor variables may be created to represent interactions of interest or time averages of past purchases, e.g. by dividing the total number of past orders or total past purchases by the total time the customer has been on file. A weighted average may be used by discounting the older purchases more than the recent purchases. It is known in data base marketing that generally the best predictors for predicting whether a customer will respond to a promotion are (1) recency of last purchase and (2) consistency of past purchases. Recency can be readily deduced from the date of last purchase. Consistency can be coded as an interaction of the indicator variables for the purchases for the last several time periods. This step is called feature engineering (see Section 6.3).

To validate this predictive model, both statistical and financial criteria can be employed. For statistical criteria, one can use mean squared prediction error or R^2 for the test set as well as residual plots, significant regression coefficients and simplicity and interpretability of the models. An example of a financial criterion is the total purchase amount of the *predicted* top 1000 buyers compared to that of the *actual* top 1000 buyers in the test set. This comparison enables us to gauge the percentage of the total of the top 1000 purchases that is captured by the predictive model.

Finally, because the percentage of respondents is very low in each data set, it may be prudent to oversample respondents in the training set when fitting the logistic regression model. This strategy helps to improve the accuracy of the estimates of the regression coefficients. Oversampling may be done by simply duplicating each respondent in the training set multiple times or by drawing a weighted bootstrap sample with higher weights on respondents. The response probabilities estimated from the resulting model for the test set will be obviously biased upward. The following formula can be applied to correct this bias. Let q be the actual proportion of respondents in the sample and suppose that the respondents are oversampled by a factor of $m > 1$, so that the proportion of the respondents in the sample becomes $q' = mq$. Let p' be the estimated response probability of a customer from the sample with oversampled respondents. Then the bias-adjusted estimated response probability p is given by

$$\ln \left(\frac{p}{1-p} \right) = \ln \left(\frac{p'}{1-p'} \right) - \ln \left[\frac{q'/(1-q')}{q/(1-q)} \right] = \ln \left(\frac{p'}{1-p'} \right) - \ln \left[\frac{m(1-q)}{1-mq} \right].$$

As an example, if $q = 0.05$ and $m = 3$ then $\ln \left[\frac{3(1-0.05)}{1-0.15} \right] = 1.21$. Thus, the odds of response should be adjusted downward by a factor of $e^{1.21} = 3.35$. For very small q, $1 - q$, and $1 - mq$ are roughly equal to 1, so the odds of response can be adjusted downward approximately by $m = 3$.

C.1 Project 1

- *Business situation*: A retail company sells upscale clothing on its website and via catalogs, which helps drive customers to the website. All customers were sent a catalog mailing in early fall 2012 and purchases made by them during fall 2012 were recorded. There is one row for each customer. The `targdol` is the response variable, which is the purchase amount during fall 2012; `targdol` $= 0$ indicates that the customer did not make a purchase. Other variables are potential predictor variables which give information about the customer as of the time of the mailing. We want to build a predictive model for responders to a similar promotion in future and how much they will buy. The purpose of the model is primarily prediction but it is also of interest to learn which are the most important predictors. The model is also intended to choose a subset of customers to be targeted in future promotions.

- *Data*: The data are in the file `Project 1 data.csv`. There are a total of 101,532 customers, of whom 9571 (9.43%) are respondents. The data are randomly divided into a training set with 50,418 observations and the remaining 51,114 into a test set (`train` is the indicator variable for training or test set: `train` $=1$ for the training set, `train` $=0$ for the test set). The definitions of the variables are as follows.

```
-----------------------------------------------------------------
Variable  Description
-----------------------------------------------------------------
targdol    dollar purchase resulting from catalog mailing
datead6    date added to file
datelp6    date of last purchase
lpuryear   latest purchase year
slstyr     sales ($) this year
slslyr     sales ($) last year
sls2ago    sales ($) 2 years ago
sls3ago    sales ($) 3 years ago
slshist    LTD dollars
ordtyr     number of orders this year
ordlyr     number of orders last year
ord2ago    number of orders  2 years ago
ord3ago    number of orders 3 years ago
ordhist    LTD orders
falord     LTD fall orders
sprord     LTD spring orders
train      training/test set indicator (1 = training, 0 = test)
-----------------------------------------------------------------
```

LTD means "life-to-date," i.e. the time since the customer purchased for the first time.

- *Hints*:

  1. Some errors in the data are as follows. If you run a histogram or frequency distribution of the `datelp6` variable among only those with `targdol > 0` you will see that, for the most part, `datelp6` equals one of two distinct dates in the calendar year. It is as if the data are binned into six-month bins. There are other inconsistencies in the data, e.g. `falord` + `sprord` is not equal to `ordhist` in about 9% of the cases. Similarly, the year of the latest purchase obtained from `lpuryear` variable and from `datelp6` variable do not always agree. Some of these errors result because when two variables measure the same thing, both are not updated.

  2. There are other inconsistencies in the data as well. For example, in a few cases the number of orders are not recorded but there are purchase amounts, date added to file does not match with date of last purchase, etc. You will need to make reasonable decisions to resolve the inconsistencies. Clearly state in your report how you resolved the inconsistencies.

C.2   Project 2

The response variable in this data set is `targamnt`. The total sample size is 106,284 of which 5698 are respondents (5.36%). The data set is divided randomly into a training set (`Project 2 train.csv`) consisting of 52,844 observations and test set (`Project 2 test.csv`) consisting of 53,440 observations. The goal of the project is to build a predictive model for `targamnt` and test it on the test data set.

The descriptions of the variables are as follows.

```
----------------------------------------
Variable  Description
----------------------------------------
recmon    Months since last order
ordcls1   5 Year Product Class 1 Orders
ordcls2   5 Year Product Class 2 Orders
ordcls3   5 Year Product Class 3 Orders
ordcls4   5 Year Product Class 4 Orders
ordcls5   5 Year Product Class 5 Orders
ordcls6   5 Year Product Class 6 Orders
ordcls7   5 Year Product Class 7 Orders
salcls1   5 Year Sales Product Class 1
salcls2   5 Year Sales Product Class 2
salcls3   5 Year Sales Product Class 3
salcls4   5 Year Sales Product Class 4
salcls5   5 Year Sales Product Class 5
salcls6   5 Year Sales Product Class 6
salcls7   5 Year Sales Product Class 7
ord185    Order Yr 1, Prom 85 (Y/N)
ord285    Order Yr 2, Prom 85 (Y/N)
ord385    Order Yr 3, Prom 85 (Y/N)
ord485    Order Yr 4, Prom 85 (Y/N)
tof       Time on File
totord    Lifetime Orders
totsale   Lifetime Sales
targamnt  Promotion Sales
----------------------------------------
```

This data set is also riddled with similar errors and inconsistencies as in Project 1 data set and similar actions need to be taken to clean the data. Specifically,

1. Most of the predictor variables are counts or amounts and are right skewed with outliers. You will need to transform them as necessary.

2. About 95% of the values of `targamnt` are 0 (the customers did not make a purchase).

3. You should consider creating new predictor variables from the given set. For example, `aoa` = average order amount = `totsale/totord` or `pr` = purchase rate = `totord/tof` may be good predictors of `targamnt` in the multiple regression model (when `targamnt` >0). (Note that `tof` measures how long someone has been a customer.) There may be other "interaction" variables that are good predictors.

C.3   Project 3

- *Business situation*: An online book seller has provided data on a sample of 33,713 customers on their purchases of books prior to 1 August 2014 when a promotional offer was made and their purchase amounts (`targamt`) in euros over the next three months in response to the offer. The total sample is divided into a training sample of 25,402 customers and a test sample 8311 customers. The goal of the project is to build a predictive model for `targamt` based on the predictor variables from the past purchase history and use the model to predict `targamt` for the test sample customers. The response rates (proportion of customers with `targamt` > 0, i.e. who bought a book) are $999/25,402 = 3.93\%$ in the training sample and also $327/8311 = 3.93\%$ in the test sample.

- *Data*: There are a four data files which can be matched by common customer id's.

  1. `Project 3 data.csv` *file*: This is the most extensive data file with the following data fields on all 33,713 customers.

     `id`: unique customer id

     `recency`: no. of days since the last order

     `frequency`: number of orders

     `amount`: total past purchase amount in euros

     `tof`: time on file

     `Fxx`: frequency of orders of books of category xx

     `Mxx`: amount of purchase of books of category xx

     The following are the categories : 1=fiction, 3=classics, 5=cartoons, 6=legends, 7=philosophy, 8=religion, 9=psychology, 10=linguistics, 12=art, 14=music, 17=art reprints, 19=history, 20=contemporary history, 21=economy, 22=politics, 23=science, 26=computer science, 27=traffic, railroads, 30=maps, 31=travel guides, 35=health, 36=cooking, 37=learning, 38=games and riddles, 39=sports, 40=hobbies, 41=nature/animals/plants, 44=encyclopedias, 50=videos, DVDs, 99=non-books

     The aggregate `frequency` and `amount` variables are the totals of `Fxx` and `Mxx` variables, respectively.

2. Project 3 orders.csv *file*: This file contains data on all 627,955 orders, which translates to an average of 18.45 orders per customer. The data fields are as follows.

   id: unique customer id

   orddate: order date

   ordnum: order number

   category: category of the book

   qty: quantity ordered

   price: price

3. Project 3 train.csv: This file has only two variables: id and logtargamt for 25,402 customers in the training set.

4. Project 3 test.csv: This file also has only two variables: id and logtargamt for 8311 customers in the test set.

Note that the log-transformation of the purchase amount is already done in both the training and test data sets.

Instead of fixing the number of top prospects, you might try to find the optimum number of top prospects by maximizing the short term profit if the profit margin is assumed to be 25% of targamt and the cost of mailing the promotional material to each prospect is assumed to be €0.5. Thus you will have to find x to maximize $0.25 \times$ (sales revenue from the top x prospects) $- 0.5 \times x$.

- *Hints*:

  1. There are the usual data errors. For example, the aggregate frequency and amount variables do not equal to the totals of Fxx and Mxx variables, respectively, for all customers. In such cases, you may set the frequency and amount variables equal to the appropriate sums.

  2. The frequency and amount variables for a few customers are much larger than those for the large majority of customers. These are probably institutional customers such as schools or libraries. Since their purchases are not likely to be responsive to promotional offers, it would be best to omit them.

  3. Another odd feature of the data is that there are 271 customers in the training set and 87 customers in the test set who have no past purchase history. The response rates for these customers in the two data sets

are more than 50% (159 out of 271 in the training set and 46 out of 87 in the test set). These are "first-time" customers who responded to the promotion at a much higher rate. They cannot be simply omitted since they constitute significant portions of respondents. On the other hand, they cannot be included in predictive modeling because they don't have any data on predictors. One way to deal with these customers is to treat them separately. In the test set, estimate the response probability for each new customer $159/271 = 58.7\%$ and the estimated purchase amount to be the average purchase amount for the new customers in the training set.

Appendix D

Statistical tables

Predictive Analytics: Parametric Models for Regression and Classification Using R, First Edition. Ajit C. Tamhane.
Companion website: www.wiley.com/go/tamhane/predictiveanalytics

Table D.1. Standard normal cumulative probabilities $\Phi(z)$

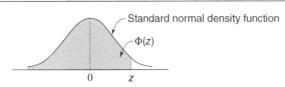

z	0.00	0.01	0.02	0.03	0.04	0.05	0.06	0.07	0.08	0.09
−3.4	0.0003	0.0003	0.0003	0.0003	0.0003	0.0003	0.0003	0.0003	0.0003	0.0002
−3.3	0.0005	0.0005	0.0005	0.0004	0.0004	0.0004	0.0004	0.0004	0.0004	0.0003
−3.2	0.0007	0.0007	0.0006	0.0006	0.0006	0.0006	0.0006	0.0005	0.0005	0.0005
−3.1	0.0010	0.0009	0.0009	0.0009	0.0008	0.0008	0.0008	0.0008	0.0007	0.0007
−3.0	0.0013	0.0013	0.0013	0.0012	0.0012	0.0011	0.0011	0.0011	0.0010	0.0010
−2.9	0.0019	0.0018	0.0017	0.0017	0.0016	0.0016	0.0015	0.0015	0.0014	0.0014
−2.8	0.0026	0.0025	0.0024	0.0023	0.0023	0.0022	0.0021	0.0021	0.0020	0.0019
−2.7	0.0035	0.0034	0.0033	0.0032	0.0031	0.0030	0.0029	0.0028	0.0027	0.0026
−2.6	0.0047	0.0045	0.0044	0.0043	0.0041	0.0040	0.0039	0.0038	0.0037	0.0036
−2.5	0.0062	0.0060	0.0059	0.0057	0.0055	0.0054	0.0052	0.0051	0.0049	0.0048
−2.4	0.0082	0.0080	0.0078	0.0075	0.0073	0.0071	0.0069	0.0068	0.0066	0.0064
−2.3	0.0107	0.0104	0.0102	0.0099	0.0096	0.0094	0.0091	0.0089	0.0087	0.0084
−2.2	0.0139	0.0136	0.0132	0.0129	0.0125	0.0122	0.0119	0.0116	0.0113	0.0110
−2.1	0.0179	0.0174	0.0170	0.0166	0.0162	0.0158	0.0154	0.0150	0.0146	0.0143
−2.0	0.0228	0.0222	0.0217	0.0212	0.0207	0.0202	0.0197	0.0192	0.0188	0.0183
−1.9	0.0287	0.0281	0.0274	0.0268	0.0262	0.0256	0.0250	0.0244	0.0239	0.0233
−1.8	0.0359	0.0352	0.0344	0.0336	0.0329	0.0322	0.0314	0.0307	0.0301	0.0294
−1.7	0.0446	0.0436	0.0427	0.0418	0.0409	0.0401	0.0392	0.0394	0.0375	0.0367
−1.6	0.0548	0.0537	0.0526	0.0516	0.0505	0.0495	0.0485	0.0475	0.0465	0.0455
−1.5	0.0668	0.0655	0.0643	0.0630	0.0618	0.0606	0.0594	0.0582	0.0571	0.0559
−1.4	0.0808	0.0793	0.0778	0.0764	0.0749	0.0735	0.0722	0.0708	0.0694	0.0681
−1.3	0.0968	0.0951	0.0934	0.0918	0.0901	0.0885	0.0869	0.0853	0.0838	0.0823
−1.2	0.1151	0.1131	0.1112	0.1093	0.1075	0.1056	0.1038	0.1020	0.1003	0.0985
−1.1	0.1357	0.1335	0.1314	0.1292	0.1271	0.1251	0.1230	0.1210	0.1190	0.1170
−1.0	0.1587	0.1562	0.1539	0.1515	0.1492	0.1469	0.1446	0.1423	0.1401	0.1379
−0.9	0.1841	0.1814	0.1788	0.1762	0.1736	0.1711	0.1685	0.1660	0.1635	0.1611
−0.8	0.2119	0.2090	0.2061	0.2033	0.2005	0.1977	0.1949	0.1922	0.1894	0.1867
−0.7	0.2420	0.2389	0.2358	0.2327	0.2296	0.2266	0.2236	0.2206	0.2177	0.2148
−0.6	0.2743	0.2709	0.2676	0.2643	0.2611	0.2578	0.2546	0.2514	0.2483	0.2451
−0.5	0.3085	0.3050	0.3015	0.2981	0.2946	0.2912	0.2877	0.2843	0.2810	0.2776
−0.4	0.3446	0.3409	0.3372	0.3336	0.3300	0.3264	0.3228	0.3192	0.3156	0.3121
−0.3	0.3821	0.3783	0.3745	0.3707	0.3669	0.3632	0.3594	0.3557	0.3520	0.3483
−0.2	0.4207	0.4168	0.4129	0.4090	0.4052	0.4013	0.3974	0.3936	0.3897	0.3859
−0.1	0.4602	0.4562	0.4522	0.4483	0.4443	0.4404	0.4364	0.4325	0.4286	0.4247
−0.0	0.5000	0.4960	0.4920	0.4880	0.4840	0.4801	0.4761	0.4721	0.4681	0.4641

Table D.1. Continued

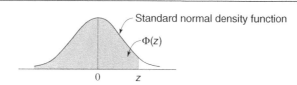

Standard normal density function

$\Phi(z)$

z	0.00	0.01	0.02	0.03	0.04	0.05	0.06	0.07	0.08	0.09
0.0	0.5000	0.5040	0.5080	0.5120	0.5160	0.5199	0.5239	0.5279	0.5319	0.5359
0.1	0.5398	0.5438	0.5478	0.5517	0.5557	0.5596	0.5636	0.5675	0.5714	0.5753
0.2	0.5793	0.5832	0.5871	0.5910	0.5948	0.5987	0.6026	0.6064	0.6103	0.6141
0.3	0.6179	0.6217	0.6255	0.6293	0.6331	0.6368	0.6406	0.6443	0.6480	0.6517
0.4	0.6554	0.6591	0.6628	0.6664	0.6700	0.6736	0.6772	0.6808	0.6844	0.6879
0.5	0.6915	0.6950	0.6985	0.7019	0.7054	0.7088	0.7123	0.7157	0.7190	0.7224
0.6	0.7257	0.7291	0.7324	0.7357	0.7389	0.7422	0.7454	0.7486	0.7517	0.7549
0.7	0.7580	0.7611	0.7642	0.7673	0.7704	0.7734	0.7764	0.7794	0.7823	0.7852
0.8	0.7881	0.7910	0.7939	0.7967	0.7995	0.8023	0.8051	0.8078	0.8106	0.8133
0.9	0.8159	0.8186	0.8212	0.8238	0.8264	0.8289	0.8315	0.8340	0.8365	0.8389
1.0	0.8413	0.8438	0.8461	0.8485	0.8508	0.8531	0.8554	0.8577	0.8599	0.8621
1.1	0.8643	0.8665	0.8686	0.8708	0.8729	0.8749	0.8770	0.8790	0.8810	0.8830
1.2	0.8849	0.8869	0.8888	0.8907	0.8925	0.8944	0.8962	0.8980	0.8997	0.9015
1.3	0.9032	0.9049	0.9066	0.9082	0.9099	0.9115	0.9131	0.9147	0.9162	0.9177
1.4	0.9192	0.9207	0.9222	0.9236	0.9251	0.9265	0.9278	0.9292	0.9306	0.9319
1.5	0.9332	0.9345	0.9357	0.9370	0.9382	0.9394	0.9406	0.9418	0.9429	0.9441
1.6	0.9452	0.9463	0.9474	0.9484	0.9495	0.9505	0.9515	0.9525	0.9535	0.9545
1.7	0.9554	0.9564	0.9573	0.9582	0.9591	0.9599	0.9608	0.9616	0.9625	0.9633
1.8	0.9641	0.9649	0.9656	0.9664	0.9671	0.9678	0.9686	0.9693	0.9699	0.9706
1.9	0.9713	0.9719	0.9726	0.9732	0.9738	0.9744	0.9750	0.9756	0.9761	0.9767
2.0	0.9772	0.9778	0.9783	0.9788	0.9793	0.9798	0.9803	0.9808	0.9812	0.9817
2.1	0.9821	0.9826	0.9830	0.9834	0.9838	0.9842	0.9846	0.9850	0.9854	0.9857
2.2	0.9861	0.9864	0.9868	0.9871	0.9875	0.9878	0.9881	0.9884	0.9887	0.9890
2.3	0.9893	0.9896	0.9898	0.9901	0.9904	0.9906	0.9909	0.9911	0.9913	0.9916
2.4	0.9918	0.9920	0.9922	0.9925	0.9927	0.9929	0.9931	0.9932	0.9934	0.9936
2.5	0.9938	0.9940	0.9941	0.9943	0.9945	0.9946	0.9948	0.9949	0.9951	0.9952
2.6	0.9953	0.9955	0.9956	0.9957	0.9959	0.9960	0.9961	0.9962	0.9963	0.9964
2.7	0.9965	0.9966	0.9967	0.9968	0.9969	0.9970	0.9971	0.9972	0.9973	0.9974
2.8	0.9974	0.9975	0.9976	0.9977	0.9977	0.9978	0.9979	0.9979	0.9980	0.9981
2.9	0.9981	0.9982	0.9982	0.9983	0.9984	0.9984	0.9985	0.9985	0.9986	0.9986
3.0	0.9987	0.9987	0.9987	0.9988	0.9988	0.9989	0.9989	0.9989	0.9990	0.9990
3.1	0.9990	0.9991	0.9991	0.9991	0.9992	0.9992	0.9992	0.9992	0.9993	0.9993
3.2	0.9993	0.9993	0.9994	0.9994	0.9994	0.9994	0.9994	0.9995	0.9995	0.9995
3.3	0.9995	0.9995	0.9995	0.9996	0.9996	0.9996	0.9996	0.9996	0.9996	0.9997
3.4	0.9997	0.9997	0.9997	0.9997	0.9997	0.9997	0.9997	0.9997	0.9997	0.9998

Source: Reprinted with permission of Pearson Education, Inc.

Table D.2. Critical values $t_{\nu,\alpha}$ for Student's t-distribution

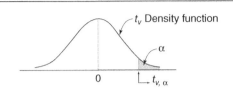

ν	α						
	.10	.05	.025	.01	.005	.001	.0005
1	3.078	6.314	12.706	31.821	63.657	318.31	636.62
2	1.886	2.920	4.303	6.965	9.925	22.326	31.598
3	1.638	2.353	3.182	4.541	5.841	10.213	12.924
4	1.533	2.132	2.776	3.747	4.604	7.173	8.610
5	1.476	2.015	2.571	3.365	4.032	5.893	6.869
6	1.440	1.943	2.447	3.143	3.707	5.208	5.959
7	1.415	1.895	2.365	2.998	3.499	4.785	5.408
8	1.397	1.860	2.306	2.896	3.355	4.501	5.041
9	1.383	1.833	2.262	2.821	3.250	4.297	4.781
10	1.372	1.812	2.228	2.764	3.169	4.144	4.587
11	1.363	1.796	2.201	2.718	3.106	4.025	4.437
12	1.356	1.782	2.179	2.681	3.055	3.930	4.318
13	1.350	1.771	2.160	2.650	3.012	3.852	4.221
14	1.345	1.761	2.145	2.624	2.977	3.787	4.140
15	1.341	1.753	2.131	2.602	2.947	3.733	4.073
16	1.337	1.746	2.120	2.583	2.921	3.686	4.015
17	1.333	1.740	2.110	2.567	2.898	3.646	3.965
18	1.330	1.734	2.101	2.552	2.878	3.610	3.922
19	1.328	1.729	2.093	2.539	2.861	3.579	3.883
20	1.325	1.725	2.086	2.528	2.845	3.552	3.850
21	1.323	1.721	2.080	2.518	2.831	3.527	3.819
22	1.321	1.717	2.074	2.508	2.819	3.505	3.792
23	1.319	1.714	2.069	2.500	2.807	3.485	3.767
24	1.318	1.711	2.064	2.492	2.797	3.467	3.745
25	1.316	1.708	2.060	2.485	2.787	3.450	3.725
26	1.315	1.706	2.056	2.479	2.779	3.435	3.707
27	1.314	1.703	2.052	2.473	2.771	3.421	3.690
28	1.313	1.701	2.048	2.467	2.763	3.408	3.674
29	1.311	1.699	2.045	2.462	2.756	3.396	3.659
30	1.310	1.697	2.042	2.457	2.750	3.385	3.646
40	1.303	1.684	2.021	2.423	2.704	3.307	3.551
60	1.296	1.671	2.000	2.390	2.660	3.232	3.460
120	1.289	1.658	1.980	2.358	2.617	3.160	3.373
∞	1.282	1.645	1.960	2.326	2.576	3.090	3.291

Source: Reprinted with permission of Pearson Education, Inc.

Table D.3. Critical values $\chi^2_{\nu,\alpha}$ for the chi-square distribution

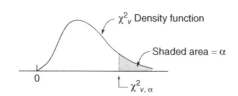

ν	.995	.99	.975	.95	.90	.10	.05	.025	.01	.005
1	0.000	0.000	0.001	0.004	0.016	2.706	3.843	5.025	6.637	7.882
2	0.010	0.020	0.051	0.103	0.211	4.605	5.992	7.378	9.210	10.597
3	0.072	0.115	0.216	0.352	0.584	6.251	7.815	9.348	11.344	12.837
4	0.207	0.297	0.484	0.711	1.064	7.779	9.488	11.143	13.277	14.860
5	0.412	0.554	0.831	1.145	1.610	9.236	11.070	12.832	15.085	16.748
6	0.676	0.872	1.237	1.635	2.204	10.645	12.592	14.440	16.812	18.548
7	0.989	1.239	1.690	2.167	2.833	12.017	14.067	16.012	18.474	20.276
8	1.344	1.646	2.180	2.733	3.490	13.362	15.507	17.534	20.090	21.954
9	1.735	2.088	2.700	3.325	4.168	14.684	16.919	19.022	21.665	23.587
10	2.156	2.558	3.247	3.940	4.865	15.987	18.307	20.483	23.209	25.188
11	2.603	3.053	3.816	4.575	5.578	17.275	19.675	21.920	24.724	26.755
12	3.074	3.571	4.404	5.226	6.304	18.549	21.026	23.337	26.217	28.300
13	3.565	4.107	5.009	5.892	7.041	19.812	22.362	24.735	27.687	29.817
14	4.075	4.660	5.629	6.571	7.790	21.064	23.685	26.119	29.141	31.319
15	4.600	5.229	6.262	7.261	8.547	22.307	24.996	27.488	30.577	32.799
16	5.142	5.812	6.908	7.962	9.312	23.542	26.296	28.845	32.000	34.267
17	5.697	6.407	7.564	8.682	10.085	24.769	27.587	30.190	33.408	35.716
18	6.265	7.015	8.231	9.390	10.865	25.989	28.869	31.526	34.805	37.156
19	6.843	7.632	8.906	10.117	11.651	27.203	30.143	32.852	36.190	38.580
20	7.434	8.260	9.591	10.851	12.443	28.412	31.410	34.170	37.566	39.997
21	8.033	8.897	10.283	11.591	13.240	29.615	32.670	35.478	38.930	41.399
22	8.643	9.542	10.982	12.338	14.042	30.813	33.924	36.781	40.289	42.796
23	9.260	10.195	11.688	13.090	14.848	32.007	35.172	38.075	41.637	44.179
24	9.886	10.856	12.401	13.848	15.659	33.196	36.415	39.364	42.980	45.558
25	10.519	11.523	13.120	14.611	16.473	34.381	37.652	40.646	44.313	46.925
26	11.160	12.198	13.844	15.379	17.292	35.563	38.885	41.923	45.642	48.290
27	11.807	12.878	14.573	16.151	18.114	36.741	40.113	43.194	46.962	49.642
28	12.461	13.565	15.308	16.928	18.939	37.916	41.337	44.461	48.278	50.993
29	13.120	14.256	16.147	17.708	19.768	39.087	42.557	45.772	49.586	52.333
30	13.787	14.954	16.791	18.493	20.599	40.256	43.773	46.979	50.892	53.672
31	14.457	15.655	17.538	19.280	21.433	41.422	44.985	48.231	52.190	55.000
32	15.134	16.362	18.291	20.072	22.271	42.585	46.194	49.480	53.486	56.328
33	15.814	17.073	19.046	20.866	23.110	43.745	47.400	50.724	54.774	57.646
34	16.501	17.789	19.806	21.664	23.952	44.903	48.602	51.966	56.061	58.964
35	17.191	18.508	20.569	22.465	24.796	46.059	49.802	53.203	57.340	60.272
36	17.887	19.233	21.336	23.269	25.643	47.212	50.998	54.437	58.619	61.581
37	18.584	19.960	22.105	24.075	26.492	48.363	52.192	55.667	59.891	62.880
38	19.289	20.691	22.878	24.884	27.343	49.513	53.384	56.896	61.162	64.181
39	19.994	21.425	23.654	25.695	28.196	50.660	54.572	58.119	62.420	65.473
40^{α}	20.706	22.164	24.433	26.509	29.050	51.805	55.758	59.342	63.691	66.766

$^{\alpha}$For $\nu > 40$, $\chi^2_{\nu,\alpha} \simeq \nu \left(1 - \frac{2}{9\nu} + z_{\alpha}\sqrt{\frac{2}{9\nu}}\right)^3$.

Source: Reprinted with permission of Pearson Education, Inc.

Table D.4. Critical values $f_{\nu 1,\, \nu 2,\, \alpha}$ for the F-distribution

$F_{\nu 1,\, \nu 2}$ Density function

Shaded area $= \alpha$

0    $f_{\nu 1,\, \nu 2,\, \alpha}$

$\alpha = 0.01$

										Degrees of freedom for numerator (ν_1)									
	1	2	3	4	5	6	7	8	9	10	12	15	20	24	30	40	60	120	∞
1	4052.0	4999.5	5403.0	5625.0	5764.0	5859.0	5928.0	5982.0	6022.0	6056.0	6106.0	6157.0	6209.0	6235.0	6261.0	6287.0	6311.0	6339.0	6366.0
2	98.50	99.00	99.17	99.25	99.30	99.33	99.36	99.37	99.39	99.40	99.42	99.43	99.45	99.46	99.47	99.47	99.48	99.49	99.50
3	34.12	30.82	29.46	28.71	28.24	27.91	27.67	27.49	27.35	27.23	27.05	26.87	26.69	26.60	26.50	26.41	26.32	26.22	26.13
4	21.20	18.00	16.69	15.98	15.52	15.21	14.98	14.80	14.66	14.55	14.37	14.20	14.02	13.93	13.84	13.75	13.65	13.56	13.46
5	16.26	13.27	12.06	11.39	10.97	10.67	10.46	10.29	10.16	10.05	9.89	9.72	9.55	9.47	9.38	9.29	9.20	9.11	9.02
6	13.75	10.92	9.78	9.15	8.75	8.47	8.26	8.10	7.98	7.87	7.72	7.56	7.40	7.31	7.23	7.14	7.06	6.97	6.88
7	12.25	9.55	8.45	7.85	7.46	7.19	6.99	6.84	6.72	6.62	6.47	6.31	6.16	6.07	5.99	5.91	5.82	5.74	5.65
8	11.26	8.65	7.59	7.01	6.63	6.37	6.18	6.03	5.91	5.81	5.67	5.52	5.36	5.28	5.20	5.12	5.03	4.95	4.86
9	10.56	8.02	6.99	6.42	6.06	5.80	5.61	5.47	5.35	5.26	5.11	4.96	4.81	4.73	4.65	4.57	4.48	4.40	4.31
10	10.04	7.56	6.55	5.99	5.64	5.39	5.20	5.06	4.94	4.85	4.71	4.56	4.41	4.33	4.25	4.17	4.08	4.00	3.91
11	9.65	7.21	6.22	5.67	5.32	5.07	4.89	4.74	4.63	4.54	4.40	4.25	4.10	4.02	3.94	3.86	3.78	3.69	3.60
12	9.33	6.93	5.95	5.41	5.06	4.82	4.64	4.50	4.39	4.30	4.16	4.01	3.86	3.78	3.70	3.62	3.54	3.45	3.36
13	9.07	6.70	5.74	5.21	4.86	4.62	4.44	4.30	4.19	4.10	3.96	3.82	3.66	3.59	3.51	3.43	3.34	3.25	3.17
14	8.86	6.51	5.56	5.04	4.69	4.46	4.28	4.14	4.03	3.94	3.80	3.66	3.51	3.43	3.35	3.27	3.18	3.09	3.00
15	8.68	6.36	5.42	4.89	4.56	4.32	4.14	4.00	3.89	3.80	3.67	3.52	3.37	3.29	3.21	3.13	3.05	2.96	2.87
16	8.53	6.23	5.29	4.77	4.44	4.20	4.03	3.89	3.78	3.69	3.55	3.41	3.26	3.18	3.10	3.02	2.93	2.84	2.75
17	8.40	6.11	5.18	4.67	4.34	4.10	3.93	3.79	3.68	3.59	3.46	3.31	3.16	3.08	3.00	2.92	2.83	2.75	2.65
18	8.29	6.01	5.09	4.58	4.25	4.01	3.84	3.71	3.60	3.51	3.37	3.23	3.08	3.00	2.92	2.84	2.75	2.66	2.57
19	8.18	5.93	5.01	4.50	4.17	3.94	3.77	3.63	3.52	3.43	3.30	3.15	3.00	2.92	2.84	2.76	2.67	2.58	2.49

Degrees of freedom for denominator (ν_2)

20	8.10	5.85	4.94	4.43	4.10	3.87	3.70	3.56	3.46	3.37	3.23	3.09	2.94	2.86	2.78	2.69	2.61	2.52	2.42
21	8.02	5.78	4.87	4.37	4.04	3.81	3.64	3.51	3.40	3.31	3.17	3.03	2.88	2.80	2.72	2.64	2.55	2.46	2.36
22	7.95	5.72	4.81	4.31	3.99	3.76	3.59	3.45	3.35	3.26	3.12	2.98	2.83	2.75	2.67	2.58	2.50	2.40	2.31
23	7.88	5.66	4.76	4.26	3.94	3.71	3.54	3.41	3.30	3.21	3.07	2.93	2.78	2.70	2.62	2.54	2.45	2.35	2.26
24	7.82	5.61	4.72	4.22	3.90	3.67	3.50	3.36	3.26	3.17	3.03	2.89	2.74	2.66	2.58	2.49	2.40	2.31	2.21
25	7.77	5.57	4.68	4.18	3.85	3.63	3.46	3.32	3.22	3.13	2.99	2.85	2.70	2.62	2.54	2.45	2.36	2.27	2.17
26	7.72	5.53	4.64	4.14	3.82	3.59	3.42	3.29	3.18	3.09	2.96	2.81	2.66	2.58	2.50	2.42	2.33	2.23	2.13
27	7.68	5.49	4.60	4.11	3.78	3.56	3.39	3.26	3.15	3.06	2.93	2.78	2.63	2.55	2.47	2.38	2.29	2.20	2.10
28	7.64	5.45	4.57	4.07	3.75	3.53	3.36	3.23	3.12	3.03	2.90	2.75	2.60	2.52	2.44	2.35	2.26	2.17	2.06
29	7.60	5.42	4.54	4.04	3.73	3.50	3.33	3.20	3.09	3.00	2.87	2.73	2.57	2.49	2.41	2.33	2.23	2.14	2.03
30	7.56	5.39	4.51	4.02	3.70	3.47	3.30	3.17	3.07	2.98	2.84	2.70	2.55	2.47	2.39	2.30	2.21	2.11	2.01
40	7.31	5.18	4.31	3.83	3.51	3.29	3.12	2.99	2.89	2.80	2.66	2.52	2.37	2.29	2.20	2.11	2.02	1.92	1.80
60	7.08	4.98	4.13	3.65	3.34	3.12	2.95	2.82	2.72	2.63	2.50	2.35	2.20	2.12	2.03	1.94	1.84	1.73	1.60
120	6.85	4.79	3.95	3.48	3.17	2.96	2.79	2.66	2.56	2.47	2.34	2.19	2.03	1.95	1.86	1.76	1.66	1.53	1.38
∞	6.63	4.61	3.78	3.32	3.02	2.80	2.64	2.51	2.41	2.32	2.18	2.04	1.88	1.79	1.70	1.51	1.47	1.32	1.00

(Continued)

Table D.4. Continued

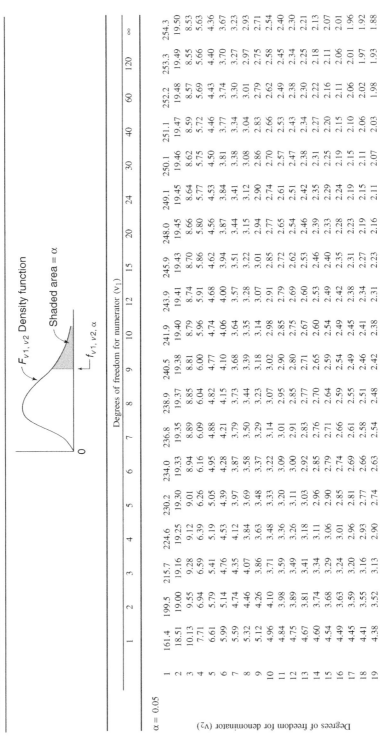

$F_{v1,v2}$ Density function

Shaded area = α

$f_{v1,v2,\alpha}$

α = 0.05

v_2	Degrees of freedom for numerator (v_1)																		
	1	2	3	4	5	6	7	8	9	10	12	15	20	24	30	40	60	120	∞
1	161.4	199.5	215.7	224.6	230.2	234.0	236.8	238.9	240.5	241.9	243.9	245.9	248.0	249.1	250.1	251.1	252.2	253.3	254.3
2	18.51	19.00	19.16	19.25	19.30	19.33	19.35	19.37	19.38	19.40	19.41	19.43	19.45	19.45	19.46	19.47	19.48	19.49	19.50
3	10.13	9.55	9.28	9.12	9.01	8.94	8.89	8.85	8.81	8.79	8.74	8.70	8.66	8.64	8.62	8.59	8.57	8.55	8.53
4	7.71	6.94	6.59	6.39	6.26	6.16	6.09	6.04	6.00	5.96	5.91	5.86	5.80	5.77	5.75	5.72	5.69	5.66	5.63
5	6.61	5.79	5.41	5.19	5.05	4.95	4.88	4.82	4.77	4.74	4.68	4.62	4.56	4.53	4.50	4.46	4.43	4.40	4.36
6	5.99	5.14	4.76	4.53	4.39	4.28	4.21	4.15	4.10	4.06	4.00	3.94	3.87	3.84	3.81	3.77	3.74	3.70	3.67
7	5.59	4.74	4.35	4.12	3.97	3.87	3.79	3.73	3.68	3.64	3.57	3.51	3.44	3.41	3.38	3.34	3.30	3.27	3.23
8	5.32	4.46	4.07	3.84	3.69	3.58	3.50	3.44	3.39	3.35	3.28	3.22	3.15	3.12	3.08	3.04	3.01	2.97	2.93
9	5.12	4.26	3.86	3.63	3.48	3.37	3.29	3.23	3.18	3.14	3.07	3.01	2.94	2.90	2.86	2.83	2.79	2.75	2.71
10	4.96	4.10	3.71	3.48	3.33	3.22	3.14	3.07	3.02	2.98	2.91	2.85	2.77	2.74	2.70	2.66	2.62	2.58	2.54
11	4.84	3.98	3.59	3.36	3.20	3.09	3.01	2.95	2.90	2.85	2.79	2.72	2.65	2.61	2.57	2.53	2.49	2.45	2.40
12	4.75	3.89	3.49	3.26	3.11	3.00	2.91	2.85	2.80	2.75	2.69	2.62	2.54	2.51	2.47	2.43	2.38	2.34	2.30
13	4.67	3.81	3.41	3.18	3.03	2.92	2.83	2.77	2.71	2.67	2.60	2.53	2.46	2.42	2.38	2.34	2.30	2.25	2.21
14	4.60	3.74	3.34	3.11	2.96	2.85	2.76	2.70	2.65	2.60	2.53	2.46	2.39	2.35	2.31	2.27	2.22	2.18	2.13
15	4.54	3.68	3.29	3.06	2.90	2.79	2.71	2.64	2.59	2.54	2.49	2.40	2.33	2.29	2.25	2.20	2.16	2.11	2.07
16	4.49	3.63	3.24	3.01	2.85	2.74	2.66	2.59	2.54	2.49	2.42	2.35	2.28	2.24	2.19	2.15	2.11	2.06	2.01
17	4.45	3.59	3.20	2.96	2.81	2.69	2.61	2.55	2.49	2.45	2.38	2.31	2.23	2.19	2.15	2.10	2.06	2.01	1.96
18	4.41	3.55	3.16	2.93	2.77	2.66	2.58	2.51	2.46	2.41	2.34	2.27	2.19	2.15	2.11	2.06	2.02	1.97	1.92
19	4.38	3.52	3.13	2.90	2.74	2.63	2.54	2.48	2.42	2.38	2.31	2.23	2.16	2.11	2.07	2.03	1.98	1.93	1.88

Degrees of freedom for denominator (v_2)

20	4.35	3.49	3.10	2.87	2.71	2.60	2.51	2.45	2.39	2.35	2.28	2.20	2.12	2.08	2.04	1.99	1.95	1.90	1.84
21	4.32	3.47	3.07	2.84	2.68	2.57	2.49	2.42	2.37	2.32	2.25	2.18	2.10	2.05	2.01	1.96	1.92	1.87	1.81
22	4.30	3.44	3.05	2.82	2.66	2.55	2.46	2.40	2.34	2.30	2.23	2.15	2.07	2.03	1.98	1.94	1.89	1.84	1.78
23	4.28	3.42	3.03	2.80	2.64	2.53	2.44	2.37	2.32	2.27	2.20	2.13	2.05	2.01	1.96	1.91	1.86	1.81	1.76
24	4.26	3.40	3.01	2.78	2.62	2.51	2.42	2.36	2.30	2.25	2.18	2.11	2.03	1.98	1.94	1.89	1.84	1.79	1.73
25	4.24	3.39	2.99	2.76	2.60	2.49	2.40	2.34	2.28	2.24	2.16	2.09	2.01	1.96	1.92	1.87	1.82	1.77	1.71
26	4.23	3.37	2.98	2.74	2.59	2.47	2.39	2.32	2.27	2.22	2.15	2.07	1.99	1.95	1.90	1.85	1.80	1.75	1.69
27	4.21	3.35	2.96	2.73	2.57	2.46	2.37	2.31	2.25	2.20	2.13	2.06	1.97	1.93	1.88	1.84	1.79	1.73	1.67
28	4.20	3.34	2.95	2.71	2.56	2.45	2.36	2.29	2.24	2.19	2.12	2.04	1.96	1.91	1.87	1.82	1.77	1.71	1.65
29	4.18	3.33	2.93	2.70	2.55	2.43	2.35	2.28	2.22	2.18	2.10	2.03	1.94	1.90	1.85	1.81	1.75	1.70	1.64
30	4.17	3.32	2.92	2.69	2.53	2.42	2.33	2.27	2.21	2.16	2.09	2.01	1.93	1.89	1.84	1.79	1.74	1.68	1.62
40	4.09	3.23	2.84	2.61	2.45	2.34	2.25	2.18	2.12	2.08	2.00	1.92	1.84	1.79	1.74	1.69	1.64	1.59	1.51
60	4.00	3.15	2.76	2.53	2.37	2.25	2.17	2.10	2.04	1.99	1.92	1.84	1.75	1.70	1.65	1.59	1.53	1.47	1.39
120	3.92	3.07	2.68	2.45	2.29	2.17	2.09	2.02	1.96	1.91	1.83	1.75	1.66	1.61	1.55	1.55	1.43	1.35	1.25
∞	3.84	3.00	2.60	2.37	2.21	2.10	2.01	1.94	1.88	1.83	1.75	1.67	1.57	1.52	1.46	1.39	1.32	1.22	1.00

(Continued)

Table D.4. Continued

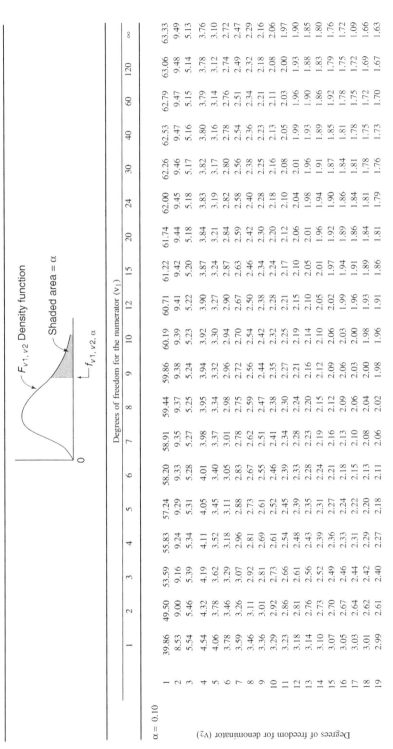

F_{v_1, v_2} Density function

Shaded area = α

$f_{v_1, v_2, \alpha}$

$\alpha = 0.10$

Degrees of freedom for the numerator (v_1)

Degrees of freedom for denominator (v_2)

v_2	1	2	3	4	5	6	7	8	9	10	12	15	20	24	30	40	60	120	∞
1	39.86	49.50	53.59	55.83	57.24	58.20	58.91	59.44	59.86	60.19	60.71	61.22	61.74	62.00	62.26	62.53	62.79	63.06	63.33
2	8.53	9.00	9.16	9.24	9.29	9.33	9.35	9.37	9.38	9.39	9.41	9.42	9.44	9.45	9.46	9.47	9.47	9.48	9.49
3	5.54	5.46	5.39	5.34	5.31	5.28	5.27	5.25	5.24	5.23	5.22	5.20	5.18	5.18	5.17	5.16	5.15	5.14	5.13
4	4.54	4.32	4.19	4.11	4.05	4.01	3.98	3.95	3.94	3.92	3.90	3.87	3.84	3.83	3.82	3.80	3.79	3.78	3.76
5	4.06	3.78	3.62	3.52	3.45	3.40	3.37	3.34	3.32	3.30	3.27	3.24	3.21	3.19	3.17	3.16	3.14	3.12	3.10
6	3.78	3.46	3.29	3.18	3.11	3.05	3.01	2.98	2.96	2.94	2.90	2.87	2.84	2.82	2.80	2.78	2.76	2.74	2.72
7	3.59	3.26	3.07	2.96	2.88	2.83	2.78	2.75	2.72	2.70	2.67	2.63	2.59	2.58	2.56	2.54	2.51	2.49	2.47
8	3.46	3.11	2.92	2.81	2.73	2.67	2.62	2.59	2.56	2.54	2.50	2.46	2.42	2.40	2.38	2.36	2.34	2.32	2.29
9	3.36	3.01	2.81	2.69	2.61	2.55	2.51	2.47	2.44	2.42	2.38	2.34	2.30	2.28	2.25	2.23	2.21	2.18	2.16
10	3.29	2.92	2.73	2.61	2.52	2.46	2.41	2.38	2.35	2.32	2.28	2.24	2.20	2.18	2.16	2.13	2.11	2.08	2.06
11	3.23	2.86	2.66	2.54	2.45	2.39	2.34	2.30	2.27	2.25	2.21	2.17	2.12	2.10	2.08	2.05	2.03	2.00	1.97
12	3.18	2.81	2.61	2.48	2.39	2.33	2.28	2.24	2.21	2.19	2.15	2.10	2.06	2.04	2.01	1.99	1.96	1.93	1.90
13	3.14	2.76	2.56	2.43	2.35	2.28	2.23	2.20	2.16	2.14	2.10	2.05	2.01	1.98	1.96	1.93	1.90	1.88	1.85
14	3.10	2.73	2.52	2.39	2.31	2.24	2.19	2.15	2.12	2.10	2.05	2.01	1.96	1.94	1.91	1.89	1.86	1.83	1.80
15	3.07	2.70	2.49	2.36	2.27	2.21	2.16	2.12	2.09	2.06	2.02	1.97	1.92	1.90	1.87	1.85	1.82	1.79	1.76
16	3.05	2.67	2.46	2.33	2.24	2.18	2.13	2.09	2.06	2.03	1.99	1.94	1.89	1.87	1.84	1.81	1.78	1.75	1.72
17	3.03	2.64	2.44	2.31	2.22	2.15	2.10	2.06	2.03	2.00	1.96	1.91	1.86	1.84	1.81	1.78	1.75	1.72	1.69
18	3.01	2.62	2.42	2.29	2.20	2.13	2.08	2.04	2.00	1.98	1.93	1.89	1.84	1.81	1.78	1.75	1.72	1.69	1.66
19	2.99	2.61	2.40	2.27	2.18	2.11	2.06	2.02	1.98	1.96	1.91	1.86	1.81	1.79	1.76	1.73	1.70	1.67	1.63

df																			
20	2.97	2.59	2.38	2.25	2.16	2.09	2.04	2.00	1.96	1.94	1.89	1.84	1.79	1.77	1.74	1.71	1.68	1.64	1.61
21	2.96	2.57	2.36	2.23	2.14	2.08	2.02	1.98	1.95	1.92	1.87	1.83	1.78	1.75	1.72	1.69	1.66	1.62	1.59
22	2.95	2.56	2.35	2.22	2.13	2.06	2.01	1.97	1.93	1.90	1.86	1.81	1.76	1.73	1.70	1.67	1.64	1.60	1.57
23	2.94	2.55	2.34	2.21	2.11	2.05	1.99	1.95	1.92	1.89	1.84	1.80	1.74	1.72	1.69	1.66	1.62	1.59	1.55
24	2.93	2.54	2.33	2.19	2.10	2.04	1.98	1.94	1.91	1.88	1.83	1.78	1.73	1.70	1.67	1.64	1.61	1.57	1.53
25	2.92	2.53	2.32	2.18	2.09	2.02	1.97	1.93	1.89	1.87	1.82	1.77	1.72	1.69	1.66	1.63	1.59	1.56	1.52
26	2.91	2.52	2.31	2.17	2.08	2.01	1.96	1.92	1.88	1.86	1.81	1.76	1.71	1.68	1.65	1.61	1.58	1.54	1.50
27	2.90	2.51	2.30	2.17	2.07	2.00	1.95	1.91	1.87	1.85	1.80	1.75	1.70	1.67	1.64	1.60	1.57	1.53	1.49
28	2.89	2.50	2.29	2.16	2.06	2.00	1.94	1.90	1.87	1.84	1.79	1.74	1.69	1.66	1.63	1.59	1.56	1.52	1.48
29	2.89	2.50	2.28	2.15	2.06	1.99	1.93	1.89	1.86	1.83	1.78	1.73	1.68	1.65	1.62	1.58	1.55	1.51	1.47
30	2.88	2.49	2.28	2.14	2.03	1.98	1.93	1.88	1.85	1.82	1.77	1.72	1.67	1.64	1.61	1.57	1.54	1.50	1.46
40	2.84	2.44	2.23	2.09	2.00	1.93	1.87	1.83	1.79	1.76	1.71	1.66	1.61	1.57	1.54	1.51	1.47	1.42	1.38
60	2.79	2.39	2.18	2.04	1.95	1.87	1.82	1.77	1.74	1.71	1.66	1.60	1.54	1.51	1.48	1.44	1.40	1.35	1.29
120	2.75	2.35	2.13	1.99	1.90	1.82	1.77	1.72	1.68	1.65	1.60	1.55	1.48	1.45	1.41	1.37	1.32	1.26	1.19
∞	2.71	2.30	2.08	1.94	1.85	1.77	1.72	1.67	1.63	1.60	1.55	1.49	1.42	1.38	1.34	1.30	1.24	1.17	1.00

Source: Reprinted with permission of Pearson Education, Inc.

Table D.5. Distribution of the Durbin-Watson statistic d: the 5% significance points of d_L and d_U (p is the number of predictor variables)

n	$p = 1$		$p = 2$		$p = 3$		$p = 4$		$p = 5$	
	d_L	d_U	d_L	d_U	d_L	d_U	d_L	d_U	d_L	d_U
15	1.08	1.36	0.95	1.54	0.82	1.75	0.69	1.97	0.56	2.21
16	1.10	1.37	0.98	1.54	0.86	1.73	0.74	1.93	0.62	2.15
17	1.13	1.38	1.02	1.54	0.90	1.71	0.78	1.90	0.67	2.10
18	1.16	1.39	1.05	1.53	0.93	1.69	0.82	1.87	0.71	2.06
19	1.18	1.40	1.08	1.53	0.97	1.68	0.86	1.85	0.75	2.02
20	1.20	1.41	1.10	1.54	1.00	1.68	0.90	1.83	0.79	1.99
21	1.22	1.42	1.13	1.54	1.03	1.67	0.93	1.81	0.83	1.96
22	1.24	1.43	1.15	1.54	1.05	1.66	0.96	1.80	0.86	1.94
23	1.26	1.44	1.17	1.54	1.08	1.66	0.99	1.79	0.90	1.92
24	1.27	1.45	1.19	1.55	1.10	1.66	1.01	1.78	0.93	1.90
25	1.29	1.45	1.21	1.55	1.12	1.66	1.04	1.77	0.95	1.89
26	1.30	1.46	1.22	1.55	1.14	1.65	1.06	1.76	0.98	1.88
27	1.32	1.47	1.24	1.56	1.16	1.65	1.08	1.76	1.01	1.86
28	1.33	1.48	1.26	1.56	1.18	1.65	1.10	1.75	1.03	1.85
29	1.34	1.48	1.27	1.56	1.20	1.65	1.12	1.74	1.05	1.84
30	1.35	1.49	1.28	1.57	1.21	1.65	1.14	1.74	1.07	1.83
31	1.36	1.50	1.30	1.57	1.23	1.65	1.16	1.74	1.09	1.83
32	1.37	1.50	1.31	1.57	1.24	1.65	1.18	1.73	1.11	1.82
33	1.38	1.51	1.32	1.58	1.26	1.65	1.19	1.73	1.13	1.81
34	1.39	1.51	1.33	1.58	1.27	1.65	1.21	1.73	1.15	1.81
35	1.40	1.52	1.34	1.58	1.28	1.65	1.22	1.73	1.16	1.80
36	1.41	1.52	1.35	1.59	1.29	1.65	1.24	1.73	1.18	1.80
37	1.42	1.53	1.36	1.59	1.31	1.66	1.25	1.72	1.19	1.80
38	1.43	1.54	1.37	1.59	1.32	1.66	1.26	1.72	1.21	1.78
39	1.43	1.54	1.38	1.60	1.33	1.66	1.27	1.72	1.22	1.79
40	1.44	1.54	1.39	1.60	1.34	1.66	1.29	1.72	1.23	1.79
45	1.48	1.57	1.43	1.62	1.38	1.67	1.34	1.72	1.29	1.78
50	1.50	1.59	1.46	1.63	1.42	1.67	1.38	1.72	1.34	1.77
55	1.53	1.60	1.49	1.64	1.45	1.68	1.41	1.72	1.38	1.77
60	1.55	1.62	1.51	1.65	1.48	1.69	1.44	1.73	1.41	1.77
65	1.57	1.63	1.54	1.66	1.50	1.70	1.47	1.73	1.44	1.77
70	1.58	1.64	1.55	1.67	1.52	1.70	1.49	1.74	1.46	1.77
75	1.60	1.65	1.57	1.68	1.54	1.71	1.51	1.74	1.49	1.77
80	1.61	1.66	1.59	1.69	1.56	1.72	1.53	1.74	1.51	1.77
85	1.62	1.67	1.60	1.70	1.57	1.72	1.55	1.75	1.52	1.77
90	1.63	1.68	1.61	1.70	1.59	1.73	1.57	1.75	1.54	1.78
95	1.64	1.69	1.62	1.71	1.60	1.73	1.58	1.75	1.56	1.78
100	1.65	1.69	1.63	1.72	1.61	1.74	1.59	1.76	1.57	1.78

Source: Durbin and Watson (1951).

References

Allison, P.D. (2008). Convergence Failures in Logistic Regression. *SAS Global Forum*, Paper No. 360-2008.

Anscombe, F.J. (1973). Graphs in statistical analysis. *The American Statistician* **27**: 17–21.

Box, G.E.P. and Cox, D.R. (1964). An analysis of transformations (with discussion). *Journal of the Royal Statistical Society: Series B* **26**: 211–252.

Chatterjee, S. and Hadi, A.S. (2012). *Regression Analysis by Example*, 5e. New York: Wiley.

Cox, D.R. (1972). Regression models and life-tables. *Journal of the Royal Statistical Society: Series B (Methodological)* **34**: 187–220.

Crowley, J. and Hu, M. (1977). Covariance analysis of heart transplant survival data. *Journal of the American Statistical Association* **72**: 27–36.

Cule, E., Vineis, P., and De Iorio, M. (2011). Significance testing in ridge regression for genetic data. *BMC Bioinformatics*. https://doi.org/10.1186/147-2105-12-372.

Draper, N.R. and Smith, H. (1998). *Applied Regression Analysis*, 3e. New York: Wiley.

Draper, N.R. and Stoneman, D.M. (1966). Testing for the inclusion of variables in linear regression by a randomisation technique. *Technometrics* **8**: 695–699.

Fisher, R.A. (1936). The use of multiple measurements in taxonomic problems. *Annals of Eugenics* **7**: 179–188.

Hald, A. (1952). *Statistical Theory with Engineering Applications*. New York: John Wiley.

Hamilton, D.J. (1987). Sometimes $R^2 > r^2_{y \cdot x_1} + r^2_{y \cdot x_2}$, correlated variables are not always redundant. *The American Statistician* **41**: 129–132.

Predictive Analytics: Parametric Models for Regression and Classification Using R,
First Edition. Ajit C. Tamhane.
© 2021 John Wiley & Sons, Inc. Published 2021 by John Wiley & Sons, Inc.
Companion website: www.wiley.com/go/tamhane/predictiveanalytics

Hardin, J.W. and Hilbe, J.M. (2012). *Generalized Linear Models and Extension*, 3e. College Station, TX: Stata Press.

Hastie, T.J., Tibshirani, R.J. (1990). *Generalized Additive Models*, London: Chapman and Hall.

Hastie, T., Tibshirani, R., and Friedman, J.H. (2001). *The Elements of Statistical Learning*. New York: Springer.

Hastie, T., Tibshirani, R., and Friedman, J.H. (2009). *The Elements of Statistical Learning*, 2e. New York: Springer.

Hastie, T., Tibshirani, R., and Wainwright, M. (2015). *Statistical Learning with Sparsity: The Lasso and Generalizations*. CRC Press.

Hochberg, Y. and Tamhane, A.C. (1987). *Multiple Comparison Procedures*. New York: Wiley.

Hoerl, A.E. and Kennard, R.W. (1970). Ridge regression: biased estimation for non-orthogonal problems. *Technometrics* **12**: 69–82.

Hosmer, D.W. and Lemeshow, S. (1989). *Applied Logistic Regression*. New York: Wiley.

James, G., Witten, D., Hastie, T., and Tibshirani, R. (2013). *An Introduction to Statistical Learning with Applications in R*. New York: Springer.

Johnson, R.A. and Wichern, D.W. (2002). *Applied Multivariate Statistical Analysis*, 5e. Upper Saddle River, NJ: Prentice Hall.

Koch, G.G., Atkinson, S.S., and Stokes, M.E. (1986). Poisson regression. In: *Encyclopedia of Statistical Sciences*, vol. **7** (ed. N.L. Johnson and S. Kotz), 32–42. New York: Wiley.

Kuiper, S. (2008). Introduction to multiple regression: how much is your car worth? *Journal of Statistics Education* **16** (3). http://www.amstat.org/publications/jse/v16n3/datasets.kuiper.html.

Kutner, M.H., Nachtsheim, C.J., Neter, J. et al. (2005). *Applied Linear Statistical Models*, 5e. New York: McGraw-Hill/Irwin.

Mallows, C.L. (1973). Some comments on C_p. *Technometrics* **15**: 661–673.

Marquardt, D.W. and Snee, R.D. (1975). Ridge regression in practice. *The American Statistician* **29**: 3–20.

McClave, J.T. and Dietrich, F.H. (1994). *Statistics*, 6e. New York: Dellen-MacMillan.

McClintock, S., Stangl, D., and Cetinkya-Rundel, M. (2014). The real secret to genius? Reading between lines. *Chance* **27**: 41–44.

McCullagh, P. and Nelder, J.A. (1989). *Generalized Linear Models*, 2e. London: Chapman & Hall.

McKenzie, J.D. Jr. and Goldman, R. (1999). *The Student Edition of MINITAB*. Reading, MA: Addison Wesley.

Messerli, F.H. (2012). Chocolate consumption, cognitive function and Nobel laureates. *New England Journal of Medicine* **367**: 1562–1564.

Mevik, B.-H. and Wehrens, R. (2007). The PLS package: principal component and partial least squares regression in R. *Journal of Statistical Software* **18** (2): 1–23.

Montgomery, D.C., Peck, E.A., and Vining, G.G. (2012). *Introduction to Linear Regression Analysis*, 5e. New York: Wiley.

Myers, R.H., Montgomery, D.C., and Vining, G.G. (2010). *Generalized Linear Models: With Applications in Engineering and the Sciences*, 1e. New York: Wiley.

Nelder, J. A. and Wedderburn, R. W. M. (1972). Generalized linear models. *Journal of Royal Statistical Society, Series. A* **135**: 370–384.

Rossi, P.H., Berk, R.A., and Lenihan, K.J. (1980). *Money, Work and Crime: Some Experimental Results*. New York: Academic Press.

Tableman, M. and Kim, J.S. (2003). *Survival Analysis Using S*. Chapman and Hall/CRC.

Tamhane, A.C. and Dunlop, D.D. (2000). *Statistics and Data Analysis: From Elementary to Intermediate*. Upper Saddle River, NJ: Prentice Hall.

Tanner, M. (1996). *Tools for Statistical Inference*, 3e. New York: Springer.

Therneu, T., Crowson, C., and Atkinson, E. (2018). Using time dependent covariates and time dependent coefficients in the Cox model. https://cran.r-project.org/web/packages/survival/vignettes/timedep.pdf (accessed 05 June 2020).

Tibshirani, R. (1996). Regression shrinking and selection via the lasso. *Journal of the Royal Statistical Society: Series B* **58**: 267–288.

Tversky, A. and Kahnemann, D. (1973). Availability: a heuristic for judging frequency and probability. *Cognitive Psychology* **5**: 207–232.

Webster, J.T., Gunst, R.F., and Mason, R.L. (1974). Latent root regression analysis. *Technometrics* **16**: 513–522.

Wold, H. (1966). Estimation of principal components and related models by iterative least squares. In: *Multivariate Analysis* (ed. P.R. Krishniah), 391–420. New York: Academic Press.

Answers to selected exercises

Chapter 2

2.6 (d) $\widehat{x}_{\text{new}} = 6.441$, 95% CI for x_{new}: $[5.829, 7.093]$.

2.8 If $r = 0.25$, then $\widehat{y} = 68''$ for $x = 64''$ and $\widehat{y} = 70''$ for $x = 72''$. If $r = 0.75$, then $\widehat{y} = 66''$ for $x = 64''$ and $\widehat{y} = 72''$ for $x = 72''$.

2.9 IBM Beta: 0.7448, Apple Beta: 1.2449, IBM SD: 0.0556, Apple SD: 0.1031, r between IBM and S&P 500: 0.5975, r between Apple and S&P 500: 0.5382.

2.10 (a) Price elasticities are as follows:

    Chuck: -1.3687, Porter House: -2.6565, Rib Eye: -1.4460.

   (b) Drops in demand: chuck: 13.69%, porter house: 26.57%, rib eye: 14.46%.

2.11 The correlations between the number of cigarettes smoked per capita and number of deaths due to different types of cancer are as follows:

    Bladder: 0.7036, Lung: 0.6974, Kidney: 0.4874, Leukemia: -0.0685.

The t-statistics are as follows:

    Bladder: $t = 6.417$, Lung: $t = 5.505$, Kidney: $t = 3.617$,

    Leukemia: $t = -0.445$.

2.12 Spearman $r_S = 0.9641$.

2.13 Lack of fit test statistic: $F = 9.28$ with 5 and 6 d.f. ($P = 0.009$).

Predictive Analytics: Parametric Models for Regression and Classification Using R, First Edition. Ajit C. Tamhane.
© 2021 John Wiley & Sons, Inc. Published 2021 by John Wiley & Sons, Inc.
Companion website: www.wiley.com/go/tamhane/predictiveanalytics

Chapter 3

3.8 $\widehat{\beta}_0 = 1.1, \widehat{\beta}_1 = 1.9$.

3.9 $\widehat{\beta}_0 = 50, \widehat{\beta}_1 = 5, \widehat{\beta}_2 = 7.5, \widehat{\beta}_3 = 2.5$.

3.10 (a) The capital and labor elasticities are 0.2076 and 0.7149, respectively.

(b) Estimated output = € 28.962 million.

(c) Test statistic for testing constant returns to scale: $t = -4.510$. Hence reject the hypothesis. SSE for the full model: 129.358, SSE for the partial model: 134.006. $F = 20.335 = t^2$. Hence reject the hypothesis.

3.11 (a) The correlation matrix and the scatter plot are shown below:

	Research_num	Faculty	PhD
Research_num	1.0000000	0.7648421	0.8174254
Faculty	0.7648421	1.0000000	0.9036829
PhD	0.8174254	0.9036829	1.0000000

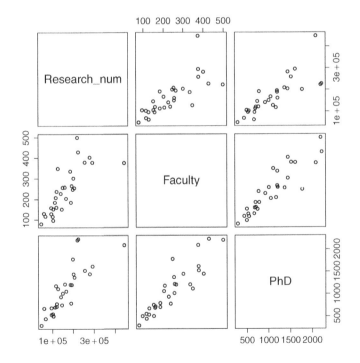

(b) Coefficients:

	Estimate	Std. Error	t value	Pr(>|t|)	
(Intercept)	23525.91	22034.47	1.068	0.2951	
Faculty	107.13	193.39	0.554	0.5842	
PhD	107.14	40.06	2.675	0.0125	*

```
Signif. codes:   0 *** 0.001 ** 0.01 * 0.05 . 0.1   1

Residual standard error: 49040 on 27 degrees of freedom
Multiple R-squared:  0.6719,    Adjusted R-squared:  0.6476
F-statistic: 27.65 on 2 and 27 DF,  p-value: 2.923e-07
```

(c) Denoting Research $= y$, Faculty $= x_1$, and PhD $= x_2$,

$$r_{yx_1|x_2} = 0.1060, r_{yx_2|x_1} = 0.4577, t_{yx_1|x_2} = 0.554, t_{yx_2|x_1} = 2.675.$$

3.12 $\widehat{\beta}_1^* = 0, \widehat{\beta}_2^* = 0.8, \widehat{\beta}_0 = 5, \widehat{\beta}_1 = 0, \widehat{\beta}_2 = 1.$

3.13 (a) $r_{yx_1|x_2} = 0.835$ and $r_{yx_2|x_1} = 0.179$.

(b) $\widehat{\beta}_1^* = 0.875$ and $\widehat{\beta}_2^* = 0.105$. Thus x_1 is a better predictor of y than x_2.

Chapter 4

4.7 (a)
$$E(R) = 10.6, \text{Var}(R) = 4.345, z = -0.192.$$

Hence the null hypothesis of independence is not rejected.

(b) The Durbin–Watson statistic $d = 1.32$. The test is inconclusive.

4.8 (a) Obs. ♯4 appears to be influential.

(b) Obs. ♯ 4 has leverage $= 0.6044$ and Cook's distance $= 0.4756$. Both influence metrics meet the criteria.

(c) Let $y = $ Woodbeam Strength, $x_1 = $ Specific Gravity and $x_2 = $ Moisture. Regression equation for full data is $\widehat{y} = 10.3015 + 8.4947x_1 - 0.2663x_2$. Regression equation if obs. ♯ 4 is dropped is $\widehat{y} = 12.4197 + 6.7992x_1 - 0.3905x_2$.

4.10 (a) The correlation matrix is

```
            x1          x2          x3          x4
x1  1.00000000  0.05230658 -0.3433818 -0.4976109
x2  0.05230658  1.00000000 -0.4315953 -0.3706964
x3 -0.34338179 -0.43159531  1.0000000 -0.3551214
x4 -0.49761095 -0.37069641 -0.3551214  1.0000000
```

(b) The VIFs are all > 150:

```
      x1        x2        x3        x4
178.2874  158.0460  257.9074  289.3750
```

4.12 (a) The correlation matrix is

```
            x1          x2          x3
x1   1.0000000   0.2236278  -0.9582041
x2   0.2236278   1.0000000  -0.2402310
x3  -0.9582041  -0.2402310   1.0000000
```

(b) The VIFs are

x1	x2	x3	x1sq	x2sq	x3sq
2.857e+06	1.0956e+04	2.0172e+06	2.502e+06	6.573e+01	1.267e+04

x1x2	x1x3	x2x3
9.803e+03	1.428e+06	2.4034e+02

(c) After centering the x_i's the VIFs are

x1	x2	x3	x1sq	x2sq	x3sq
375.25	1.741	680.28	1762.6	3.164	1156.7

x1x2	x1x3	x2x3
31.04	6563.3	35.61

Chapter 5

5.5 For ridge regression $\lambda = 0.132$. The coefficients are

```
(Intercept)     cylinders displacement    horsepower
 23.4459184    -0.6787924    -0.4898495    -1.7635131
      weight   acceleration
  -3.8984293    -0.1428772
```

$R^2 = 0.7072$ and $R^2_{adj} = 0.7034$ for this model.
For lasso regression $\lambda = 0.183$. The coefficients are

```
(Intercept)     cylinders displacement    horsepower
 23.44591837  -0.59980032  -0.02950645  -1.58002666
      weight   acceleration
  -4.43164443    0.00000000
```

$R^2 = 0.7070$ and $R^2_{adj} = 0.7032$ for this model.

5.6 For ridge regression $\lambda = 1.012$. The coefficients are

```
(Intercept)           x1            x2           x3      I(x1^2)
 35.1934280     5.9334447     2.4148470   -4.4992594    1.8807948
   I(x2^2)     I(x3^2)           x1:x2         x1:x3        x2:x3
 -0.4916187  -0.3268384     -2.2685298   -0.6905291    1.0116200
```

$R^2 = 0.9908$ and $R^2_{adj} = 0.9769$ for this model.
For lasso regression $\lambda = 0.301$. The coefficients are

```
(Intercept)            x1              x2             x3        I(x1^2)
 36.490640386  10.391413556    2.489902205   -0.001679847   0.905709655
     I(x2^2)        I(x3^2)           x1:x2          x1:x3          x2:x3
 -0.592090866   0.000000000   -3.235891614    0.000000000    0.000000000
```

$R^2 = 0.9939$ and $R^2_{adj} = 0.9847$ for this model.

5.7 (a) Four PCs are needed.

(b) Proportion of the variance in Fat explained $= R^2 = 56.46\%$.

5.8 (a) Two LVs are needed.

(b) Four LVs are needed.

(c) Peak loadings are at highest wavelengths around 1650 nm. Peak coefficients are also around 1650 nm.

Chapter 6

6.2 $F = 10$ with 2 and 20 d.f., significant at $\alpha = 0.01$.

6.4 (a) Complete the table.

(b) $R^2_{adj,p}$ criterion: Choose model $\{x_1, x_3\}$, C_p criterion: Choose model $\{x_1, x_3\}$, AIC_p criterion: Choose model $\{x_1, x_3\}$.

(c) The first variable to enter the model: x_3 with $F_{in} = 13.67$.

(d) The second variable to enter the model: x_1 with $F_{in} = 4.6$. Partial correlation $r_{yx_1|x_3} = \pm 0.462$

(e) Partial F of x_3 equals 11.8. So x_3 can't be removed.

(f) Partial F of x_2 equals 1.0. So x_2 can't be added.

6.5 (a) $r_{yx_2|x_1} = 0.568, r_{yx_3|x_1} = 0.318$. Enter x_2 in the model.

(b) Since x_1 and x_2 are almost uncorrelated, while x_1 and x_3 are highly correlated.

(c) $F = 14.29 > f_{1,30,0.05} = 4.17$. So add x_2 to the model.

6.7 Model 1 SSE $= 0.6046$, Model 2 SSE $= 0.6103$. Thus Model 1 SSE is slightly smaller. But Model 1 has one nonsignificant coefficient for Cruise, whereas all coefficients in Model 2 are highly significant. Model 1 has 12 predictors while Model 2 has only 10.

Chapter 7

7.6 (a)
$$\ln \widehat{\psi} = 1.558, \widehat{\mathrm{Var}}(\widehat{\psi}) = 0.4906, z = 2.224,$$

which is significant at $\alpha = 0.05$.

(b) $z = 2.346$.

7.8 (a) The fitted logistic response model is shown below.

```
Coefficients:
            Estimate Std. Error z value Pr(>|z|)
(Intercept)  3.81944    1.83518   2.081   0.0374 *
Days        -0.08648    0.04322  -2.001   0.0454 *
---
Signif. codes:  0 *** 0.001 ** 0.01 * 0.05 . 0.1   1

(Dispersion parameter for binomial family taken to be 1)

    Null deviance: 32.601  on 23  degrees of freedom
Residual deviance: 27.788  on 22  degrees of freedom
AIC: 31.788

Number of Fisher Scoring iterations: 4
```

(b) $[0.4249, 0.9914]$,

(c) The optimum threshold: $p^* = 0.40$ with CCR $= 0.625$.

$$\text{Sensitivity} = 0.7857, \text{Specificity} = 0.4000,$$
$$\text{Precision} = 0.6471, \text{Recall} = 0.7857, F_1 = 0.710.$$

7.9 (a) 95% CI on the odds ratio: $[1.006, 1.066]$.

(b) The odds of coronary disease for this male are 1.160. The probability is $p = 0.5370$.

7.10 First table:

$$\text{CCR} = 0.90, \text{Sensitivity} = 0.90, \text{Specificity} = 0.90,$$
$$\text{Precision} = 0.50, \text{Recall} = 0.90, F_1 = 0.543.$$

Second table:

CCR = 0.90, Sensitivity = 0.90, Specificity = 0.90,
Precision = 0.083, Recall = 0.90, F_1 = 0.152.

7.12 (a) The output using "Preterm" as the reference category is shown below.

```
Coefficients:
                Estimate Std. Error t-value Pr(>|t|)
2:(intercept) -1.085419   2.521756 -0.4304 0.666889
3:(intercept) -1.862345   2.657872 -0.7007 0.483496
2:Nutrition    0.015536   0.018762  0.8281 0.407625
3:Nutrition    0.037131   0.020222  1.8362 0.066328 .
2:Alcohol     -1.060329   0.821668 -1.2905 0.196891
3:Alcohol     -2.113373   0.975271 -2.1670 0.030238 *
2:Smoking     -0.627292   0.857542 -0.7315 0.464474
3:Smoking     -2.717454   0.998979 -2.7202 0.006524 **
2:Age1         1.006349   1.025369  0.9815 0.326370
3:Age1        -2.361864   1.504796 -1.5696 0.116518
2:Age3         0.474632   1.075975  0.4411 0.659128
3:Age3        -0.740482   1.329040 -0.5572 0.577421
---
Signif. codes:  0 *** 0.001 ** 0.01 * 0.05 . 0.1   1

Log-Likelihood: -42.534
McFadden R^2:  0.22997
Likelihood ratio test: chisq = 25.406 (p. value = 0.0046268)
```

The confusion matrix is as follows.

```
testpred
    1  2  3
1   4  6  3
2   4  9  4
3   2  3 16
```

CCR = 56.86%.

(b) The output for ordinal regression is shown below.

```
Coefficients:
            Value Std. Error t value
Nutrition  0.02954    0.01384  2.1350
Alcohol   -1.62122    0.64720 -2.5050
Smoking   -1.80054    0.61526 -2.9264
Age1      -1.37253    0.74461 -1.8433
Age3      -0.54337    0.79352 -0.6848
```

```
Intercepts:
      Value   Std. Error  t value
1|2   0.5350  1.8214       0.2937
2|3   2.5610  1.8523       1.3826
```

```
Residual Deviance: 91.30409
AIC: 105.3041
```

The confusion matrix is as follows.

```
   test.prob
     1  2  3
 1   5  5  3
 2   3 10  4
 3   2  0 19
```

$$\text{CCR} = 66.67\%.$$

7.13 (a) The output for nominal regression is as follows.

```
Coefficients:
                 Estimate Std. Error  t-value Pr(>|t|)
1:(intercept)    0.242514   0.791882   0.3062 0.759414
2:(intercept)    1.035712   0.715321   1.4479 0.147646
1:PB            -0.225043   0.106426  -2.1146 0.034468 *
2:PB            -0.299698   0.098346  -3.0474 0.002308 **
1:HIST           1.064827   0.630023   1.6901 0.091001.
2:HIST           1.655380   0.526786   3.1424 0.001676 **
---
Signif. codes:  0 *** 0.001 ** 0.01 * 0.05 . 0.1   1
```

```
Log-Likelihood: -181.62
McFadden R^2:  0.06256
Likelihood ratio test: chisq = 24.241 (p. value = 7.1462e-05)
   testpred
       0   1   2
 0 106   0   4
 1  37   0   4
 2  48   0   7
```

$$\text{CCR} = 54.85\%.$$

(b) The fitted ordinal regression model is

```
Coefficients:
        Value Std. Error  t value
PB    -0.2626    0.07728   -3.398
HIST   1.3724    0.42023    3.266
```

```
Intercepts:
     Value   Std. Error t value
0|1 -1.3439  0.5746      -2.3389
1|2 -0.5102  0.5699      -0.8952

Residual Deviance: 363.2292
AIC: 371.2292
```

The confusion matrix is the same as that obtained for nominal logistic regression, so CCR is the same.

Chapter 8

8.3 (a) $L_1 = 2.722 < L_2 = 3.000$. Classify x to Group 2.

(b) $LD = 3.889, LD_1 = 5.333, LD_2 = 3.000, \overline{LD} = 4.167$. Classify x to Group 2.

8.4 (a) Because the coefficients of the LDF for CHD are > those of the LDF for NCHD.

(b) $L_{CHD} = 25.649, L_{NCHD} = 25.664$. $\hat{p}_{CHD} = 0.5038, \hat{p}_{NCHD} = 0.4962$. Since $\hat{p}_{CHD} > \hat{p}_{NCHD}$, classify observations as CHD.

(c) $d_{NCHD} = 12.935, d_{CHD} = 13.414$. Since $d_{NCHD} < d_{CHD}$, classify observations as NCHD. Euclidean distance does not take into account the covariance matrix.

8.5 (a) Fisher's LDFs:

```
Coefficients of linear discriminants:
                   LD1          LD2
Petal_width  -2.8104603 -2.83918785
Petal_length -2.2012117  0.93192121
Sepal_length  0.8293776 -0.02410215
Sepal_width   1.5344731 -2.16452123

Proportion of trace:
   LD1    LD2
0.9912 0.0088
```

(b) The three probabilities are

```
$posterior
              1          2          3
1 1.962381e-19 0.9993753 0.0006246577
```

Chapter 9

9.3 (1) Simple linear regression without transformation: SSE = 123.5302

(2) Simple linear regression with square-root transformation: SSE = 123.0247

(3) Poisson regression: SSE = 117.3472

9.4 (a) Unweighted Poisson regression output.

```
Call:
glm(formula = Count ~ Day + Time + Road + Light + Weather
        + Traffic_Control, family = poisson, data = traffic)

Deviance Residuals:
    Min      1Q   Median      3Q      Max
 -61.305  -3.476  -0.584   1.395   78.294

Coefficients:
                          Estimate Std. Error  z value Pr(>|z|)
(Intercept)               7.444902   0.008396  886.766 < 2e-16 ***
DayWeekend               -0.974418   0.006492 -150.103 < 2e-16 ***
TimeMidday                0.028733   0.007604    3.779 0.000158 ***
TimeMorning              -0.281834   0.008258  -34.128 < 2e-16 ***
TimeNight                -0.335986   0.008388  -40.055 < 2e-16 ***
RoadOther                -2.213312   0.010808 -204.777 < 2e-16 ***
RoadWet                  -1.331213   0.007423 -179.334 < 2e-16 ***
LightDawn/Dusk           -2.048168   0.015995 -128.053 < 2e-16 ***
LightDaylight             0.805835   0.006502  123.930 < 2e-16 ***
LightUnknown             -2.131293   0.016597 -128.416 < 2e-16 ***
WeatherOther             -2.890320   0.014077 -205.327 < 2e-16 ***
WeatherPoor Visibilty    -5.437777   0.049077 -110.802 < 2e-16 ***
WeatherRain/Snow         -1.691636   0.008188 -206.597 < 2e-16 ***
Traffic_ControlNo Control 0.316084   0.005933   53.275 < 2e-16 ***
Traffic_ControlUnknown   -2.866373   0.019446 -147.404 < 2e-16 ***
---
Signif. codes:  0 *** 0.001 ** 0.01 * 0.05 . 0.1   1

(Dispersion parameter for poisson family taken to be 1)

    Null deviance: 655308  on 1151  degrees of freedom
Residual deviance: 162264  on 1137  degrees of freedom
AIC: 165864

Number of Fisher Scoring iterations: 7
```

(b) Weighted Poisson regression output:

```
Call:
glm(formula = Count ~ Day + Time + Road + Light + Weather + Traffic_Control,
    family = poisson, data = traffic, weights = Weight)
```

```
Deviance Residuals:
     Min       1Q    Median       3Q       Max
-130.483   -6.058   -0.976    2.372   147.782

Coefficients:
                           Estimate Std. Error  z value Pr(>|z|)
(Intercept)                7.398366   0.004095 1806.558  < 2e-16 ***
DayWeekend                -0.974418   0.004196 -232.223  < 2e-16 ***
TimeMidday                 0.014997   0.003697    4.057 4.97e-05 ***
TimeMorning               -0.235902   0.003950  -59.728  < 2e-16 ***
TimeNight                 -0.468529   0.004229 -110.797  < 2e-16 ***
RoadOther                 -2.226426   0.005321 -418.392  < 2e-16 ***
RoadWet                   -1.321563   0.003619 -365.144  < 2e-16 ***
LightDawn/Dusk            -1.968025   0.007829 -251.392  < 2e-16 ***
LightDaylight              0.909690   0.003246  280.215  < 2e-16 ***
LightUnknown              -2.108363   0.008330 -253.116  < 2e-16 ***
WeatherOther              -2.918760   0.006980 -418.164  < 2e-16 ***
WeatherPoor Visibilty     -5.444327   0.024087 -226.025  < 2e-16 ***
WeatherRain/Snow          -1.683637   0.003992 -421.730  < 2e-16 ***
Traffic-ControlNo Control  0.305629   0.002900  105.392  < 2e-16 ***
Traffic-ControlUnknown    -2.884356   0.009565 -301.567  < 2e-16 ***
---
Signif. codes:  0 *** 0.001 ** 0.01 * 0.05 . 0.1   1

(Dispersion parameter for poisson family taken to be 1)

    Null deviance: 2701803  on 1151  degrees of freedom
Residual deviance:  651854  on 1137  degrees of freedom
AIC: 665102

Number of Fisher Scoring iterations: 7
```

9.5 (a) The output of exponential regression:

```
> fit1=glm(Cost~factor(Merit)+factor(Class)+log(Insured)+log(Premium)
      +log(Claims), data=claims, family=Gamma)
> summary(fit1,dispersion=1)

Call:
glm(formula = Cost ~ factor(Merit) + factor(Class) + log(Insured) +
    log(Premium) + log(Claims), family = Gamma, data = claims)

Deviance Residuals:
    Min       1Q    Median       3Q      Max
-1.19580  -0.23223  -0.02970   0.04726   0.86208

Coefficients:
                   Estimate Std. Error z value Pr(>|z|)
(Intercept)       0.0323898  0.0241604   1.341   0.1800
factor(Merit)1   -0.0008813  0.0014481  -0.609   0.5428
factor(Merit)2   -0.0018777  0.0021674  -0.866   0.3863
factor(Merit)3   -0.0026610  0.0014418  -1.846   0.0649.
factor(Class)2    0.0026665  0.0032454   0.822   0.4113
```

```
factor(Class)3   0.0019428   0.0011941    1.627   0.1037
factor(Class)4   0.0069292   0.0047478    1.459   0.1444
factor(Class)5   0.0036654   0.0050445    0.727   0.4675
log(Insured)    -0.0074169   0.0108500   -0.684   0.4942
log(Premium)     0.0160141   0.0123882    1.293   0.1961
log(Claims)     -0.0090759   0.0056629   -1.603   0.1090
---
Signif. codes:  0 *** 0.001 ** 0.01 * 0.05 . 0.1   1

(Dispersion parameter for Gamma family taken to be 1)

    Null deviance: 48.7047  on 19  degrees of freedom
Residual deviance:  4.0303  on  9  degrees of freedom
AIC: 349.14

Number of Fisher Scoring iterations: 5
```

(b) The output of gamma regression is the same as above except the dispersion parameter estimate is 0.4060507.

Chapter 10

10.8 (a) The Kaplan–Meier curves are shown below.

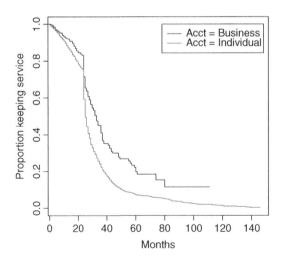

(b) Logrank $\chi^2 = 35.6$ on 1 d.f.

10.10 The strongest predictors are prevcatmerch, prevcattravel, prefood, prebonus, baselen.

10.11 When the time dependent covariate employed is included in the model, we see that Aid is now even less significant with the p-value nearly doubled ($p = 0.1260$).

Index

Predictive Analytics: Parametric Models for Regression and Classification Using R,
First Edition. Ajit C. Tamhane.
© 2021 John Wiley & Sons, Inc. Published 2021 by John Wiley & Sons, Inc.
Companion website: www.wiley.com/go/tamhane/predictiveanalytics

Printed and bound by CPI Group (UK) Ltd, Croydon, CR0 4YY